The Calculus Reader, Part I

Textbook for first semester calculus

Preliminary Edition

David A. Smith and Lawrence C. Moore

Duke University

Acknowledgments:

We acknowledge with gratitude the assistance of members of the mathematics faculty of the North Carolina School of Science and Mathematics (NCSSM). This book was written under the auspices of *Project CALC: Calculus As a Laboratory Course*, a joint project of Duke University and NCSSM, with the support of the National Science Foundation. Opinions expressed in this work are those of the authors and are not necessarily endorsed by the National Science Foundation.

This book is "work in progress." We welcome comments, criticisms, and/or suggestions from any reader. Please send your responses via e-mail to *das@math.duke.edu* or via ordinary mail to

Project CALC   or   Mr. Charles Hartford
Department of Mathematics   D. C. Heath and Co.
Duke University   College Division
Durham, NC 27706   125 Spring Street
     Lexington, MA 02173

This preliminary edition was produced in EXP on IBM PS/2, Sun 386i, and Cordata 386SX computers with output to a Hewlett-Packard Series III LaserJet printer. Other software used in the production includes MathCAD (versions 2.0, 2.5, 3.0, 3.1), Windows 3.0, Paintbrush, Converge, SmartScan (with the Complete Half Page Scanner/400), Arts & Letters, Bridge Math Utilities, Hijaak, and Quattro.

TABLE OF CONTENTS

Chapter 2.   Rates of Change: Models of Growth

Chapter 3. Initial Value Problems

Chapter 5.   Applications of Euler's Method

Chapter 6.   Periodic Motion

What is This Book About?

Most math textbooks are about "answers" (and how to get them). This book is about QUESTIONS — and what to do about them. Just like the world around us, this book has far more questions than answers. Indeed, our questions *are* the questions of the world around us. Almost everything of importance in our world is moving or changing, and **calculus is the mathematical language of motion and change.**

To give you some idea of the importance of our subject, we will pose some sample questions that calculus might help us answer. You won't find the answers in the back of the book; indeed, answers that fit neatly in books are seldom real solutions to real problems. Here we go.

Are we in the midst of a global population explosion? If so, what resources will we exhaust first: food, fuel, or terrestrial space? As a response to such a crisis, should we colonize outer space? If so, what does it take to do that, and how do we go about it? Can people survive in large numbers on the moon or Mars? Can we move enough of them there to make any difference? If so, what are the scientific, engineering, economic, political, sociological, theological, and biomedical problems we have to solve? How do we solve them?

Suppose we find there *is* a population crisis, but there is *no* viable solution to the problem of space colonization. What problems do we have to solve to continue our existence in relative peace on Earth? Population control? Waste management? Pollution control? Technological advances in computers, consumer goods, weapons, communications? Arms control or reduction? Management of international relations? Peace through strength or strength through peace? Economic growth or economic stability?

Suppose there is *no* impending population explosion — population may be self-limiting. What then? Will we see world population "level off" at some stable number? If so, how big can we expect that number to be?

Would its sheer size lead us to grapple with a host of other problems, such as extreme scarcity of resources and drastically lowered standard of living?

If there is no leveling-off point, will there be oscillations in the population level? If so, will these be wild swings between very high and very low levels, or will they be modest variations at manageable levels? If the latter (which would suggest that population problems need not be high on our priority scale), what are the *important* problems of a society and a world that appear to be changing ever more rapidly?

These are challenging questions about *change*, more precisely about the *rates* at which dynamic quantities change and about the consequences we can determine from those rates. Calculus provides us the conceptual framework and many of the computational tools for the quantitative and qualitative study of rates of change, and *that's* what this book is about.

Why Study Calculus?

Each semester we ask our beginning calculus students why they are taking the course. Here are a few of the most common answers to this question:

- It's required for my major.
- I have always had a mathematics course, and this was the next one in line.
- My parents said I had to take it.
- I like mathematics.
- Everyone says mathematics is important; I just felt that I ought to do it.
- Calculus is central to understanding the development of philosophy and science in the last three centuries; without a thorough grasp of this fundamental branch of mathematics, one cannot be considered an educated person.

Well, honesty compels us to admit that no one has actually given the last response, but our hope springs eternal.

Why do you, your major department, and/or your parents think that mathematics (calculus, in particular) is important? In large part, this rests on the belief that much of what we experience in the world around us can be *understood* — that our experience can predict what will happen in the future or explain what happened in the past. In other words, mathematics helps us *solve problems*.

Calculus is particularly important for solving problems involving change, problems involving limiting behavior, problems involving stasis in the midst of change. Calculus is essential for understanding the motion of atomic particles, automobiles, satellites, galaxies. Calculus is basic to the study of flow in rivers, currents in the oceans, and air over airplane wings.

What is This Course About?

Often a problem comes to us in the form of data: An object falls through the air; we observe data consisting of distances fallen at, say, ten different times. Can we tell how far the object had fallen at some time other than those at which we made the observations? Can we predict how a similar object will fall in the future? In this case, the theory comes to us from physics; the language of the theory is a mixture of English and calculus, and the calculations necessary to answer the questions require the same mix.

This course concentrates on the use of calculus to solve problems. It might seem that *any* calculus course ought to do that, but not all do. Traditional courses emphasize pencil-and-paper calculations, often without revealing how these calculations are used to solve real problems. In many cases, the calculations take on a life of their own and are performed mindlessly, without any purpose or application in mind.

Calculators and computers can now do much of the numeric, algebraic, and graphic manipulations that are the focus of a traditional course. These tools empower us to solve problems involving "messy" data and large numbers of computations, exactly the kind of problems we find in the real world. We can also use these tools to experiment with different ways of attacking problems and thereby obtain the intuition that comes from experience. For these reasons, the heart of this course is the *laboratory experience*.

Three paragraphs back we said, "the language of the theory is a mixture of English and calculus, and the calculations necessary to answer the questions require the same mix." What has English — reading and writing — to do with mathematics? We want to solve problems; solving problems requires deciding what should be done, executing the calculations, and interpreting the results. The environment for this intellectual activity is language, English in our case. Until you can describe what you have done, why you did it, and what it means, you have not solved the problem. For this reason, our course is a *writing course*; we expect you to write up the laboratory experiments and the in-class and take-home projects on which you will work.

How to Succeed in Calculus While Really Trying

Clearly, this textbook,[1] your calculator, the lab computers, and your instructor are important resources for learning about calculus and the art of problem solving. There is another resource just as important as those already mentioned: your fellow students. You will do in-class, take-home, and laboratory projects in teams. You will be expected to work with your teammates, to talk about what you are doing, to explain your ideas and insights to each other. Most of the time, each team will submit just one written report for each project. In the course of this work, you will find your fellow students an excellent source of help for understanding the course in general. Whatever your question, it is likely somebody else in the class has considered it already and has some ideas for an answer. The key here is to talk to one another. When you do not understand why one thing follows from another, say so; when you do not see the evidence to support a conclusion, say so.

Learning is a cooperative, not a competitive, activity. We are about to embark on a great cooperative adventure: learning calculus. *Bon voyage!*

---

[1]When you encounter a footnote, ignore it the first time you read the sentence. Then go back and read the sentence again with the footnote as a parenthetical insertion. Some of the most important information appears in footnotes. Why is important information relegated to footnotes? Because stating it at that point in the text (even parenthetically) would break into an even more important flow of ideas. *This* footnote contains important information, which, however, has little to do with the sentence it interrupts. *Some* footnotes are merely frivolous.

CHAPTER 1

Relationships

1.1 Functions and Relations

As students of calculus, nothing is more important to us than *functions*.[1]
Thus, in this introductory chapter, we deal with one big question:
WHAT'S A FUNCTION?

No doubt you already know an answer to the "big question" from your
previous study of mathematics. In fact, we will use your prior knowledge as
our starting point for discussion and refinement of this central concept. Of
course, only *you* can provide that prior knowledge.

Exercise 1.[2] *In the space below, write your present answer to the big question. You
may use definitions, examples, symbols, or whatever you think appropriate for
explaining to someone else what a function is. In the process, you should
acquire a sharper focus on your explanation.*

---

[1]You may well find other things more important in other contexts, e.g., social
justice, Mozart, baseball.

[2]We mean *Exercise* literally. Reading is a mostly passive activity, like watching
television. You may occasionally interrupt your TV watching (during commercials, say)
by physical exercise (walking to the refrigerator or the bathroom). We will frequently
interrupt your reading with a request that you "move something": your brain, the keys
on your calculator, and/or a pencil. Read this book with a pencil or pen in hand, and
use it to make comments in the margins. After some practice, the pen or pencil
becomes a tactile reminder of your active role as reader and learner. A highlighter is
not as useful for this; you need to add your part to the dialogue, not merely distill ours.

The word "function," like all the other words we shall use in this course, belongs to the English language. As such, it has a definition — or several:

> **function** (fŭngk′shen) *n.* **1.** The action for which a person or thing is particularly fitted or employed. **2. a.** Assigned duty or activity. **b.** Specific occupation or role: *in his function as attorney.* **3.** An official ceremony or formal social occasion. **4.** Something closely related to another thing and dependent on it for its existence, value, or significance: *Growth is a function of nutrition.*[3]

Do you see any point of contact between that dictionary definition and what you wrote in response to Exercise 1? Perhaps not — and therein lies the root of a fundamental difficulty in the study of mathematics. Math books and math teachers seem to use words from the English language, but often with meanings that seem arbitrary and unrelated to common usage. Actually, as we shall see, definition **4.** is rather close to the meaning we shall establish in this chapter. If you wrote something like that, give yourself a pat on the back. If nothing you wrote looks like that, don't despair — by the end of this chapter we will have connected your prior experience with the concept of "function" as the word will be used in this course.

You may have noticed that the title of this chapter is not "Functions" but "Relationships." The key word in definition **4.** is "related," and we will establish our meaning for "function" within the more general context of "relationships."

Variables and Data

Think for a moment about each of the following questions:

- The United States has a serious dropout problem. What is the relationship between state expenditures on teacher salaries and high school graduation rates?

- High blood pressure in adults is linked to weight; is there a similar relationship for children or adolescents?

- Will the world be seriously overpopulated in 20 years?

---

[3]From *The American Heritage Dictionary*, Second College Edition, Houghton Mifflin, 1985.

These questions are different in many respects, but answering each requires the collection, organization, and interpretation of data. Each requires analysis of the relationship between *two variables*. In the first case, those variables are "state expenditure on teacher salaries" and "high school graduation rate." In the second case (both parts), they are "blood pressure" and "weight." What are they in the third case? Write your answer here: _______________

Sometimes one variable actually has a "causative" effect on another. For example, we expect that blood pressure in adults of the same height depends in some way on weight, but not necessarily the other way around. Similarly, we expect that crop yield depends on amount of rainfall, but not the other way around. Other times there is a "relationship" between the variables, but it is *not* one of *cause and effect*. For example, there is certainly a relationship between time and the population of the world — at any given time, some number is the actual population — but we would probably not consider either of these variables to be a "cause" of the other.[4] Moreover, sometimes there is no relationship at all between the two variables; for example, we do not expect there to be a relationship between the distance from a student's home to college and her height. We may not know what to expect in the case of state expenditures and graduation rates — a relationship is possible, but we don't know; indeed, an (unstated, but necessary) part of the question is to determine whether in fact there is any relationship at all.

To determine whether there is a relationship between two variables, we must analyze *pairs* of data — each pair consisting of a value of the first variable and a "corresponding" value of the second variable. Sometimes these data are gathered from a well-designed, carefully controlled scientific experiment, as might be the case for a study of blood pressure or crop yields. Other times we want to analyze data that already exist in the world around us, such as census data on populations.

____

[4]Even though "time" does not *cause* "population," we still say "population *depends on* time" or "population is *a function of* time" (in the sense of definition 4. of function); each of these phrases abbreviates the slightly longer expression, "at any given time, some number is the actual population." Mathematical usage of words often coincides with "common" usage, but *not always*. (By the way, this note answers the question in the previous paragraph whose answer you wrote at the end of that paragraph.)

The September 7, 1987, *U.S. News and World Report* claims that "spending heavily on teachers doesn't always yield a bumper crop of graduates." The comment is followed by the data provided in Table 1. Study this list of paired data to determine whether you agree with the statement made by the magazine.

	Spending	Graduation Rate		Spending	Graduation Rate
Alaska	$8,842	67.1%	California	$3,751	65.8%
New York	$6,299	62.7%	Iowa	$3,740	86.5%
Wyoming	$6,229	74.3%	Maine	$3,650	78.6%
New Jersey	$6,120	77.3%	Texas	$3,584	63.2%
Connecticut	$5,532	80.4%	New Mexico	$3,537	71.9%
Dist. of Columbia	$5,349	54.8%	North Carolina	$3,473	70.3%
Massachusetts	$4,856	76.3%	Nebraska	$3,437	86.9%
Delaware	$4,776	69.9%	New Hampshire	$3,386	75.2%
Pennsylvania	$4,752	77.2%	Indiana	$3,379	76.4%
Wisconsin	$4,701	84.0%	Missouri	$3,345	76.1%
Maryland	$4,659	77.7%	Louisiana	$3,237	54.7%
Rhode Island	$4,574	67.6%	North Dakota	$3,209	86.1%
Vermont	$4,459	83.4%	South Dakota	$3,190	85.1%
Hawaii	$4,372	73.8%	Georgia	$3,167	62.6%
Minnesota	$4,241	90.6%	Kentucky	$3,107	68.2%
Oregon	$4,236	72.7%	South Carolina	$3,005	62.4%
Kansas	$4,137	81.4%	West Virginia	$2,959	72.8%
Colorado	$4,129	72.2%	Tennessee	$2,842	64.1%
Montana	$4,070	82.9%	Arkansas	$2,795	75.7%
Florida	$4,056	61.2%	Arizona	$2,784	64.5%
Illinois	$3,980	74.0%	Oklahoma	$2,701	71.1%
Michigan	$3,954	71.9%	Alabama	$2,610	63.0%
Virginia	$3,809	73.7%	Idaho	$2,555	76.7%
Washington	$3,808	74.9%	Mississippi	$2,534	61.8%
Ohio	$3,769	76.1%	Utah	$2,455	75.9%
Nevada	$3,768	63.9%			

Table 1. Public School Spending per Pupil and High School Graduation Rates

How do we get information out of the numbers in Table 1? (Did you actually *read* the table, or did you just skip to this paragraph?) As presented, the data are difficult to interpret. To find information hidden in the numbers, we need some way to organize the data so we can see meaning without getting lost in a horde of numbers. Then maybe we can decide whether or not "graduation rate" is related to "educational spending."

Scatter Plots

One way to display the general relationship between two variables is to make a "scatter plot," that is, a graph in a rectangular coordinate system of all the pairs of data. In Figure 1 we show a computer-generated scatter plot of the fifty-one data points listed in Table 1. Making such a graph is easy if there are not too many points; for this many points, it would be tedious to make the plot by hand. In an early lab session, you will learn how to make scatter plots with a computer or a graphing calculator.

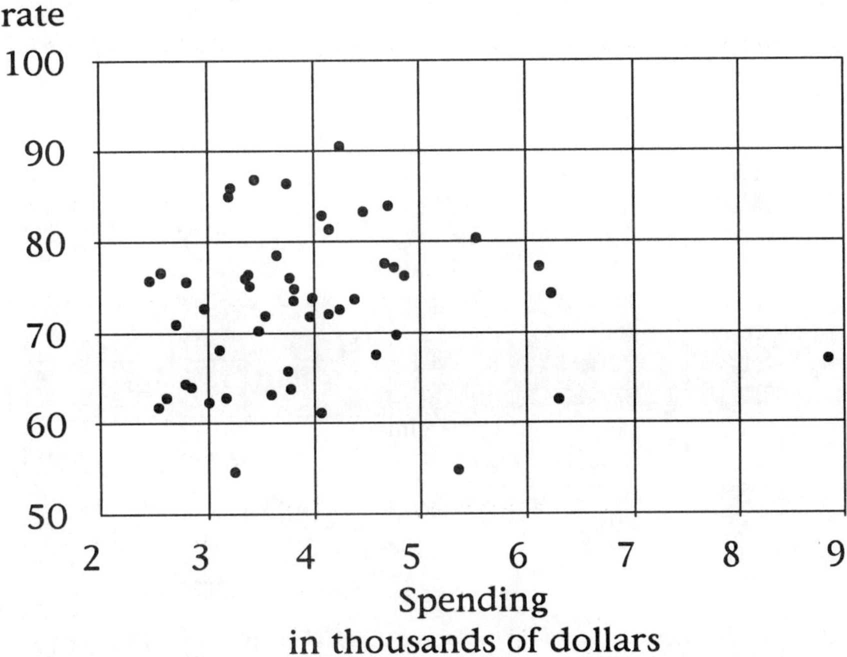

Figure 1. Graduation Rate versus Public School Spending.

For each state, the pair consisting of a spending amount and a graduation rate is represented by one point in the plane. State spending is plotted on the horizontal axis, and graduation rate is plotted on the vertical axis. For example, the pair representing Montana is ($4070, 82.9\%$). [Find the plotted point you think represents Montana, and circle it.] When making a scatter plot, it really does not matter which variable is plotted on which axis. If we suspect one variable to be dependent on the other, however, we usually plot the dependent variable on the vertical axis.

Exercise 2. *(a) Study the scatter plot in Figure 1. Do you agree with the U.S. News and World Report claim? Does there appear to be any relationship between state spending and graduation rate?*

(b) Are you surprised by the point that lies far to the right of the others? Which state does that point represent?

(c) Which display do you find easier to interpret, the table or the graph?[5]

The data in Table 2 were gathered by a team of two students in a physics lab; each used an ultrasonic motion detector to measure, at a number of closely spaced times, the distance a falling object had traveled. Times that appear twice in the table were selected by both students, and each selected some times not used by the other. The scatter plot for this data is shown in Figure 2.

Time (in sec)	Distance (in cm)	Time (in sec)	Distance (in cm)
.16	12.1	.57	150.2
.24	29.8	.61	182.2
.25	32.7	.61	189.4
.30	42.8	.68	220.4
.30	44.2	.72	254.0
.32	55.8	.72	261.0
.36	63.5	.83	334.6
.36	65.1	.88	375.5
.50	124.6	.89	399.1
.50	129.7		

Table 2. Falling body data.

Figure 2. Scatter plot of falling body data.

Exercise 3. *(a) Do you think there should be a relationship between time and distance for an object falling in a gravitational field?*

(b) Does the scatter plot in Figure 2 confirm or deny a relationship?

(c) Does the data agree or disagree with what you think should happen? If there are discrepancies with what you think should happen, how do you think those discrepancies might arise in the process of data gathering?

---

[5]Hints and answers for all exercises embedded in the text (and many of the exercises at ends of sections and chapters) are collected at the back of the book. You should always work out your own answers before consulting ours or looking for a hint. However, if you need help, you may find it there. If not, don't stay frustrated — ask another student or your instructor.

The two examples considered thus far are extreme cases of scatter plots in opposite directions — *lots of scatter* and *not very much scatter.* The data in Table 3 will provide us an "in between" example; the table shows the total numbers of rebounds, assists, personal fouls committed, and points scored by individual basketball players on two well-known teams during the 1984-85 basketball season.

Duke University

Player	Rebounds	Assists	Total Fouls Committed	Total Points Scored
Alarie	158	49	78	492
Amaker	69	184	55	253
Anderson	20	0	12	15
Bilas	186	14	96	312
Bryan	8	2	6	12
Dawkins	141	154	64	582
Henderson	98	51	69	317
King	50	22	34	51
Meagher	130	46	85	241
Nessley	18	3	15	16
Strickland	36	15	37	108
Williams	29	2	15	47

University of North Carolina, Chapel Hill

Player	Rebounds	Assists	Total Fouls Committed	Total Points Scored
Brust	4	3	2	2
Daugherty	349	77	112	623
Daye	3	2	1	6
Hale	119	168	102	349
Hunter	36	32	34	120
Martin	199	44	103	347
Morris	5	0	0	8
Peterson	65	57	31	198
Popson	87	19	66	211
Roper	1	0	2	0
K. Smith	92	235	54	444
R. Smith	19	7	18	65
Wolf	158	58	79	274

Table 3. Basketball Statistics[6]

---

[6]Provided by the Sports Information Offices at Duke University and the University of North Carolina at Chapel Hill.

Exercise 4. (a) Study the "fouls" and "points" columns in Table 3. Do you think there is a relationship between the number of personal fouls that a basketball player commits in a season of play and the number of points he scores?

(b) In the scatter plot shown in Figure 3, we plot (for the two teams combined) personal fouls on the horizontal axis and points scored on the vertical axis. Now do you think there is a relationship? Write your answer here: ___________________

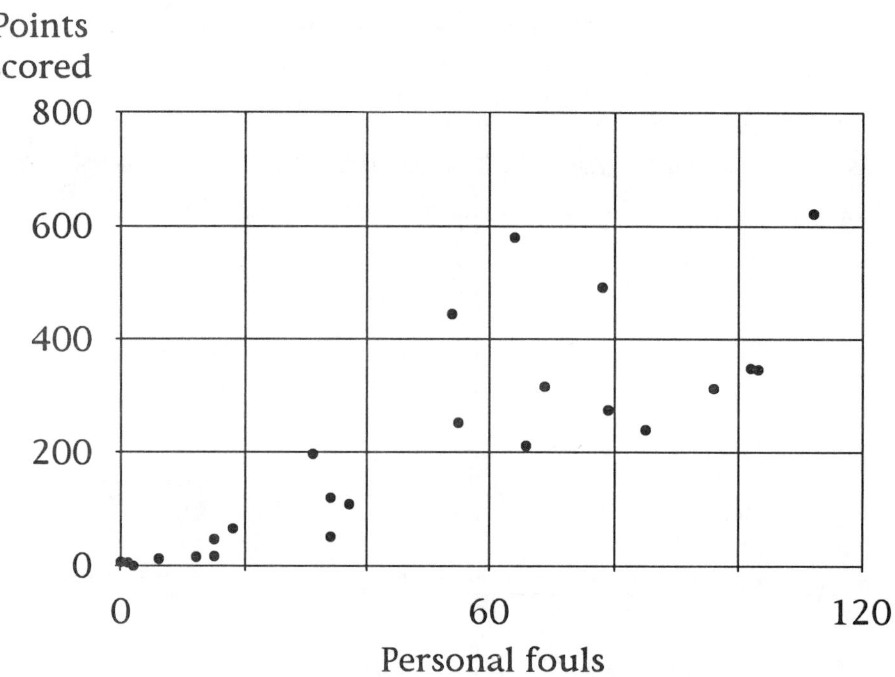

Figure 3. Points Scored versus Personal Fouls.

Exercise 5. (a) What is the general "shape" of the plot in Figure 3?

(b) How strong is the relationship between these variables? How well could you predict the points scored by a player committing 25 personal fouls?

(c) In general, if you were given a value for one variable, how confident would you be in predicting a value for the other variable?

(d) Are there any points that appear to "stand out" from the rest, that is, that do not seem consistent with the other observations? If so, what explanation would you offer for this apparent inconsistency?

Exercise 6. (a) How do you feel about the variables in this example? Does a change in one variable cause the other to change, or is this a situation where they

simply change together as the result of some underlying factor?
(b) Would you expect the coach to encourage his players to commit lots of fouls in order to be sure of scoring many points? If not, what underlying factors could be influencing both of our variables?

Mathematical Models

When a relationship between varying quantities is suggested by a scatter plot, we may want to describe it mathematically. We describe the relationship by finding an equation that summarizes the way the two variables are related; such an equation is a **"mathematical model."** A good model simplifies the phenomenon it represents and gives us the ability to predict. If we can find an equation of a line or curve that closely "fits" a scatter plot, we can focus on the important characteristics of the relationship between the variables without the clutter of a scatter plot. We can also use this equation to predict the values of one variable for specific values of the other variable. Sometimes we use the model to "interpolate," or estimate values among observed values; sometimes we use the model to "extrapolate," or predict values outside the region of observations.

We illustrate this "model" concept with a simpler example than the basketball data. Figure 4 is adapted from a newspaper article[7] about advertising expenditures by major drug manufacturers. In this figure, the "scatter plot" (with no scatter at all) takes the form of a *bar chart*. The top of each bar represents a pair of related numbers: a year and a sales figure for that year.

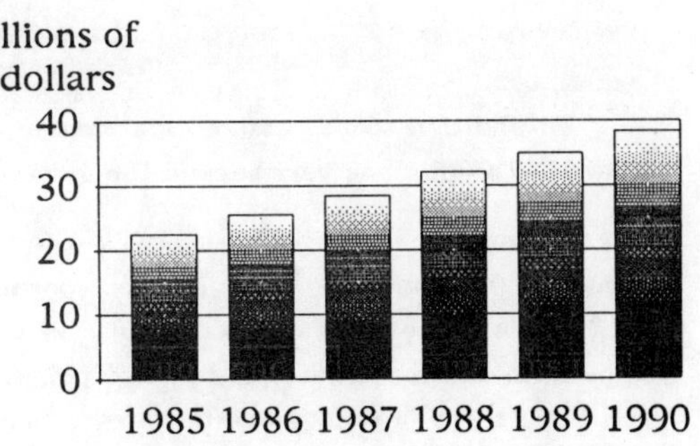

Figure 4. Sales by the pharmaceutical industry in North Carolina.

---

[7]"Drug makers spend money to make money," by Jim Barnett, *The News and Observer* (Raleigh, NC), April 14, 1991. The 1990 figure is estimated.

Exercise 7. *(a) In Figure 4, draw a straight line from the top of the "1985" bar to the top of the "1990" bar. (You decide whether you want to use left corner, right corner, or center for the top of each bar.) Notice that this line passes pretty close to the corresponding points on the tops of all the other bars.*

(b) Find an equation for your line of the form $y = mx + b$, where x stands for "year" and y stands for "sales (in billions of dollars)."

(c) Use your equation to estimate the sales by North Carolina drug companies in each of the years 1986, 1987, 1988, and 1989, and compare your "model" estimates with the graphical data in Figure 4.

(d) Now use your equation to estimate the sales in 1991, i.e., to predict the next data point beyond the time at which the article was written.

Our estimate for the sales by North Carolina drug companies in 1991 is 41.2 billion dollars. Whether you agree with that estimate or not depends not only on the correctness of your calculations, but also on some assumptions you had to make.[8] Specifically, you had to decide how accurately you could read sales numbers from the bar chart, and then you had to decide what those numbers were. Once you had your first and last points, you had to correctly calculate an equation for the line. (That's *not* an assumption.) Finally, you had to correctly substitute each of the years for x and calculate each of the corresponding y's.[9]

Have you noticed that none of the exercises so far have clearly determined "right" answers? That's often the way mathematics works in the real world. There are answers that are more sensible and answers that are less sensible and answers that are clearly *wrong* — but not always

---

[8]In addition to *our* assumption that a straight line model would fit the data well enough to predict one year beyond the given data.

[9]If you had difficulty with any of those steps, stop right now and straighten out whatever went wrong. What are the coordinates of the two points determining your line? Do the y-coordinates make sense in terms of the graph? How do you calculate slope from the coordinates of the two points? Once you know the slope m, how do you find b? You don't have the luxury of substituting $x = 0$, because that is not a year in which we know anything about drug sales in N. C. However, you can substitute either of your known pairs of x and y to get an equation in which b is the only unknown. Once you have sensible values for m and b, you should be able to do the rest of the arithmetic with your calculator.

answers that are clearly right. Your job is to find answers that are *sensible* — and to be prepared to defend those answers.

Exercise 8. Figure 5 shows data on the percentage of U. S. citizens among those earning doctorates in the mathematical sciences at universities in the U. S. Estimate from the graph the percentage in each year from 1978-79 on. If the trend in the 1980's continued for another decade, what would the percentage be in 1999-2000?

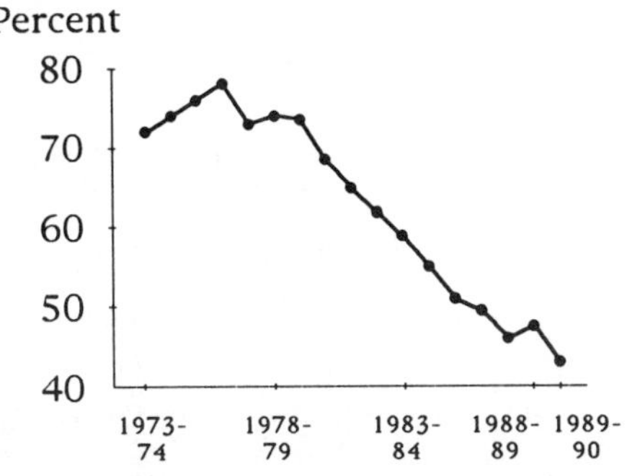

Figure 5. Percentage of U. S. citizens earning doctorates in the mathematical sciences.[10]

Functions as Mathematical Models

We are all familiar with scientific formulas (e.g., $E = mc^2$) that express relationships between variables; often these formulas are used to make predictions. But it is not only scientists who use knowledge about relationships this way. In order for each of us to make reasonable decisions, we must know certain relationships and use them to make informed choices. For instance, we know that stopping distance (for a car) depends on (or "is a function of") our driving speed, and we use this information to decide on a safe following distance as we drive.

In relationships such as those just described, it may be that one of the quantities in the relationship *determines* the value of the other; the second quantity then "depends on" the first. When this is true, the quantity that depends on the other is called the "dependent" variable; the other is called the "independent" variable. For example, since a safe following distance depends on driving speed, "driving speed" is the independent variable and "following distance" is the dependent variable.

---

[10]From the 1990 Annual Survey of the American Mathematical Society and the Mathematical Association of America, as reported in *Focus*, January-February, 1991.

In other situations, "dependency" of one variable on the other is not so obvious. For instance, the amount people are willing to pay for bonds actually determines the effective rate of return on the bonds. At times the rate might determine the price, but the price also determines the rate. We might declare *either* variable to be "independent" and the other "dependent" as we set up a description of this relationship.

We can display or describe the relationship (if any) between two quantities or variables in many different ways. Here are some of the ways that should already be familiar:

(1) *List all the pairs of values that occur together.* As we have seen, it is often difficult to see the characteristics of the relationship in this form.

(2) *Make a table or chart that displays some or all of the pairs in the relationship.* Again, it may be difficult to see properties of a relationship in a table, but tables are often useful, especially when we already know that a meaningful relationship exists. For example, your doctor probably has a chart that shows ideal weights for various heights. Each April, wage earners use a tax table to determine from their taxable income how much income tax they owe. Some store clerks use a table to find the amount of sales tax determined by the purchase price.

(3) *Give a formula or equation that relates the two variables.* For some relationships, formulas that express how the variables are related are already known. For example, sales tax is *defined* by a simple percentage-of-the-price formula. (But sales clerks often find the table easier to use.) For other relationships, data values can be gathered and used to determine a "mathematical model" of the relationship.

(4) *Draw a graph in a rectangular coordinate system.* As noted already, we usually graph the dependent variable on the vertical axis and the independent variable on the horizontal axis. Such a graph can reveal many of the properties of the relationship and thereby help us to understand it.

You will spend a fair amount of time in this course studying graphs of relationships between two variables. These graphs may be scatter plots, bar charts, or line graphs of data, as in the "Scatter Plots" section. Or they may be graphs of formulas, whether derived from known relationships or from "fitting" formulas that approximate data. Or they may be "conceptual pictures" of relationships for which we know neither data nor formulas. We illustrate this third possibility now.

Exercise 9. *On the axes below, label and sketch a graph to represent the relationship between the number of customers checking out at a grocery store and the average amount of time each person must wait in a check-out line. Does the number of customers determine the waiting time, or is it the other way around? Which is the independent variable in this relationship? What characteristics of the relationship can you determine even without specific information about the time required for a customer to complete the check-out process? As the number of people waiting to check out increases, what can you say about their average waiting time?*

If you assumed that the number of check-out lines does not change, the shape of your graph should indicate an "increasing" relationship; as the number of customers increases, the waiting time also increases. We do not know the coordinates of any particular points on the graph, nor do we know the precise steepness of the graph. Nevertheless, we can represent an important characteristic of the relationship by the fact that the graph is increasing.

Relations

The technical term for the relationships we have considered so far — such as weight paired with blood pressure, time paired with distance fallen, or income paired with tax rate — is **"relation."** We use this term whenever values of one variable are paired with values of another variable, regardless of the nature of the relationship between them. Thus, the technical term "relation" is actually *broader* than the intuitive concept of "relationship between two variables," in the sense that "relation" applies whenever values of two variables are paired, whether or not the two variables appear to be "related."

Let's consider another relation — that between the price you pay for an airplane ticket and the distance you fly. It would seem reasonable that the cost of an airplane ticket should depend on the distance you plan to fly. However, it often happens that two tickets for flights of, say, 500 miles (even on the same plane!) have very different prices. Given a value for the independent variable (distance), we cannot confidently determine the value of the would-be dependent variable (price). The price of a ticket is not *uniquely* determined by the length of the flight; other factors — the choice of airline, the sizes of the airports, the day of the week, the class of service, the date of the ticket purchase, and special deals or incentives — all influence the ticket price.

In some relations, knowing the value of the independent variable guarantees that there is a *unique* value for the dependent variable. Whenever this is true, the relation is called a **"function."** In the airplane ticket relation, we are *not* guaranteed a unique value for the dependent variable, so this relation is *not* a function. On the other hand, your fitted line in Exercise 7 has a formula of the form $y = mx + b$ (where m and b are whatever numbers you calculated to make the line fit the data); for each value of the independent variable x (a year), there is a *unique* value for the dependent variable y (an actual or predicted sales figure, according to your "linear" model).

We hasten to add that it is *not* having a formula such as $y = mx + b$ that makes the relationship in Exercise 7 a function; rather *it is the pairing of values of the independent variable with* **unique** *values of the dependent variable.* Indeed, that exercise involves *another* function for which we have *no* formula: the pairing of a year with the *actual* sales in that year. It is not very likely that the drug companies in North Carolina would have combined sales year after year that neatly and exactly fit *any* simple formula.

It is useful to think of a function as a *process* that associates each permissible first coordinate with a *unique* second coordinate — a "pairing process," if you like, with the special condition that the second value in each pair is uniquely determined by the first. In this course, it is important to think about functions as being "dynamic," as "doing something" to an x-value to get a y-value.[11]

Exercise 10. *Consider the relationship between* the amount of money earned on a part-time job *and* the number of hours worked.

(a) Identify the two quantities that vary, and decide which should be represented by an independent variable and which by a dependent variable.

(b) On the axes provided, sketch a reasonable graph to show a possible relationship between the two variables. (You should be concerned only with the basic shape of the graph, not with particular points.)

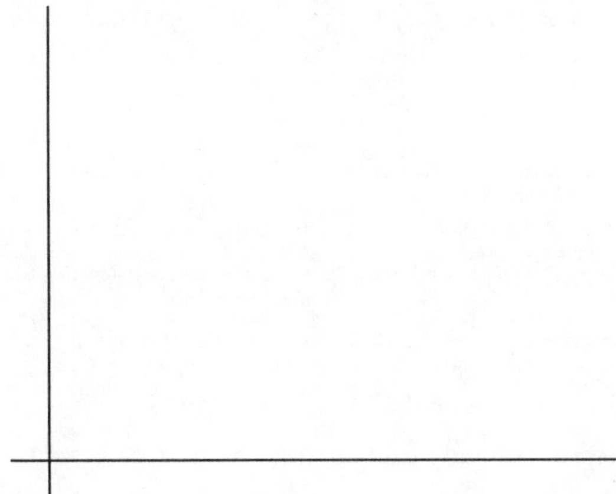

(c) Write a sentence or two to justify the shape and behavior of your graph.

---

[11]It will often be convenient to use variable names other than x and y, as we shall see. They are used here as "generic" variable names.

Exercise 11. Consider the relationship between the postage on a first class letter *and* the weight of the letter.

(a) Identify the two quantities that vary, and decide which should be represented by an independent variable and which by a dependent variable.

(b) On the axes provided, sketch a reasonable graph to show a possible relationship between the two variables. (You should be concerned only with the basic shape of the graph, not with particular points.)

(c) Write a sentence or two to justify the shape and behavior of your graph.

Exercise 12. Consider the relationship between the number of people absent from class *and* the day of the academic year.

(a) Identify the two quantities that vary, and decide which should be represented by an independent variable and which by a dependent variable.

(b) On the axes provided, sketch a reasonable graph to show a possible relationship between the two variables.

(c) Write a sentence or two to justify the shape and behavior of your graph.

Exercise 13. For each of the relations in Exercises 10, 11, and 12, determine whether the relation is or is not a function, and explain why you think it is or is not. For each relation that is not a function, can you think of additional conditions you could impose on the description that would make it a function?

The time has come to clarify a point about "dependence" that often causes confusion — indeed, that may have affected some of the decisions you made while thinking about exercises 10, 11, and 12. As we said earlier, we are sometimes interested in "cause and effect" relationships between variables — but *not always!* Often we need only a "descriptive" relationship in which each value of the "independent" variable is associated with (but need not "cause") a single value of the other variable. In this case, the "other" variable is still called "dependent," and the relation is still a function.

We have a good example in Exercise 3, which dealt with data on a falling object. Think for a moment about the physical position of the object at each instant during its fall. Certainly, at every such instant, there is a definite distance that the object has fallen — whether or not we can measure that distance. Thus, "distance fallen" is a function of "time," even though readings on a stopwatch are certainly not the "cause" of the fall or of the occurrence of a particular distance.

The physical "distance fallen" function is different from the data reported in Exercise 3 in at least two respects. First, there are only a finite number of data points; our experimenters could not take a measurement at literally *every* time during the fall, because time changes "continuously." Second, because there were two experimenters, and they chose some times in common, there are tabulated times that do not determine *unique* distances. (Check this: Find and circle the entries in Table 2 where the same time is paired with *different* distances.) Thus, in this case, the data describe a relation that is not a function, even though it closely approximates a "real" (physical) function.

Does this clarification about "cause and effect" affect any of your answers to Exercises 10, 11, or 12? Go back and check to be sure.

Exercises

14. *Consider the relationship between* the temperature of an ice-cold drink left in a warm room *and* a period of time.

(a) Identify the two quantities that vary, and decide which should be represented by an independent variable and which by a dependent variable.

(b) Sketch a reasonable graph to show a possible relationship between the two variables.

(c) Write a sentence or two to justify the shape and behavior of your graph.

15. *Consider the relationship between* the number of hours of daylight *and* the day of the year.

(a) Identify the two quantities that vary, and decide which should be represented by an independent variable and which by a dependent variable.

(b) Sketch a reasonable graph to show a possible relationship between the two variables.

(c) Write a sentence or two to justify the shape and behavior of your graph.

16. *Consider the relationship between* the cost of a pizza *and* the diameter of the pizza.

(a) Identify the two quantities that vary, and decide which should be represented by an independent variable and which by a dependent variable.

(b) Sketch a reasonable graph to show a possible relationship between the two variables.

(c) Write a sentence or two to justify the shape and behavior of your graph.

17. *For each of the relations in Exercises 14, 15, and 16, determine whether the relation is or is not a function, and explain why you think it is or is not. For each relation that is not a function, can you think of additional conditions you could impose on the description that would make it a function?*

1.2 Functions of Time

You're driving along the highway, impatient to reach your destination, because you fear you may be late for an important interview (or your cousin's wedding or your favorite sporting event or a rock concert — whatever). You look at the dashboard clock, and you look at the speedometer. You have 28 minutes to go, your destination is still 25 miles away, and you have to park the car. Are you going fast enough?

If this were a course in probability, we might want to discuss the related question of what might actually happen if you decide to go faster — for example, what are the odds that you will get caught speeding, and is the risk worth it? For now, we want to focus on the two familiar devices just mentioned: clock and speedometer. They represent a *function*: Given a time on the clock, there is a simultaneous speedometer reading (whether you are reading it or not) that gives the speed of the car at that instant of time[12] — and there is *just one* such reading. In mathematical language, "speed is a function of time."

Exercise 1. (a) *What other device on the dashboard of a car expresses some quantity as a function of time? What quantity is so expressed?*
(b) *Is time a function of speed? Why or why not?*

Every ten years, the U. S. Census Bureau attempts to count the population of the United States; the results of these efforts are shown in Table 4, which expresses (counted) population as a function of time. The same information is presented graphically in Figure 7 as a "line graph" — which is just a scatter plot with the points connected by line segments.

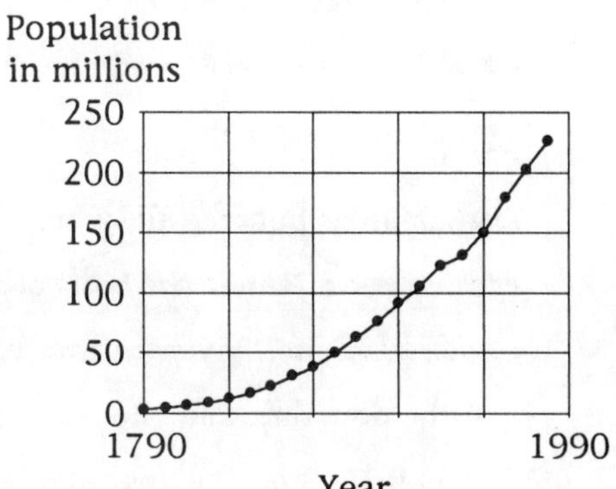

Figure 6. U. S. Census data (see Table 4).

---

[12]Strictly speaking, this assumes that your clock knows the year, the date, and the difference between A.M. and P.M. If it doesn't, the statement will still be true in any consecutive 12-hour block of time.

Census Date	Population (millions)	Census Date	Population (millions)
1790	3.939	1890	62.948
1800	5.308	1900	75.996
1810	7.240	1910	91.972
1820	9.638	1920	105.711
1830	12.866	1930	122.775
1840	17.069	1940	131.669
1850	23.192	1950	150.697
1860	31.443	1960	179.323
1870	38.558	1970	203.185
1880	50.156	1980	226.546

Table 4. U. S. Census Data, 1790–1980.

Our speedometer and population examples again illustrate the essential features of a function: (1) We have a collection of numbers (*times* in both examples) that constitute allowable "inputs" to the function.   This collection is called the **"domain"** of the function. (2) We have a "rule" for associating with each input a *unique* "output" or "result" or "value" of the function. In the speedometer example, the domain consists of all the times on a twelve-hour digital clock (at least — see Note 12), and the rule is "Associate with a time the corresponding speedometer reading."   In the population table, the domain consists of the 20 numbers (census dates) 1790, 1800, 1810, 1820, 1830, 1840, 1850, 1860, 1870, 1880, 1890, 1900, 1910, 1920, 1930, 1940, 1950, 1960, 1970, 1980, and the rule is "Read off the population number next to the date."   The domain is the same for the graph in Figure 6, and the rule for output is to read the height of each point on the graph.

Many, but certainly not all, of the functions we will study will be functions of time. In both of the examples just given, the time variable is "discrete," i.e., given a time value, there is a *next* value (except at the end of the domain), and there is a *previous* value (except at the beginning of the domain).   For the digital clock, time moves in steps of one minute (or perhaps one second, if your clock is that precise); for the population data, time moves in steps of 10 years.

You may prefer to think of time as moving "continuously," i.e., with no identifiable "next" or "previous" time after or before any given instant. An appropriate model for continuous time is the electric wall clock with hands rather than digits, sometimes called an "analog" clock. Indeed, some automobile dashboard clocks *are* analog clocks, and the speedometer *always* has a reading, whether or not the digital clock has just recorded a new time, so we may think of speed as a function of *continuous* time. Since the speedometer is not likely to be a perfectly accurate measuring device (could any such device possibly exist?), there is a slightly different, highly theoretical, function that represents the "real" speed of the car as a function of continuous time.

Let's think again about our population example in the same way. On one hand, we know that the U. S. Census is not a perfectly accurate measurement. On the other hand, we can imagine that there is *some* exact U. S. population at any instant of continuous time.[13] Unlike a "real" speed, which may or may not change "continuously," the exact, "real," highly theoretical, and unobservable population *cannot* change continuously. (Why not?)

These dichotomies — discrete vs. continuous, "real" vs. observable or measurable, exact vs. approximate — will be frequently recurring themes throughout our study of the interaction between mathematics and the world around us.

All three of these dichotomies interact with and overlap a fourth, that of *"calculational" vs. "conceptual" functions.* The functions we "observe," whether as tabulated data, graphical relationships, or measurements, can never be *known* at more than a handful of inputs, and the calculations we do with these functions can never be *exact.* On the other hand, we can often *model* those functions by mathematical formulas, about which we know (or at least can hope to know) "everything." (You did this yourself in Exercise 7, and you may have done so in Exercise 8.) You can easily be seduced into believing that these formulas are the most important objects of study, that they represent "exactness" or "truth," and that everything else

---

[13]While it may seem obvious that there should be an exact population at each instant of time (whether it can be measured or not), one can argue that no such thing exists. Is there an exact instant that an individual is born or dies? Is there an exact instant that an immigrant enters the country or an emigrant leaves?

is, at best, "approximate." In fact, the true relationships of science, social science, and engineering are very seldom described exactly by formulas. Nevertheless, formulas that approximate these realities are often convenient models, because we can study and manipulate the formulas to gain insights that might remain obscured by messy reality.

Here is an example that harks back again to Exercise 3: We observe a falling object, and, at a number of selected times, we record its position. When we plot the recorded data (as in Figure 4), we see something very much like a quadratic relationship: $s = c \cdot t^2$, where s is the distance fallen at time t. We will soon see that the *formula* implies that the velocity of the object (which is the same as speed, in this case) is proportional to time: $v = 2c \cdot t$. And it further implies that the acceleration is constant: $a = 2c$. Could that be right?

Well, one of Newton's "laws" of motion[14] says that the acceleration of the object is proportional to the force acting on it, so if the quadratic model of position is correct, the force must be constant. The forces acting on our object include gravitation, retardation due to falling through air, and perhaps wind currents; none of these is constant, and the combination of them is not likely to be. However, we can rationalize the approximate fit to a simple formula by arguing that, close to the surface of the Earth, the gravitational force is *approximately* constant, and the other forces may be so small as to have a negligible effect on our observed position data. Thus, under certain circumstances, this simple model provides our initial insight into objects moving in a gravitation field. Furthermore, knowing where we made simplifying assumptions adds insight about where to look when we need a more accurate model that will account for other factors, such as air resistance and wind.

---

[14]For "law," read "very perceptive observation." As is the case with any model of the real world, the laws of motion (and other physical phenomena) are approximations to the ways physical objects behave, not exact descriptions of reality.

Exercises

2. *Ziggy, Tom Wilson's lovable cartoon character, tells us: "Time is just Nature's way of keeping everything from happening all at once." Explain how that statement is related to the concept of function of time.*

3. *In Exercises 10-17 in Section 1.1, you constructed functional relationships from the following six pairs of variables:*

    (i) money earned, number of hours worked;

    (ii) first-class postage, weight of letter;

    (iii) number of absentees, day of year;

    (iv) temperature of a drink left in a warm room, time;

    (v) number of hours of daylight, day of year;

    (vi) cost of pizza, diameter of pizza.

    Which of your constructed functions are functions of time, and which are not? Explain.

4. *Table 5 shows annual rainfall recorded in central North Carolina over an 11-year period.*

Year	1982	1983	1984	1985	1986	1987	1988	1989	1990	1991	1992
Rainfall (in.)	44.35	47.23	46.27	38.17	36.95	42.09	37.66	54.15	37.55	35.46	43.18

Table 5. Rainfall recorded at Raleigh-Durham International Airport.[15]

(a) Plot the rainfall data as a function of time.

(b) What was the average annual rainfall for this 11-year period?

(c) The 30-year average annual rainfall for central North Carolina is about 42 inches. Was this recent 11-year period unusually dry, unusually wet, or about average?

---

[15]Source: National Weather Service, cited in "Rain ought to leave the Triangle with 1992," by Wade Rawlins, *The News and Observer* (Raleigh, NC), Jan. 1, 1993.

1.3 Words

> Words, words, words, words!
> I get words all day through,
> First from him, now from you!
> Is that all you blackguards can do?
>
> Eliza Doolittle in *My Fair Lady*

By now, you may share some of Eliza's anger and frustration. The outburst was directed at her would-be suitor, Freddie, but she was really angry at her chief tormentor, Professor Henry Higgins (the "him" in the lines quoted above), who was trying to mold her in his image of an English gentlelady by freeing her from the straitjacket of a Cockney dialect. We consciously intend to achieve the unintended result of Professor Higgins' efforts: by freeing our students from their linguistic straitjackets, to enable them to become independent thinkers.[16]

Thus, we consider it vitally important that you learn and use correctly the language of mathematics, that is, of quantitative reasoning. We have devoted these many pages of words (with very few mathematical symbols) to examination of the words "relation" and "function" precisely because these words (especially the latter) represent the concepts that will be our principal objects of study in this course.

We have a small confession to make. When we quoted *The American Heritage Dictionary* on the definition of "function," we left out something: its so-called "math" definition. Here is the missing part of the quotation:

> **function** (fŭngk′shen) *n.* **5.** *Math.* **a.** A variable so related to another that for each value assumed by one there is a value determined by the other. **b.** A rule of correspondence between two sets such that there is a unique element in one set assigned to each element in the other.

---

[16]Lerner and Loew's musical, *My Fair Lady*, was based on *Pygmalion*, a play by G. B. Shaw, which in turn was based on a Greek myth whose central character was Pygmalion, a sculptor who fell in love with one of his statues that was brought to life by Aphrodite, the goddess of love and beauty. The theme is a common one that you may encounter in many other courses; you are part of an ancient tradition that runs through thousands of years of literature and art.

Perhaps you see why we left this out earlier. First, in stark contrast to the other four definitions, these two statements are rather difficult to read and interpret. Why would a lexicographer write four straightforward definitions for different meanings of a word and then apparently obfuscate the technical meaning? Second, this appears to be *two* quite different definitions for only one concept — or do mathematicians use the one word "function" for two rather different concepts?

We suspect the question of the lexicographers' motives has a simple answer: They did not really "understand math," so they looked in several math books (whose authors are not reluctant to state definitions, apparently with authority) and found almost as many definitions. They selected two of these definitions and copied them verbatim, not understanding any more about the mathematical meaning of "function" than before.

In particular, the dictionary definition does not tell us that **5.a.** and **5.b.** actually define the *same* concept, or that the concept is merely an abstraction of definition **4**. We trust that you have figured all that out from our examples and exercises and from the labs and classroom activities associated with this chapter.

Exercise 1. *(a) In definition **5.a.**, two variables are mentioned, and each is referred to twice. Circle those four references in the definition, and label each as "independent" or "dependent."*

*(b) Definition **5.b.** refers to "one set" and "the other"; which is the domain? Circle the appropriate words in the definition, and write in "domain" above them.*

Now it is time to review your own description of "function" on page 1. Have we made a connection yet? If not, our failure to connect may result from your use of formulas to define, describe, or exemplify functions. You may have noticed that hardly any formulas have turned up yet. We will see lots of formulas later — when they serve some useful purpose — and you will have many opportunities to use all those wonderful things you have learned about algebra and trigonometry. For now, the point we want to make is this: "Function" and "formula" are not synonyms! And "equation" is not a synonym for either "function" or "formula."

You don't need a math book to sort out the meanings of these words; a dictionary will do. We have given you the dictionary definition of "function"; look up and fill in the definitions of the other two in the spaces provided here. (Each will have several definitions; select the one labeled "Math.")

formula

equation

While you have the dictionary at hand, also look up the mathematical definition of "expression." This word may not be part of the vocabulary you learned in previous mathematics courses, but its definition may be one that you routinely confuse with "function" or "formula" or "equation." In fact, "formula" and "expression" are closely related; little harm results from confusing them. "Equation" is a very special kind of formula or expression, one that asserts the *equality* of two quantities — each of which may also be described by a formula or expression.

None of this (equation, formula, expression) has any direct connection with *function*. However, the "rule" associating values of a dependent variable with values of an independent variable *might* be expressed by an equation or a formula or an expression. This *almost never* happens in the "real world." However, as we have seen already, reality often can be "modeled" by formulas, and the functional rules embodied in those formulas may turn out to be convenient and/or useful *approximations* to reality.

Thus, we will move freely among (a) the "real" (but not completely knowable) functions that relate varying quantities in physics, biology, chemistry, or economics, (b) the "observed" functions represented by data sampled from measurements of the "real" variables, and (c) mathematical "model" functions that approximate one or both of the other types. Only in this last category will we see any formulas. Indeed, that is the *purpose* of formulas: to give us approximate models of reality that are subject to manipulations that can, in turn, help us to understand reality.

1.4 Historical Background (Optional Reading)

Mental constructs, such as the concept of "function," change through history as scholars get a better grip on important ideas. The meanings of the words used to represent these concepts also change, and for the same reason. The following text is quoted from *An Introduction to the History of Mathematics* by Howard Eves[17]; the footnotes are added.

> "The concept of function, like the notions of space and geometry, has undergone a marked evolution, and every student of mathematics encounters various refinements of this evolution as his studies progress from the elementary courses of high school into the more advanced and sophisticated courses of the college postgraduate level.
>
> "The history of the term *function* furnishes another interesting example of the tendency of mathematicians to generalize and extend their concepts. The word *function*, in its Latin equivalent, seems to have been introduced by Leibniz[18] in 1694, at first as a term to denote any quantity connected with a curve, the radius of curvature, and so on. Johann Bernoulli,[19] by 1718, had come to regard a function as any expression made up of a variable and some constants, and Euler,[20] somewhat later, regarded a function as any equation or formula involving variables and constants. This latter idea is the notion of a function formed by most students of elementary mathematics courses. The Euler concept remained unchanged until Joseph Fourier (1768-1830) was led, in his investigations of heat flow, to consider so-called trigonometric series.[21] These series involve a more general type of relationship between variables than had previously been studied, and, in an attempt to furnish a definition of function broad enough to encompass such

---

[17]H. Eves, *An Introduction to the History of Mathematics, 5th Edition*, Saunders College Publishing, 1983, page 462.

[18]Gottfried Wilhelm Leibniz (1646-1716), one of the two principal inventors of the Calculus. The other was Isaac Newton (1642-1727), who was also responsible for Newton's laws of motion — see Section 1.2. Many of Leibniz' and Newton's important contributions will appear later in this course.

[19]1667-1748; a follower of Leibniz who greatly extended the power and applicability of calculus.

[20]Leonhard Euler (1707-1783), a student of Johann Bernoulli, described by Eves as "far and away the most prolific writer in the history of [mathematics]" (ibid., p. 328). The work of Euler will play a very important role throughout this course.

[21]We will touch on these ideas of Fourier in the second semester of this course. He wrote: "The deep study of nature is the most fruitful source of mathematical discovery."

relationships, Lejeune Dirichlet (1805-1859)[22] arrived at the following formulation: A **variable** is a symbol that represents any one of a set of numbers; if two variables x and y are so related that whenever a value is assigned to x there is automatically assigned, by some rule or correspondence, a value to y, then we say y is a (single-valued) **function** of x. The variable x, to which values are assigned at will, is called the **independent variable**, and the variable y, whose values depend upon those of x, is called the **dependent variable**. The permissible values that x may assume constitute the **domain of definition** of the function, and the values taken on by y constitute the **range of values** of the function.

"The student of mathematics usually meets the Dirichlet definition of function in his introductory course in calculus. The definition is a very broad one and does not imply anything regarding the possibility of expressing the relationship between x and y by some kind of analytic expression; it stresses the basic idea of a relationship between two sets of numbers."

So, you see, our purpose so far in this chapter has been to induce you to move from an 18th century understanding of function (that of Euler) to a 19th century understanding (that of Dirichlet) — but with some sense of purpose for doing so.

---

[22]A student of, and eventually successor at Göttingen University, to the great genius, Carl Friedrich Gauss (1777-1855). Some of the lesser accomplishments of Gauss (mostly from his youth) will also appear in this course.

1.5 Functions as Objects

Doing versus Being

So far, we have stressed the *process* interpretation of function: A function is a rule for "doing something" to each number in a certain set of numbers (the "domain") to get some particular result. Put in an x, get out a corresponding y. For each time t, find a distance s. Process each allowable input to produce the corresponding output. For each entry in column 1, find the corresponding entry in column 2. For each x-coordinate, go up to the graph, then over to the y-axis to find the corresponding height. Each such function rule contains one or more active verbs that collectively specify an action to be carried out on each number in the domain.

Now we turn our attention to consideration of functions as *objects*. You should think of the grammarian's meaning of the word "object": the noun or "substantive" that receives the action of the verb. Functions will often be the objects we "do things to." That is, the operations of calculus will be ones we carry out on functions to produce other functions. Thus, it is important to understand a function as a single object that can itself be the "input" to some process.

The object interpretation of function is important because the "answers" to our questions, the "solutions" to our problems, will often turn out to be functions or collections of functions, not just numbers or particular values of variables. Here's a non-mathematical example: Suppose you ask an expert on farm animals how to determine the sex of a chicken. That problem has an answer; we don't know what it is, but we know it's not a number! Whatever the answer is, it is something like a function. Specifically, your expert will describe a procedure you can apply to any member of the domain (chickens) to produce a presumably unique output, either "male" or "female."

Whatever a function is as an object, it is obviously a more complicated object than a single number or a single variable or even a small collection of these things — the sorts of things you are used to seeing as "answers." On the other hand it is *not* a more complicated object than a graph; indeed, many graphs (not all) are just pictorial representations of the very objects we have called functions. Thus, if you start from your mental image of "graph" as an object, you won't be too far away from having a mental image of "function" as an object. The hard part will be to connect the geometric image with algebraic properties of functions.

Keep in mind the characteristic that separates functions from relations in general: the fact that each first coordinate is paired with exactly one second coordinate. We can use that characteristic property to decide whether a particular graph is or is not a representation of a function.

Exercise 1. *(a) For each of the graphs in Figure 7, decide whether the graph is or is not the graph of a function. If it is not, state explicitly what keeps it from being the graph of a function.*

(b) In one complete sentence, state in your own words a simple geometric way to decide whether a graph is or is not the graph of a function. [Write your sentence on this page for future reference.]

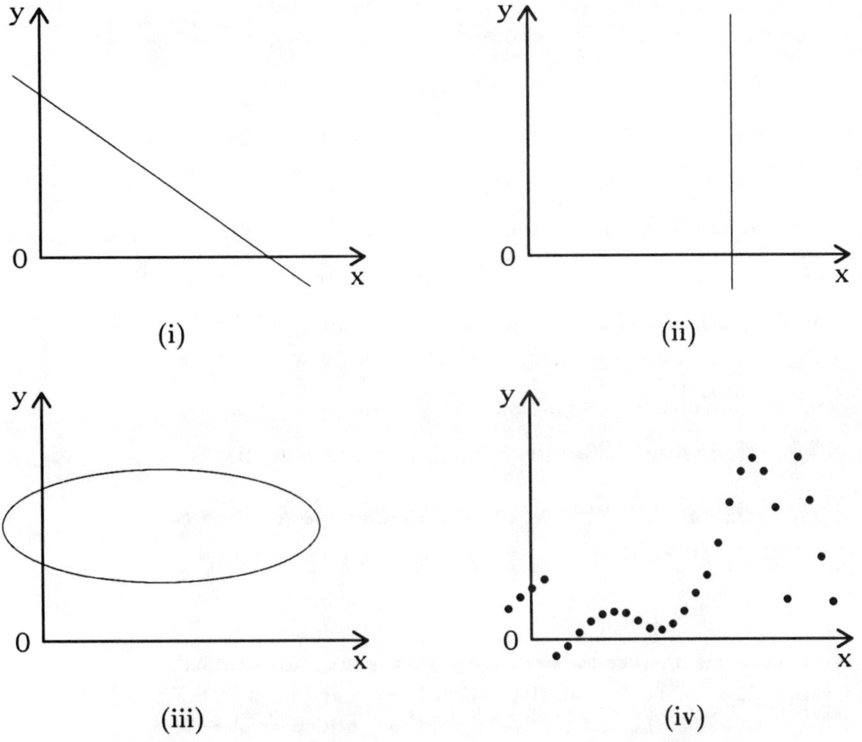

Figure 7. Which graphs represent functions?

In order to discuss algebraic properties of the objects called functions, we need some notation. It is easy to talk about "processes" in words, but it gets very awkward to do algebraic operations with words alone; thus we abbreviate words and phrases to single symbols. In a given discussion, each function is usually abbreviated to a single symbol, such as f (for function) or g (next letter after f) or ϕ (phi, the closest thing to f in the Greek alphabet). The independent (input) and dependent (output) variables for a given function are also abbreviated to single characters, such as t for time and s for position.[23] For example, the entire sentence, "The position of a falling body is a function of the time it has been falling," may be abbreviated to   "$s = f(t)$."[24]   The input to a function is inserted in parentheses following the function name, and the combined symbol is read "f of t."

Additive Functions

Here's an example of an algebraic question about functions (an important question!) for which the answer has to be a whole class of functions: What functions are "additive"? A function is called **"additive"** if what it does (as a process) to the *sum* of any two inputs is the *sum* of the corresponding outputs. In symbols, a function f is additive if $f(a + b) = f(a) + f(b)$ for every pair of numbers a and b in its domain. For example, consider the function whose rule is "multiply by 5." In symbols, that function is defined by the formula $f(x) = 5x$. We can easily calculate that

$$f(a + b) = 5(a + b) = 5a + 5b = f(a) + f(b),$$

so the "multiply-by-5" function is indeed additive.

Exercise 2. For each of the functions tabulated below, decide whether the function is additive or not. If you suspect it is not, choose particular numbers a and b

---

[23]You may be curious about why mathematicians and physicists use "s" to stand for position or distance, rather than "p" or "d." In the time of Newton and Leibniz, the language of intellectual discourse was Latin, and the Latin word for position or place is *situs*. That root survives in English words such as "site" and "situation."

[24]This notation is one of Euler's many contributions — see note 19.

in the domain of the function, and calculate $f(a+b)$ and $f(a)+f(b)$; if these two numbers turn out to be different, you have shown that the function is not additive. If they turn out to be the same, try two other numbers a and b. On the other hand, if you suspect the function is additive, try to show that by an algebraic calculation like the one above. For future reference, circle the appropriate description next to each function, and write your reason for your conclusion in the "Reason" column.

Function	Conclusion		Reason
(a) $f(x) = -2x$	additive	not additive	
(b) $f(x) = -2x + 7$	additive	not additive	
(c) $f(x) = x^2$	additive	not additive	
(d) $f(x) = \frac{2}{x}$	additive	not additive	
(e) $f(x) = \sqrt{x}$	additive	not additive	
(f) $f(x) = \log x$	additive	not additive	

Exercise 3. (a) Describe the largest class of functions you can think of that you know for sure are all additive. How do you know for sure?

(b) Would you describe additive functions as "relatively common" or "relatively rare"?

The additivity condition, $f(a+b) = f(a) + f(b)$, is an example of a **"functional equation,"** that is, an equation involving an unknown function that might or might not be "satisfied" by any particular function. Your work on the two previous exercises was an attempt to separate functions that do satisfy this equation from functions that don't.

Multiplicative Functions

A function is called **"multiplicative"** if it satisfies the functional equation $f(ab) = f(a)f(b)$, that is, if what it does to any product is to produce the product of its results on the individual factors. Is the "multiply-by-5" function multiplicative? Let's try two numbers in its domain, say $a = 2$ and $b = 3$. Then $f(a) = 5 \cdot 2 = 10$, and $f(b) = 5 \cdot 3 = 15$, so

$f(a)f(b) = 150.$ On the other hand, $f(ab) = f(6) = 30,$ so this function is definitely not multiplicative.

Exercise 4. *For each of the functions tabulated below, decide whether the function is multiplicative or not. For future reference, circle the appropriate description next to each function, and write your reason for your conclusion in the "Reason" column.*

Function	Conclusion		Reason
(a) $f(x) = -2x$	multiplicative	not multiplicative	
(b) $f(x) = -2x + 7$	multiplicative	not multiplicative	
(c) $f(x) = x^2$	multiplicative	not multiplicative	
(d) $f(x) = \frac{2}{x}$	multiplicative	not multiplicative	
(e) $f(x) = \sqrt{x}$	multiplicative	not multiplicative	
(f) $f(x) = \log x$	multiplicative	not multiplicative	

Exercise 5. *(a) Describe the largest class of functions you can think of that you know for sure are all multiplicative. How do you know for sure?*

(b) Would you describe multiplicative functions as "relatively common" or "relatively rare"?

Here's another important functional equation: $f(ab) = f(a) + f(b).$ Functions that satisfy this equation "turn products into sums," a useful thing to do, because sums are usually easier to work with than products. From your previous experience with algebra, you should know at least one such function. What is it? [Hint: Look again at Exercise 24.] Do you know more than one such function?

What   about   functions   that   "turn   sums   into   products": $f(a + b) = f(a)f(b).$ Do you know any of those? [Hint: The operation of turning sums into products "undoes" the operation of turning products into sums. You know a function that does the latter; what undoes the effect of that function?]

Symmetries: Odd and Even Functions

Other important functional equations are those that describe "symmetries" of functions that happen to have symmetries.

A function has **"even symmetry"** (or is an **"even function"**) if it satisfies the functional equation $f(-a) = f(a)$. For example, the "squaring" function, $f(x) = x^2$, has this symmetry, because

$$f(-a) = (-a)^2 = a^2 = f(a).$$

Similarly, a function has **"odd symmetry"** (or is an **"odd function"**) if it satisfies $f(-a) = -f(a)$. For example, the "multiply-by-5" function, $f(x) = 5x$, has this symmetry, because

$$f(-a) = 5(-a) = -5a = -f(a).$$

Power functions with even and odd powers provide additional simple examples — as well as a reason for the names "even" and "odd." We have just seen that the squaring function is even; in a similar manner, if $f(x) = x^3$, then

$$f(-a) = (-a)^3 = -a^3 = -f(a),$$

so the "cubing" function is odd.

Exercise 6. In Figure 8, we have (twice) sketched half the graph of a function, the half for positive values of the independent variable.

(a) In the left picture, assume the function is even, and sketch the rest of the graph. [For a typical point (a, b) on the graph, we have indicated the locations of the points $(-a, b)$, $(a, -b)$, and $(-a, -b)$. Think about which of these points must lie on the graph of an even function. Apply the same reasoning to all the other points on the graph.]

(b) In the right picture, assume the function is odd, and sketch the rest of the graph.

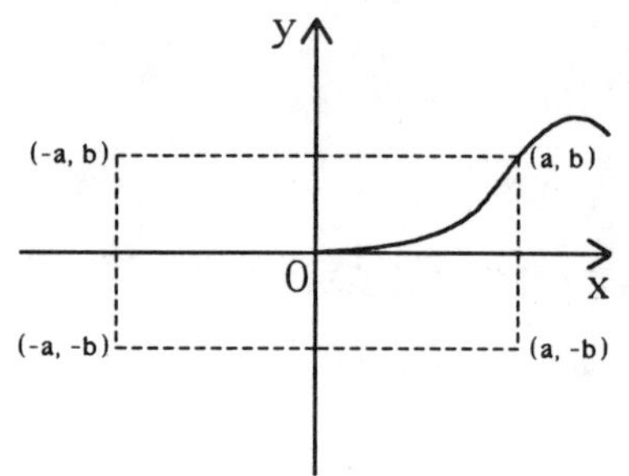 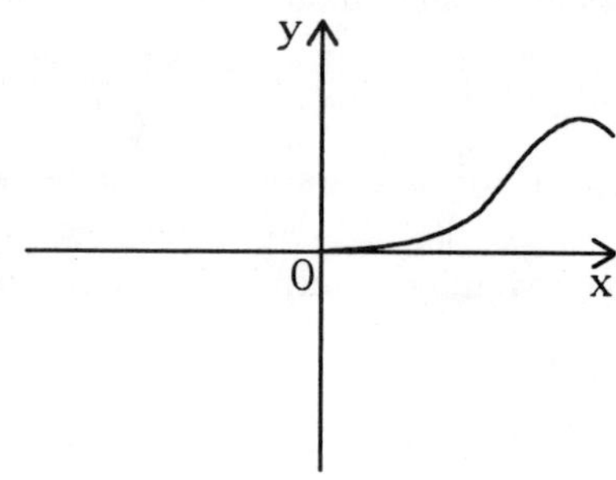

Figure 8. Make the left graph even and the right graph odd.

Exercise 7. Which of the following functions is even, which is odd, and which is neither? Circle the appropriate answer in each case.

(a) $f(x) = x^3 + 4x$ *even*   *odd*   *neither*

(b) $f(x) = x^3 + 4x^2$ *even*   *odd*   *neither*

(c) $f(x) = |x|$ *even*   *odd*   *neither*

(d) $f(x) + g(x)$, where f and g are both even   *even*   *odd*   *neither*

(e) $f(x)g(x)$, where f and g are both odd   *even*   *odd*   *neither*

(f) $f(x) + g(x)$, where f is even and g is odd   *even*   *odd*   *neither*

(g) $f(x)g(x)$, where f is even and g is odd   *even*   *odd*   *neither*

The Algebra of Functions

The function descriptions in Exercise 7 suggest some possibilities for building functions from other functions. Indeed, there is a natural way to apply the operations of algebra (addition, subtraction, multiplication, division, extraction of roots, absolute value, and so on) to function-objects to produce new functions. For example, we can describe the function in part (a) of Exercise 7 as the *sum* of the "cubing" function and the "multiply-by-4" function. In parts (d) and (f), the function $f(x) + g(x)$ is the sum of the functions $f(x)$ and $g(x)$. The function in parts (e) and (g) is the *product* of the functions $f(x)$ and $g(x)$. Similar descriptions apply to differences, quotients, and other combinations of functions.

Combining functions with algebraic operations is not just an abstract game — it's a powerful conceptual tool for expressing relationships. Consider, for example, the three functions represented graphically in Figure 9: Exports, Imports, and Trade balance, each as a function of time. The figure is adapted from a front-page newspaper article[25] reporting a monthly trade deficit that was the smallest in five years. Notice carefully the artist's device of using a positive vertical scale in *both* directions; if we attach the conventional negative signs below zero, the bottom graph represents the *negative* of the Imports function.   Either way, Figure 9 represents graphically the relationship

$$\text{Exports} - \text{Imports} = \text{Trade balance.} \tag{1}$$

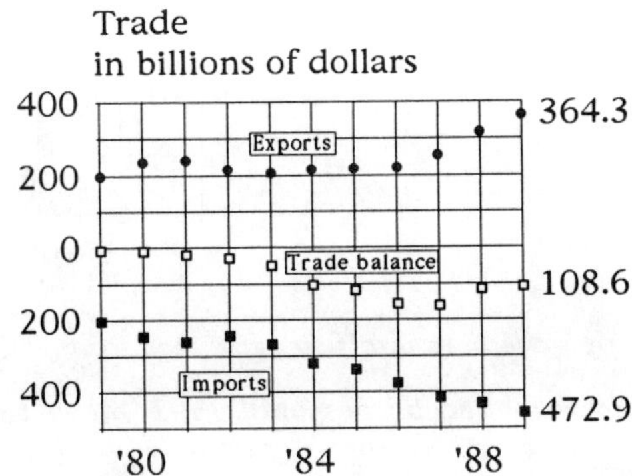

Figure 9. The merchandise trade balance over a decade.

In any given year, equation (1) states something obvious about subtracting two numbers to get a third; for example,

$$\text{Exports}(1989) - \text{Imports}(1989) = 364.3 - 472.9$$

$$= -108.6 = \text{Trade balance}(1989).$$

But by applying the subtraction to *functions* rather than numbers, we can express the entire relationship in Figure 9 by a single, simple formula.

---

[25] "U. S. trade deficit narrows" by Robert D. Hershey, Jr., *The News and Observer* (Raleigh, NC), February 17, 1990.

Exercises

8. *Figure 10 shows the market shares of U. S., Japanese, and European companies for "wafer steppers," devices that are important for manufacturing semiconductors.*[27]

(a) The figure shows the graphs of three different functions. What is the relationship among these three functions?

(b) What conclusions can you draw from the figure?

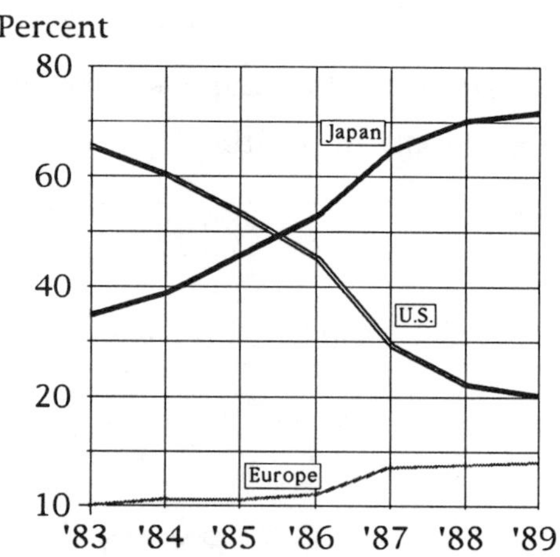

Figure 10. Market shares for wafer steppers: "X" marks the spot.

9. *Table 6 shows the number of Democratic governors in office each year from 1980 through 1993. All the other governors were Republicans until 1991, when two independent governors were inaugurated.*

Year	1980	1981	1982	1983	1984	1985	1986	1987	1988	1989	1990	1991	1992	1993
No.	31	27	27	34	35	34	34	26	26	28	29	27	28	30

Table 6. Numbers of Democratic governors.[28]

(a) Let $D(t)$ be the function of time represented by Table 6, and let $R(t)$ be the function whose value at time t is the number of Republican governors. What is the relationship between D and R, at least up to 1991?

(b) What is the domain of each of the functions D and R?

(c) Make a table of values of $R(t)$.

(d) Plot both functions, $D(t)$ and $R(t)$, on the same coordinate axes. What do you observe about these graphs, at least up to 1991?

---

[27]Adapted from "U.S. letting its technology slip away" by Andrew Pollack, *The News and Observer* (Raleigh, NC), December 17, 1989.

[28]Source: National Governors' Association, cited in *USA Today*, November 16, 1992, page 4A.

10. *Table 7 shows the revenues and net income of AMR Corporation, parent company of American Airlines, over the period 1985-1992.*

Year	Revenues (billions)	Net income (millions)
1985	6.1	346
1986	5.9	275
1987	7.2	201
1988	8.7	440
1989	10.2	415
1990	11.6	-50
1991	13.0	-220
1992	14.4	-935

Table 7. Revenues and net income of AMR Corp.[29]

(a) Plot each of these functions of time on separate coordinate axes.

(b) What would you predict about AMR's revenues in 1993?

(c) What were AMR's total revenues from 1985 through 1992?

(d) What was AMR's net income over the same period of time?

(e) Make up another question that can be answered from this data; answer it.

[29]Source: AMR Corp., cited in "American posts record loss," by Jim Barnett, *The News and Observer* (Raleigh, NC), Jan. 21, 1993.

11. *Table 8 shows numbers of highway fatalities and numbers of miles driven in North Carolina from 1986 to 1992.*

Year	Deaths	Miles driven (millions)
1986	1645	52.9
1987	1595	54.6
1988	1580	57.7
1989	1460	60.8
1990	1375	62.7
1991	1351	64.9
1992	1231	NA

Table 8. Highway deaths and miles driven in North Carolina.[30]

(a) Let $D(t)$ and $M(t)$ be the functions whose values are given in the second and third columns of Table 8, each as a function of time represented by the first column. Plot the functions D and M on separate coordinate systems.

(b) Estimate the missing entry in the Miles column, and add that point to your plot.

(c) Let $P(t)$ stand for the number of deaths per million miles. How is P related to D and M? Add a column to the table for $P(t)$.

(d) Plot the function $P(t)$. What conclusions do you draw?

(e) Make up another question that can be answered from this data; answer it.

(f) What do think might be the cause(s) of declining highway deaths in N. C.?

---

[30]Source: N. C. Highway Patrol, cited in "State's traffic deaths drop," by Allison Salerno, *The News and Observer* (Raleigh, NC), Jan. 1, 1993.

1.6 Inverse Functions

In Section 1.5 we stressed the importance of function-as-object because we will often want to do algebraic (and other) operations to functions. Here we single out one of those operations for special attention, namely the operation of "inverting" or "undoing" a function. As with the other concepts discussed in this chapter, this will be a frequently recurring theme throughout the course. We begin with an example that you may find of personal interest.

Car Payments

A car salesperson advises potential customers that the monthly payment on a new car depends on four factors: the price of the car, the size of the down payment, the annual interest rate, and the term of the loan. These quantities are related (for reasons we will see later) by the formula

$$p \; = \; \frac{(P - D) \, r \, (1 + r)^n}{(1 + r)^n - 1} \, , \tag{2}$$

where

p is the monthly payment,

P is the price of the car,

D is the down payment,

r is the monthly interest rate (as a decimal fraction), and

n is the number of months required to pay back the loan.

The interest rate and the duration of the loan are usually determined by the seller, and the down payment is agreed upon by the buyer and seller. Since these three quantities do not vary, the monthly payment p and the price P of the car are the only true variables in the formula. The price of the car determines the monthly payment, so price is usually thought of as the independent variable and the monthly payment is a function of the price.

Suppose you want to buy a $15,000 car, and you have $2400 for a down payment. The annual interest rate, because the dealer is trying move cars off the lot, is a very favorable 3.9% (which is equivalent to 0.325% per month), and the term of the loan will be 48 months. Using formula

(2), the salesperson determines that your monthly payment will be $283.93. Being a cautious consumer, you have your calculator in hand, and you check the salesperson's calculation. (Do it!) Thus, the pair of numbers (12000, 283.93) is one of the (price of car, monthly payment) pairs that make up the function defined by formula (2) when $D = 2400$, $r = 0.0075$, and $n = 48$.

Unfortunately, you cannot afford a monthly payment of $283.93. Your income and your other expenses limit you to a monthly payment of $230, so you will have to shop around to find a car you can afford. In this situation, the monthly payment is no longer the independent variable. Now the price you can afford depends on the monthly payment you can afford, and the roles of independent and dependent variables have been reversed. In effect, there is a new function involved in this problem, namely the pairing relation (monthly payment, price of car). This second function has a special — and obvious — relationship to the first function; the two functions are said to be "inverses" of each other.

Exercise 1. Substitute $p = 230$ in formula (2), along with the other known values, and solve for P to find out how expensive a car you can afford.

Exercise 2. Don't substitute $p = 230$ in formula (2), or the other known values; solve for P to find a formula for the inverse function, i.e., for P as a function of p (with D, r, and n still assumed to be constants).

Exercise 3. (a) When you substitute the known values for D, r, and n in formula (2), the resulting formula expresses p as a linear function of P; what is the slope of the linear function?

(b) When you solved for P in the previous exercise, you should have found that P is also a linear function of p. When you substitute the known values for D, r, and n in that formula, what slope do you get?

(c) What is the relationship between the two slopes in parts (a) and (b)?

Representing Functions and Their Inverses

A calculator is a useful tool for exploring the concept of inverse functions. Suppose you are doing some calculation, and the number 37.958127 is on your calculator display. If you were to accidentally press the square root key, the new display would be 6.16101671804. How could you retrieve your original display? It should be clear that pressing the x^2 key will achieve this result, since squaring "undoes" the effect of taking a square root. This sequence of events can be understood in terms of inverses. The square root key acts as a function that pairs the input 37.958127 with the output 6.16101671804. The number 6.16101671804 is then the input to a second function that is the inverse of the first; this second function squares its input, and so 6.16101671804 is paired with 37.958127. [Check the calculations in this paragraph on your own calculator.]

If the pairs of numbers that constitute a function are known explicitly (for example, by being listed in a table), it is very easy to display explicitly the inverse function: Just write all the pairs in reverse order (or interchange the columns of the table). We illustrate this with the drug-sales function that you studied in Exercise 7 in section 1.1. If you tabulated the data given graphically in Figure 7, your table might look like this:

x (Year)	y (Sales)
1985	22
1986	25
1987	28
1988	31
1989	35
1990	38

If you wanted to answer a question such as, "In what year did the sales pass $30 billion?", you could read the table backwards, or you could actually construct the inverse function:

y (Sales)	x (Year)
22	1985
25	1986
28	1987
31	1988
35	1989
38	1990

The graphs of a function and its inverse have a special relationship: For each point (a,b) on the graph of f, there is a point (b,a) on the graph of the inverse of f. The points (a,b) and (b,a) are "mirror images" in the line $y = x$ (see Figure 11). Thus, the graph of f can be reflected about the line $y = x$ to obtain the graph of the inverse of f. Notice that this reflection takes the y-axis to the x-axis and *vice versa*; that is, it interchanges x- and y-coordinates.

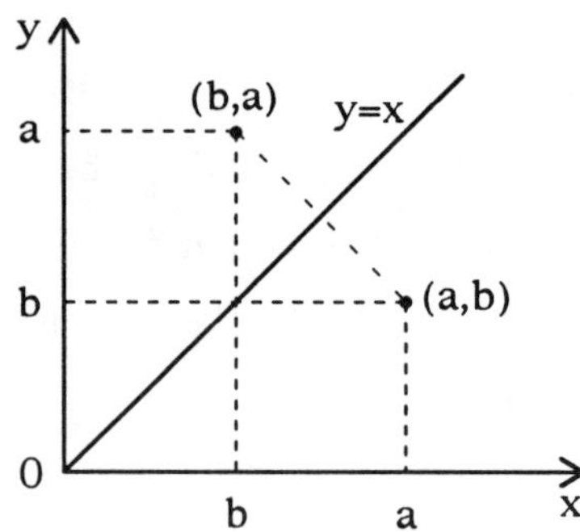

Figure 11. Mirroring in the line $y = x$.

We illustrate this mirroring with an entire figure (not just a point) by repeating Figure 7 along with its reflection (Figure 12); check that plotted points in the reflected graph match up with the second table above.

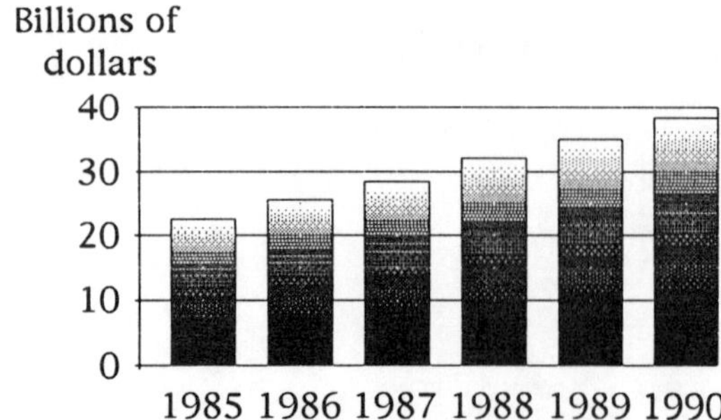

Figure 7R. Sales by the pharmaceutical industry in North Carolina.

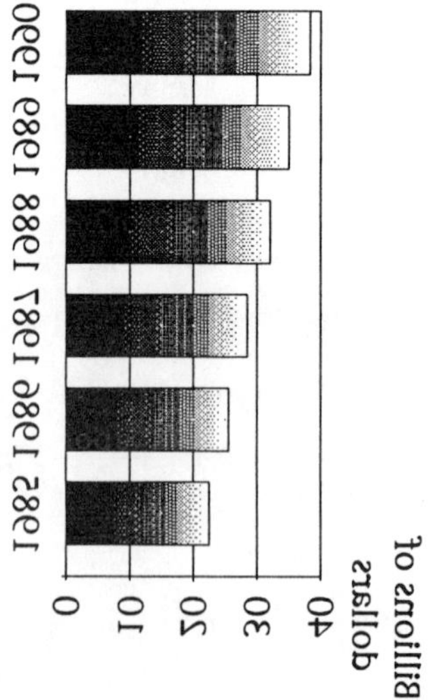

Figure 12. Reflection of Figure 7 about the line $y = x$.

Exercise 4. In Exercise 7 of section 1, you calculated the slope of the line representing the drug-sales function in Figure 7. What is the slope of the inverse function? How are these two slopes related?

Following the same geometric idea for inverting functions, we show in Figures 13 and 14 the graphs of $y^2 = x$ and $y = x^2$. Each of these equations is obtained from the other by interchanging x and y; each of the graphs is obtained from the other by reflecting in the line $y = x$, which amounts to the same thing. However, these figures reveal a new complication, suggested by the distinction between the dashed and solid parts of the two graphs. The upper (solid) part of the graph of $y^2 = x$ is also the graph of the (positive) square root function, $y = \sqrt{x}$. The inverse of this function is *not* the entire squaring function, but only the part shown with a solid curve in Figure 13. That is, the inverse of the square root function is the function defined by $y = x^2$, $x \geq 0$. Since the square root function has only positive numbers for its *outputs*, its inverse has only positive numbers for its *inputs*.[31]

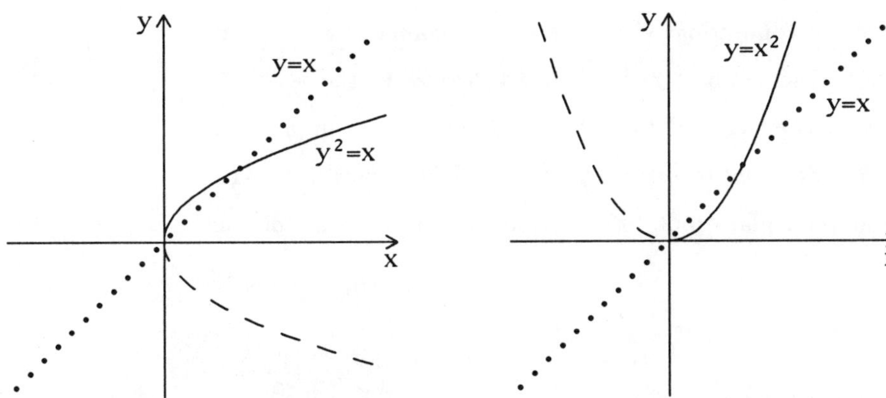

Figure 13. Positive and negative square roots.

Figure 14. Squaring positive and negative numbers (mirroring Figure 13 in the line $y = x$).

Exercise 5. Write an explicit formula for the function represented by the dashed-line curve in Figure 13. What is the inverse of this function? Where does it appear in Figure 14?

---

[31] The square root function also has only positive numbers for its *inputs*, but that fact is not related to the point we are making now.

Suppose we focus now on the *squaring* function in Figure 14. The domain for that function consists of all real numbers x, positive, negative, and in between. (What's in between?) That is, the entire graph in Figure 14, solid and dashed, is the graph of the squaring function. What's the inverse of this function? Well, if we invert all the pairs of numbers that represent points on the graph, we get all the points of the graph in Figure 13 — solid *and* dashed. That's a perfectly good *relation*, but it's not a function. (Why? Refer back to your criterion in Exercise 1 of Section 1.5.) That's the new complication: The inverse of a function may not be a function!

However, we can usually produce a function whose inverse is also a function if we restrict the domain in some appropriate way — such as considering the squaring function only for positive numbers. Your calculator already knows that, as we will see when we study its various inverse functions.

Inversion of functions has a standard notation, one that you have probably already used in connection with inverse trigonometric and inverse logarithmic functions: The symbol f^{-1} is used to represent the inverse of the function f. We hasten to add that you must interpret this notation with care: It looks like it *ought* to mean $\frac{1}{f}$, but it clearly *does not*. After all, the squaring and the square root functions are not *reciprocals* of each other.

Exercise 6. (a) If $f(x) = x^2$, what function is $\frac{1}{f}$?

(b) If $f(x) = \sqrt{x}$, what function is $\frac{1}{f}$?

(c) If $f(x) = x^3$, what function is $\frac{1}{f}$?

(d) If $f(x) = x^3$, what function is f^{-1}?

Logarithms

We take up now a very important function inversion related to the two functional equations we left dangling mysteriously in Section 1.5 at the end of the subsection on "Multiplicative Functions": We asked what functions "turn sums into products" and what functions "turn products into sums." Whatever the answers are, you should recognize now that the verbal descriptions of these two classes of functions are somehow "inverses" of each other. More specifically, if a given function is in the first of these classes, its inverse (if it turns out to be a function) will be in the second class — and *vice versa*.

You should know, from your previous study of algebra, at least one function that turns products into sums — that's the whole *point* of "logarithms," and you know from the function keys on your calculator that "log" is a function. You know "properties" of logarithms, one of which is

$$\log AB = \log A + \log B, \tag{3}$$

where A and B are any numbers that have logarithms. (What numbers have logarithms?) Equation (3) simply says that "log" is a function that satisfies the functional equation for turning products into sums. Perhaps more important for what we will do in the next chapter is the logarithmic "property" of turning exponents into factors:

$$\log A^B = B \log A. \tag{4}$$

But what's a "logarithm"?

There are, of course, many different "log" functions, one for each allowable "base." We recall the definition of **"logarithm base b,"** more or less as it appeared in your algebra or precalculus book. In deference to traditional textbook style, we even put a box around it:

$$\boxed{y = \log_b x \quad \text{if and only if} \quad x = b^y.} \tag{5}$$

What you should observe in this definition is that "taking the logarithm (base b)" and "exponentiation (base b)" are *inverse processes*; each undoes what the other does. Thus, if x is calculated as b raised to the y power ("exponentiation base b" applied to y), then y is recovered from x by taking the logarithm with the same base. Similarly, if y is calculated from x by taking a logarithm, then x is recovered from y by raising the base to the y power. The *meaning* of "take a logarithm" is essentially "undo exponentiation."

Exercise 7. (a) Check this with your calculator, using base 10. Enter 10, then a number y of your choice, and raise 10 to the y power. Then press the $\boxed{\log}$ or $\boxed{\log_{10}}$ key, and observe whether you get y back.
(b) Now enter a (positive) number x, and calculate $\log_{10}$ of it. Use the resulting number as an exponent, with 10 as base, and see whether you get x back.
(c) Repeat parts (a) and (b) with several other values of y and x.

Exercise 8. For any given base b, there are two pairs of numbers x and y that we can easily see will satisfy relationship (5). One pair is $x = 1$ and $y = 0$. What's the other? For future reference, write your answer here: ________________

Exercise 9. The fact that "base 10 logarithm" is defined as the inverse of "base 10 exponentiation" suggests that "base 10 exponentiation" may be an answer to the question about functions that turn sums into products. Show that this is the case. [Hint: Recall what you have learned about "properties" of exponents.]

In Figure 15, we show graphs of $y = 10^x$ and $y = \log_{10}x$. What do you observe about these graphs?

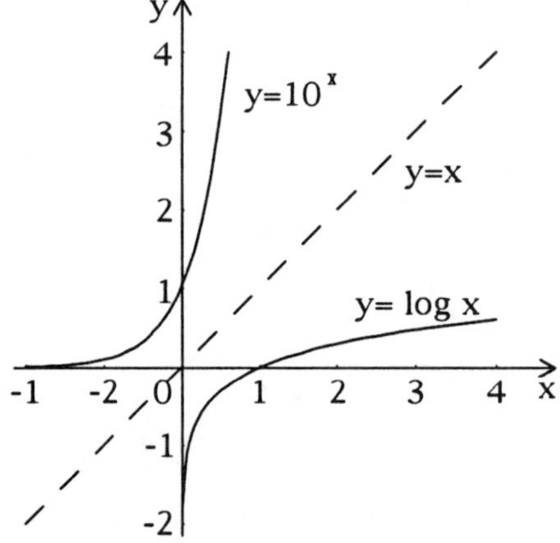

Figure 15. Exponentiation and logarithm base 10.

1.7 What's Significant About a Digit?[32]

You might wonder about this question when you are asked in this course (or others) to report your numerical answers to, say, "four significant digits."

Here's a true story about a student, a calculator, and a calculus exam. We will call the student "Anna" (not her real name); both the calculator and the exam shall remain nameless.

A question on the exam asked Anna to estimate the rate of change, on a very short interval starting at 0.5, of a function f known only by a calculator button[33] that had never been discussed in class.[34]

Being a good student, Anna knew exactly what to do. First, she chose her "very short interval" to be $[0.5, 0.501]$. Given that choice, the *rate of change* of the function over the interval can be calculated as the change in function values divided by the length of the interval:

$$\frac{f(0.501) - f(0.5)}{0.501 - 0.5} \; .$$

Anna's calculator told her that $f(0.5)$ was 0.523598775598 — which she wrote down as 0.524 — and that $f(0.501)$ was 0.524753861551 — which she recorded as 0.525. Next, she calculated the rate of change:

$$\frac{f(0.501) - f(0.5)}{0.501 - 0.5} = \frac{0.001}{0.001} = 1.$$

Actually, to her credit, Anna wrote the final step as " ≈ 1 " — she knew this was only an estimate, and that was all the problem asked for. However, by her calculator procedure, she forced the answer to have less accuracy than was available. If the problem had specified an accuracy, her answer might have been *wrong*.

---

[32]Based on an article with the same title by D. A. Smith in *The College Mathematics Journal* 20 (1989), pp. 136-139.

[33]It happened to be the "inverse sine" button, but that's not important for the story.

[34]In the next chapter, you will see that one of the important ideas in Calculus I is estimation of "instantaneous rate of change" of a function by calculating an "average rate of change" over a very short interval. The actual question asked for an estimate of the instantaneous rate of change at 0.5.

What was wrong with Anna's procedure? There was no need for her to write down, round off, or re-enter either of the function values. The calculator's value for $f(0.501) - f(0.5)$ (with two presses of the function key and one of "minus") is 0.001155085952, so the end of the calculation might have been

$$\frac{0.001155}{0.001} = 1.155,$$

an answer that differs from Anna's by about 15%.[35] In this case, there was no way Anna could have known exactly what she was trying to estimate,[36] but she *should* have known (more importantly, *you* should know) that sloppy use of meaningful digits can lead to wrong answers.

In this section we will explore some examples and exercises on "significance" and then abstract a test by which you can decide whether you are responding appropriately to assignments that call for numerical answers — which means *most* assignments.

There are at least these three reasons to be concerned about significance of digits: (a) the accuracy of available data; (b) the finite precision of your calculating device; and (c) loss of significance due to the way we manipulate the numbers. (Anna's calculation is in the third category.) We comment briefly on each of these and provide examples of each.

Accuracy of Measurements

In section 1.2, we pointed out several dichotomies by which we can sort our *functions* into various conceptual categories. For example, the position of a moving object is a "real" function of time that has an "exact" value at every instant of time — but we may "know" only an "observable" approximation to this exact value and only at selected instants of time. That is, we may have only a finite list of numbers, such as the distance numbers in Table 2. Those numbers are necessarily limited in significance

---

[35]This answer is also a much better estimate of the instantaneous rate of change of the function in question, which happens to be 1.15470.

[36]That's why we sometimes have to *estimate*!

by the precision of the device by which we made the measurements, and it would be silly to expect that we can ever gain additional significance by doing calculations on the numbers.

Here's an example. Suppose a population of 240,000,000 people grows by 2% per year for three years. What will the population be at the end of the three years? If P_0 stands for the starting population (i.e., $P_0 = 240,000,000$), then the population at the end of the first year will be $P_0 + .02 P_0$ or $P_0 \times 1.02$. Each year the population will get multiplied again by 1.02, so at the end of three years the population will be $P_0 \times (1.02)^3$. When we substitute 240,000,000 for P_0 and multiply by $(1.02)^3$, our calculator says the new population is 254,689,920. (Check this with your own calculator.) But the original "data" had only *two significant digits* ("2" and "4") — or maybe *three* (we can't be sure about the first "0") — and we can't possibly know more precise information about population at a later date simply by doing calculations with an abstract model.[37] Thus, the best "answer" to our question is "about 250,000,000" — an answer with the same number of significant digits as we started with. It would not be *wrong* to say "about 255,000,000," as long as we understand that there is no solid reason to *believe* the second "5". It definitely *is* wrong to report 254,689,920 as the answer, because that implies you know the population to a high degree of accuracy, and you certainly *don't*.

Moral: Calculation steps *never* add significant digits, so don't always believe all the digits you see on your calculator.

Exercise 1. Suppose the population of 240,000,000 continues to grow at 2% per year for 10 years. What would the population be at the end of that time?

---

[37]A substantial part of Chapter 2 will be devoted to "natural growth" models. In addition to the problem of how accurately we know (or can know) the starting population, there is another problem of how accurately the model can track the growth of a real population. That accuracy is also seldom better than two significant digits.

Accuracy of Calculating Devices

Some of our important functions are the mathematical models of reality that are defined by formulas — usually "continuous" formulas that give "exact" values for every input number, but which are at best approximations to the "reality" being modeled. Reality aside, it may be important to know something about "exact" values — and our ability to know those values may be limited by the accuracy of our calculator or computer.

Here's an example. If you ask your calculator for a value of $\log_{10}(3)$, it might respond with 0.4771212547 or .47712125472. (Try your own calculator — you may get some other response.[38]) The "leading zero" in the first answer is *not* significant; in fact, leading zeros are *never* significant. Whether you do or do not put a zero in front of a decimal fraction is a matter of style and clarity. The first answer has 10 significant digits (count them), and the second has 11. Actually, the second answer confirms for us that all 10 digits of the first answer are correct, but we have to *trust* the manufacturer of the second calculator that the 11-th digit is correct — or check it against some more accurate source. *Neither* of these answers is the exact value of $\log_{10}(3)$, because exact representation of that number as a decimal expansion would require *infinitely many* digits.

Moral: Finite machines (calculators and computers) cannot produce more significant digits than they have been programmed to produce. In particular, they cannot produce an exact numerical answer if that answer requires an infinite decimal expansion.

Without some clever programming, none of our calculators can give us more than 10 or 12 or perhaps 15 significant digits of a number such as $\log_{10}(3)$. There are some computer programs capable of "arbitrarily high precision" arithmetic; we asked one[39] for a 20-significant-digit value for $\log_{10}(3)$, and it responded 0.47712125471966243729. This shows us that (a) the 11-th decimal place is *not* "2", but (b) the 11-place calculators

---

[38]Our first answer is from a Casio calculator, and the second is from both Texas Instruments and Hewlett-Packard calculators.

[39]*Derive.*

responded with a value *correctly rounded* to 11 significant digits.

Exercise 2. *(a) Suppose you are working on a problem whose answer is* $\log_{10}(3)$, *and you have been asked to report your answer to* 7 *significant digits. What do you write down?*

(b) Give a 12-significant-digit approximation to $\log_{10}(3)$.

Loss of Significance

The information we get from our calculator or computer may be *corrupted* — in many different ways — by the sequence of operations we choose to perform. Sometimes this loss of significance is preventable, sometimes not. We have already given an example of *preventable* loss of significance in our tale of Anna and the exam problem.

Moral (from Anna's tale): Don't discard digits in an intermediate result. The only time you should round off is at the *end* of your calculation.

Exercise 3. *(a) Find an approximation to* $\pi/\sqrt{2}$ *that is as accurate as your calculator can provide. Write down all the digits. How many digits in your answer do you believe are correct?*

(b) Write down a five-significant-digit approximation to $\pi/\sqrt{2}$.

(c) Use the approximations $\pi \approx 3.14$ *and* $\sqrt{2} \approx 1.414$ *to calculate a decimal approximation to* $\pi/\sqrt{2}$. *How many digits of the answer are significant?*

Moral: You should not expect more significant digits in any answer than are in the *least* accurate input to the calculation.

Now a loss of significance that is *not* preventable. If you ask a 10-significant-digit calculator[40] for a value of π, it will respond 3.141592654, which is correctly rounded to nine decimal places. If you ask the same calculator for a value of $\frac{355}{113}$, you will see 3.14159292 — almost the same number.[41] What happens if we subtract the smaller number (π) from the larger $\left(\frac{355}{113}\right)$? If we were doing it by hand, the calculation would look like this:

---

[40]For example, a Casio.

$$3.141592920$$
$$- \ 3.141592654$$
$$\overline{0.000000266}$$

It looks like our answer has only *three* significant digits! And our calculator confirms that by reporting 2.66×10^{-7} as the answer from the subtraction. Subtraction of nearly equal numbers can be a "significance killer"; in particular, subtraction of two 10-significant-digit numbers that agree in the first seven digits produces an answer that has only three significant digits.

Moral: Watch out for "disastrous cancellations." If you can't arrange your work to avoid them, at least be aware that your numbers have fewer significant digits as a result.

We note in passing that Anna's subtraction of the nearly equal numbers 0.525 and 0.524 resulted in a cancellation that was "disastrous" for her: From numbers with only three significant digits, she ended up with an answer that had only one significant digit. However, if she had prevented the preventable part of the problem — loss from 12 down to three significant digits — it wouldn't have mattered much that she had an unpreventable loss of two digits.

Exercise 4. *(a) Use your calculator to evaluate the expression* $\sqrt{x^2 + 1} - x$ *at* $x = 42,709$. *How many digits of the answer are significant?*

(b) Evaluate the same expression at $x = 1,342,709$. *How many digits of the answer are significant?*

(c) The "disastrous cancellation" in parts (a) and (b) actually is avoidable, *but only by transforming the expression to another form. Rewrite the expression in part (a) by "rationalizing the numerator":*

$$\sqrt{x^2 + 1} - x \ = \ \frac{\sqrt{x^2 + 1} - x}{1} \ \times \ \frac{\sqrt{x^2 + 1} + x}{\sqrt{x^2 + 1} + x} \ .$$

Simplify the expression on the right. Evaluate the resulting expression at each of the numbers in parts (a) and (b). How many significant digits do you think each of your answers has?

---

[41] There is another "significance" issue here that is *not* the point of this example. If you count carefully, you will see that this last answer has only eight decimal places — nine significant digits. But we know the calculator carries 10 significant digits. A reasonable (and correct) deduction is that the ninth decimal place must be 0. Calculators often drop "trailing zeros" *even when those digits are significant.*

How to Tell Whether a Digit is Significant

The significance of digits in an approximation (e.g., in a calculator or computer answer) has to be measured relative to an "exact" value that is being approximated. Regardless of where the decimal point is located, we ignore 0's to the left of the first non-zero digit; they function only as place holders and do not tell us anything about digits of the number being approximated. Starting from the first non-zero digit on the left, we ask of each digit in turn how well it "matches" the corresponding digit in the "exact" number. We will say it "matches" if the *error* in the approximation is "less than 5 in the *next* place." If we are checking for a match in the "thousands" place, the "next" place, reading from left to right, is the "hundreds" place.

For example, if the *exact* number being approximated is 1,342,709, then 1,340,000 is a "3-significant-digit" approximation,[42] because the error (2709) is less than 5000. Similarly, 1,341,624 is also a "3-significant-digit" approximation, for exactly the same reason. (The fact that it has other non-zero digits is irrelevant; its error is less than 5000 but not less than 500.) The approximation 1,342,000 is yet another "3-significant-digit" approximation, but *not* a "4-significant-digit" approximation, because its error (709) is not less than 500. On the other hand, 1,343,000 *is* a "4-significant-digit" approximation, because its error $(\,|\,1{,}343{,}000 - 1{,}342{,}709\,| = 291)$ *is* less than 500. Thus, from the exact number, 1,342,709, we can report a 4-significant digit approximation (1,343,000) by *correct rounding* in the fourth place from the left.

Here is the formal definition of significance:

(a) Leading zeros are *never* significant.

(b) The k-th decimal digit (reading from left to right) of an approximation y to a number x is **significant** if it is not a leading zero, and $|\,x - y\,| < 0.5 \times 10^{-k}$.

(c) The approximation y to x has **n significant digits** if, after discarding leading zeros, the first n of its digits are

---

[42]Notice that trailing zeros are sometimes necessary, even when they are not significant.

significant in the sense defined in (b).

We will abbreviate the phrase "four significant digits" by "**4SD**," and similarly for other numbers of SD's. We provide some more examples to illustrate the meaning of SD, and then we ask you to check your understanding with some exercises.

A calculator approximation to $\pi/40$ is $0.078539816\ldots$. A 4SD approximation is 0.07854. Note that this requires *five* decimal places. Significance is not directly connected with numbers of decimal places.

A calculator approximation for $20013/10006$ is $2.00009994\ldots$. A 4SD approximation is 2.000. If you are asked for a 4SD answer, it does not make sense to report this answer as "2" or even as "2.0". The first suggests either an *exact* answer or a 1SD answer — only the context could make it clear which you intended. The second definitely suggests 2SD. Thus, sometimes trailing zeros are not only significant but are needed to convey that significance.

Exercises

5. *To four decimal places (5SD), π is 3.1416, and $\sqrt{2}$ is 1.4142. If you divide these approximations, how many significant digits of $\pi/\sqrt{2}$ will you get? (Don't guess — do the calculation to be sure.)*

6. *(a) Write down all the digits of your calculator's approximation to $9\pi/5$.*

   (b) Write down a 4SD approximation to $9\pi/5$.

   (c) Explain why 5.66 is not a correct 3SD approximation to $9\pi/5$. That is, explain why it is not legitimate to do repeated rounding. What is the correct 3SD approximation?

7. *Suppose your bank pays quarterly interest on savings accounts at an annual rate of 7%. That is, the interest added at the end of each quarter is calculated at a rate of $0.07/4 = 0.0175$.*

   (a) If you deposit $160 at the start of a year, how much money will be in your account at the end of the year?

   (b) How many significant digits are there in your answer?

SUMMARY

We started this chapter with "the big question": WHAT'S A FUNCTION? We have now answered that question in many different ways.

First, through a multitude of examples of functional relationships, we have tried to persuade you that "function" is linguistically distinct from "formula." Formulas often *do* define functions, but not always. On the other hand, functions may be defined by

- formulas,
- graphs (e.g., line graphs, scatter plots, bar charts),
- data tables,
- verbal descriptions,
- conceptual relationships,
- physical, biological, chemical (and other) relationships,

and perhaps in other ways as well. Functions are everywhere, and the *concept* is a powerful tool for dealing with the complexity we see all around us.

What is a function? It's a pairing of the values of one varying quantity with the values of another varying quantity in such a way that each value of the first variable is paired with *exactly one* value of the second variable. We can (and do) express that idea in terms of "inputs" and "outputs," in terms of "first column" and "second column," in terms of "x-coordinates" and "y-coordinates," and many other ways.

The *act* of pairing is a *process*, and it is often important to view functions dynamically, that is, as processes "doing something" to the input values to produce the output values. On the other hand, a function (that is, the collection of all the paired values) is an *object*, and thus we can "do things" to functions as well. In particular, functions can be treated as *algebraic* objects (albeit more complicated ones than the familiar numbers, literal constants, and variables). We can combine functions by the algebraic operations of addition, subtraction, multiplication, and division. We can form new functions from old ones by, for example, taking negatives, reciprocals, and square roots. And we have the very important operation of *taking the inverse* of a function, that is, of acting on the function *as object* to reverse the function *as process.*

As we will do with many topics in this course, we have viewed inversion of functions three ways: *symbolically, numerically,* and *graphically.* Each of the three ways is important, and we review them here.

Symbolic inversion. If a function is defined by a formula, $y = f(x)$, then inversion means solving this equation in the two variables x and y for x. The result will have the form $x = g(y)$, where g is (we hope) a function — in which case, g is the inverse of f. If we want to represent inputs to g by x and outputs by y, then we can interchange the symbols x and y. The algebraic condition for whether g is a function or not is whether each y (output from f) corresponds to a unique x (input to f).

The algebraic calculation of an inverse function may be "conceptual" — which means we know nothing more explicit than what we just said in the previous paragraph. Or it may be "actual" — that is, we may actually *do* the algebra to solve for x in terms of y. Sometimes that's not too hard, but, in general, solving equations for variables that don't appear linearly can be very hard work; indeed, it may be *impossible*, even if there is good reason to believe an inverse function exists. To put it bluntly, a function defined by a formula may have an inverse for which *no formula exists* (in any reasonable sense).

Numerical inversion. This means at least two different (but closely related) things. In the car-loan and calculator-square-root examples, we did numerical calculations to find single values of x when we knew the values of y. That's often easy — with the help of a calculator. And when it's not easy, it's almost always *possible* — often with the help of a computer. (This is a subject we shall take up later in the semester.) On the other hand, a function may be "known" only as a table of data. In that case, inversion is *trivial* — just interchange the columns in the table.

Graphical inversion. This is almost as easy as inverting a data table: Just "flip" the graph over the line $y = x$. When you draw graphs with a computer or calculator, you should see just how easy this is.

We learned in this chapter that all of the following may all be "modeled" or represented by formulas:

- relations between real-world variables that definitely *are not* functions, and
- relations that *may not be* functions, and
- relations that *definitely are* functions but have no apparent formulas.

So far, all of our examples but one have used *linear* functions as models; the exception was the physical data on a dropped object, for which the position function appeared to be modeled by a quadratic function. Much of this course will be about ways to model "real-world" functions that lead to computable solutions of problems and to insights about the varying phenomena.

In our study of functions, we have encountered a number of dichotomies, trichotomies, and other distinctions — some subtle, some obvious — that will color almost everything we do in this course.

We can classify our functions by a trichotomy: *real, observed, model.* These functions have the characteristics summarized in the following table.

<u>Type</u>	<u>Characteristics</u>
Real	Conceptual
	Not completely knowable
	Not expressible by formulas
Observed	Sampled from real relations
	Finite in extent or number
	Not expressible by formulas
Model	Approximate fit to real or observed functions
	Expressible by formulas (usually)
	Amenable to exact calculation (often)

We can classify our variables by a dichotomy: *discrete* or *continuous.* Some varying quantities "jump" from one value the next, like a digital clock; those variables are discrete. They may have only a finite number of distinct values, or they may have infinitely many. Other varying quantities change "smoothly," in such a way that there is never a "next" or a "previous" value. For example, we usually think of time proceeding in this way, and we model this change by the second hand on an analog clock.

The two clock types highlight the fact that we often model a single varying quantity (time, in this case) by *both* discrete and continuous variables. Our ability to move easily between discrete and continuous models will give us a broad range of conceptual and computational tools for representing and attacking the problems posed in this course.

Our final section of the chapter offered some cautions about the use of digital devices — calculators and computers — as tools for representing numerical values as they arise in this course. Most such devices present every "answer" with the same number of digits, whether or not those digits are meaningful. It is *your* responsibility — not your calculator's — to keep track of significant digits and to avoid the errors that arise from using either too few or too many digits at various points in your problem-solving processes.

Perhaps the most important message of this chapter is that all of this effort is *for something*. We are going to deal with (and solve!) problems that you will recognize as being important in a variety of different ways. In the process, we will see the power of mathematical abstraction to isolate essential features of a problem, to clear away irrelevant clutter, to lead to recognition of or reduction to a problem whose solution is already known. Our first step in that direction has been to come to grips with the simple, but extremely powerful, concept of FUNCTION.

Practice with Calculations

*Exercises in this category include (a) calculations that can be done by machines and (b) practice on important topics from courses that precede calculus. You need to develop **facility** with both categories — not because such calculations are a central feature of the course, but because they should not frustrate you or keep you from concentrating on the more important parts of the course by occupying a lot of your time or by leading to lots of mistakes. Even though routine calculations can be done quickly and accurately by computer or calculator, you need to develop judgment about when to use a machine and when not to, and you need to know how to tell when you might have pressed the wrong button. These skills are acquired and sharpened by **practice**.*

You should expect to see exercises like these as some portion of your homework assignments, quizzes, and tests.

1. Plot the points $(1,1)$ and $(-1,-2)$. Draw a line through these points. (Each block is one unit square.)

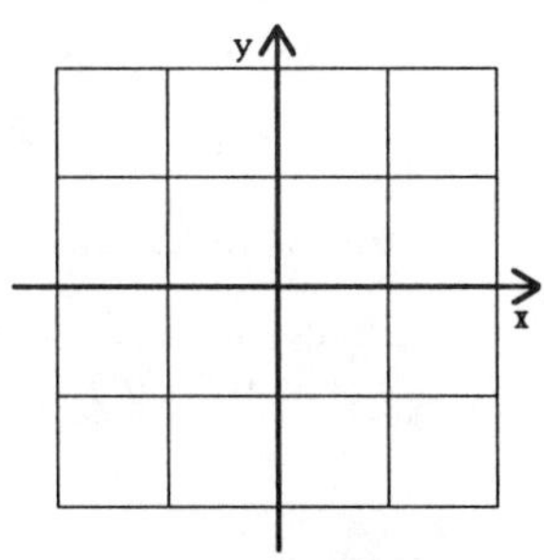

2. Plot the points $(1.5, 2)$, $(0.5, 0)$, and $(-0.5, -2)$. How many lines can you draw through pairs of these points?

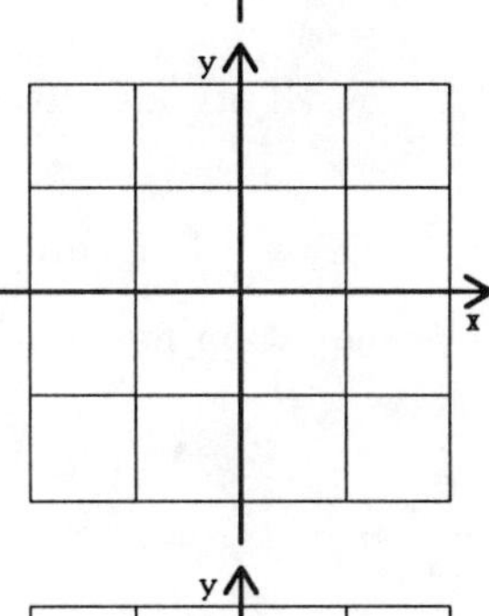

3. Draw a line through the origin and the point $(-2, 1)$.

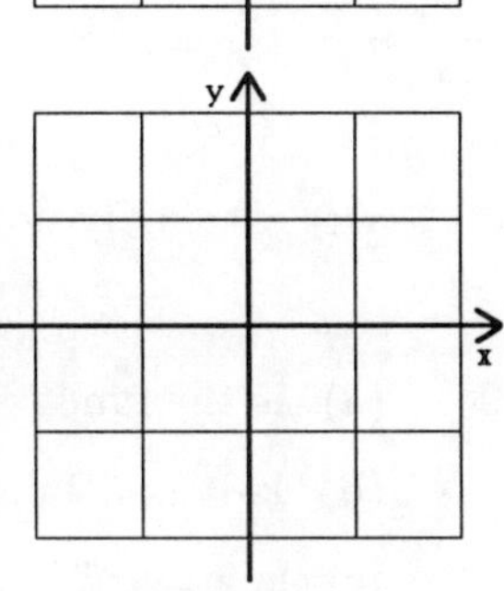

4. Plot the points $(-2, 1)$, $(1, -0.5)$, and one other so that the three points are vertices of a right triangle. Draw the triangle.

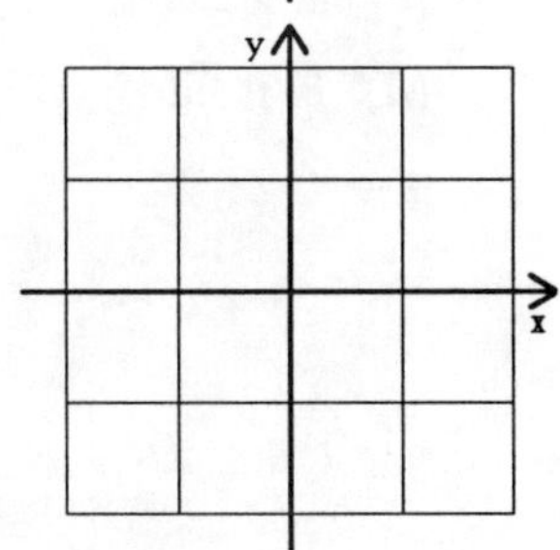

5. Draw a line through the origin with slope $\frac{1}{2}$. (Each block is one unit square.) Find an equation for your line.

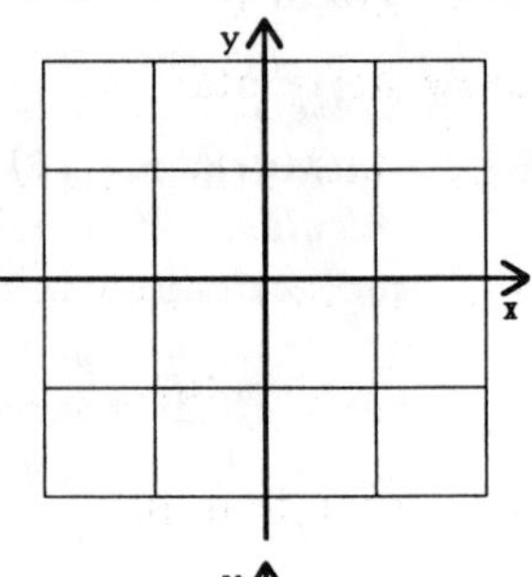

6. Draw a line with slope $\frac{3}{2}$ through the point $(-1, -0.5)$. Find an equation for your line.

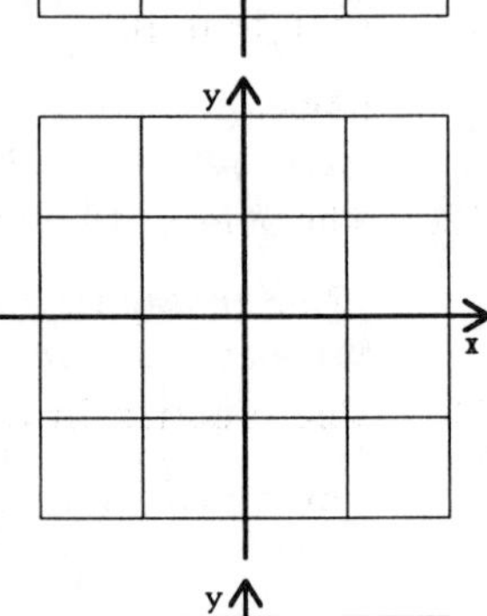

7. Draw a line with slope $-\frac{1}{2}$. Find an equation for your line.

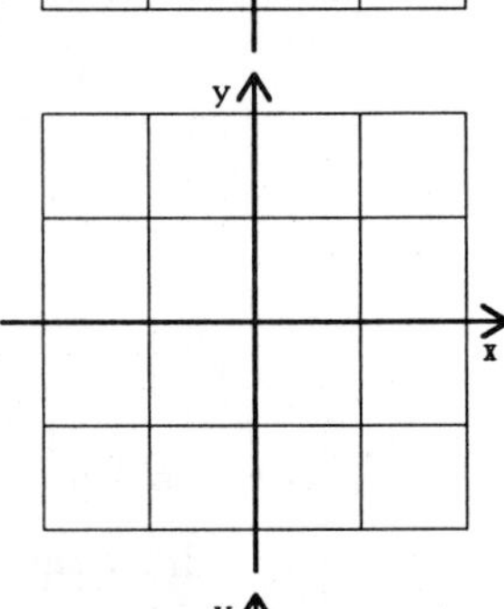

8. Draw a line through the points $(-1, -0.5)$ and $(1.5, 2)$. Find an equation for your line.

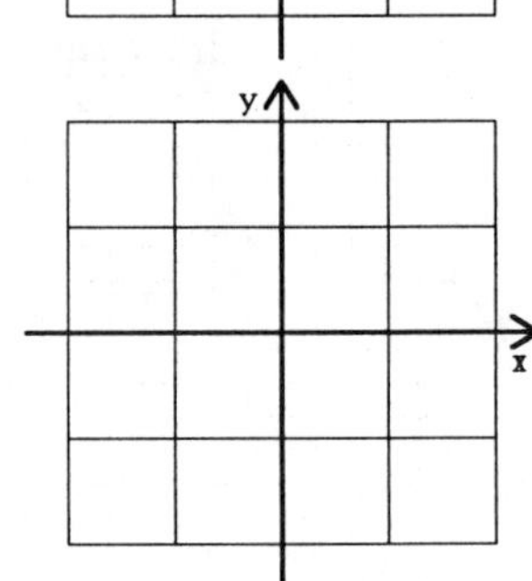

9. (a) Is the function $f(x) = x^2 + 7$ additive? Why or why not?
    (b) Is it multiplicative? Why or why not?
    (c) Is it even? Why or why not?
    (d) Is it odd? Why or why not?

The "graph paper" below has 50 blocks in each direction. Use a straightedge to draw horizontal and vertical axes through the center of the grid. You will use this grid for the next four exercises.

10. Draw the line through the points $(2.8, 2)$ and $(-3.7, 5)$. Find an equation for this line.

11. Plot four points on the line whose equation is $4x - 3y = 9$. Draw in the line.

12. The line in the previous exercise is the graph of a function; what function? (Find a way to describe the function in words and/or symbols.)

13. The function in the previous exercise has an inverse function. Describe it in words and/or symbols. Draw its graph.

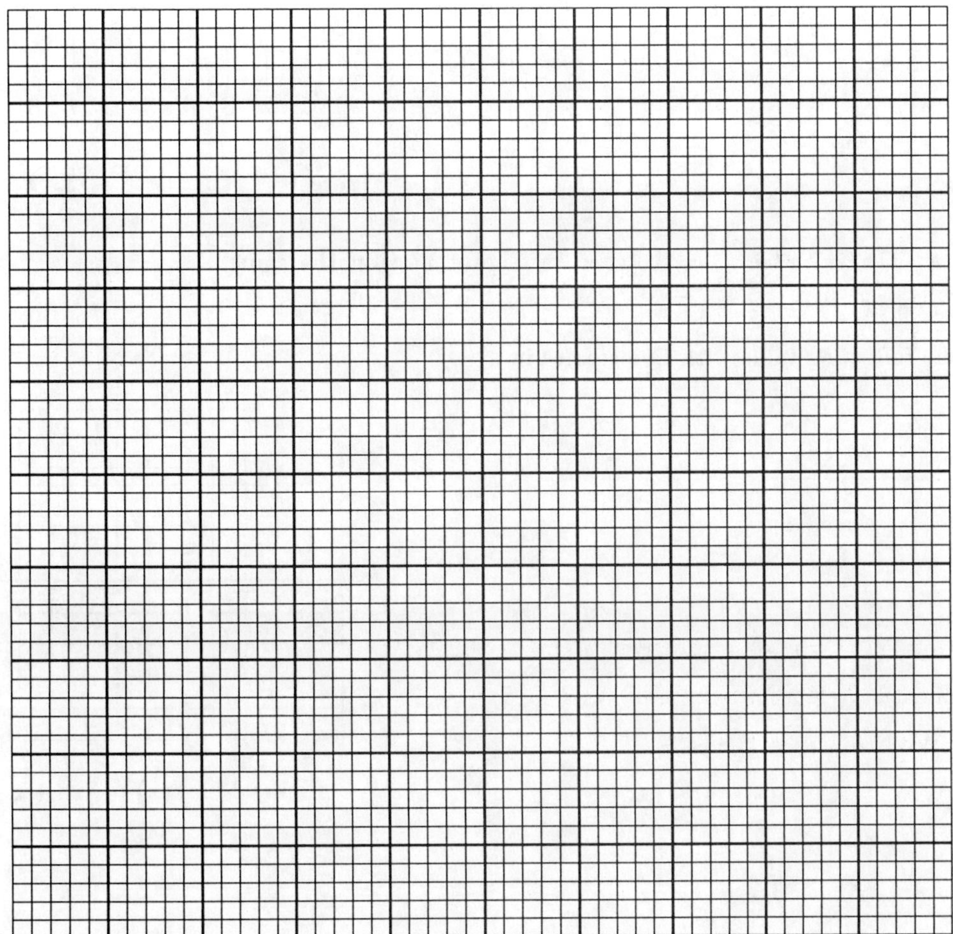

14. Some of the values of a function $f(x)$ are given in the following table:

x	0	1	3	6	8
$f(x)$	0	2	3	4	5

Sketch a graph of the inverse of the function f.

15. (a) The function $f(x) = x^{2/3} + 3$, $x \geq 0$, is sketched in Figure 16; on the same axes, sketch a graph of $f^{-1}(x)$.

(b) Find a formula for f^{-1}.

(c) Complete the following table with five entries in each column:

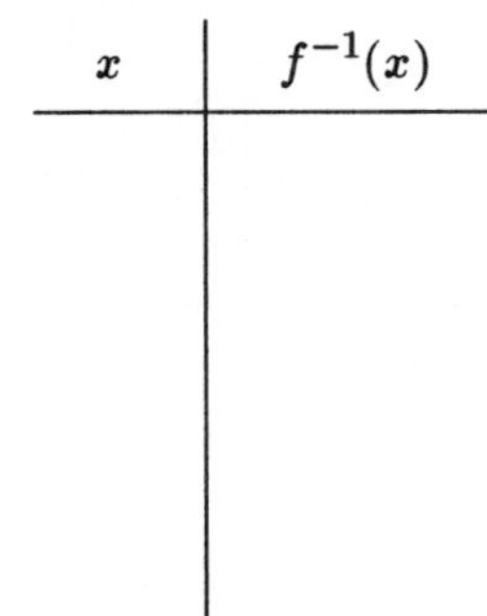

x	$f^{-1}(x)$

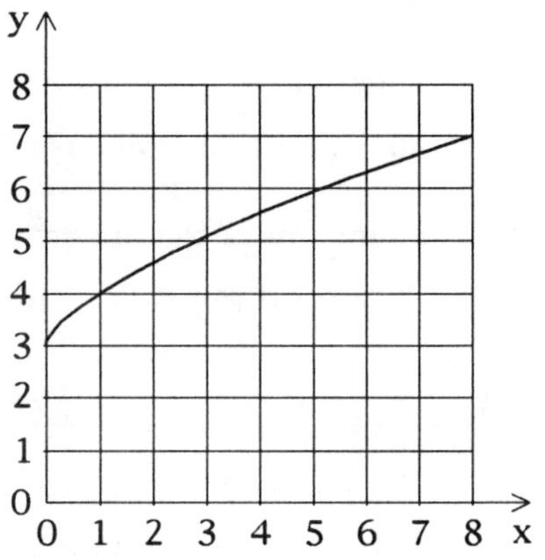

Figure 16. $f(x) = x^{2/3} + 3$, $x \geq 0$.

16. (a) The function $f(x) = \frac{1}{8}x^3 + x$ is sketched in Figure 17; on the same axes, sketch a graph of $f^{-1}(x)$.

(b) Complete the following table with five entries in each column:

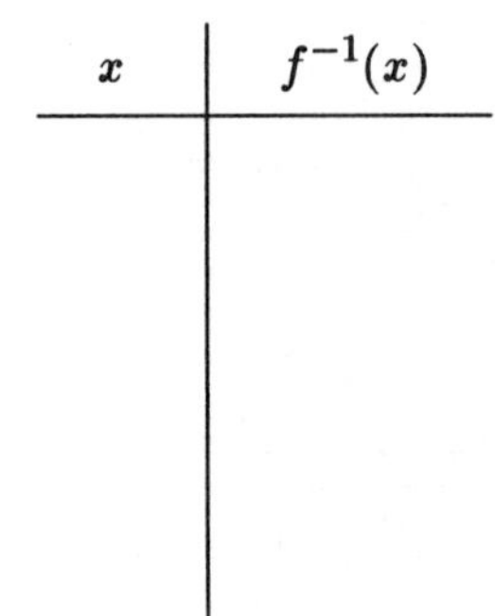

x	$f^{-1}(x)$

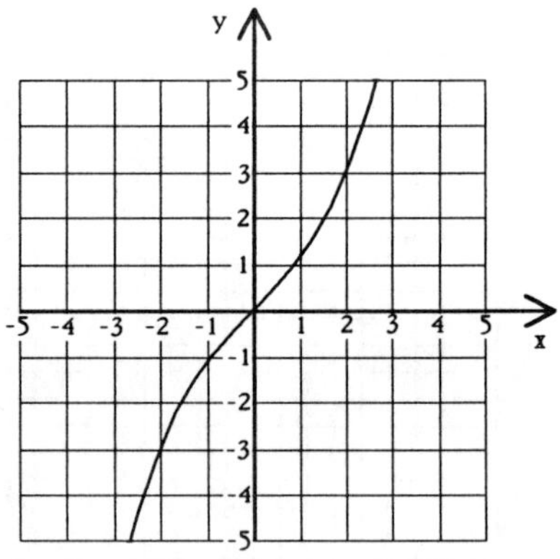

Figure 17. $f(x) = \frac{1}{8}x^3 + x$.

17. (a) A function $f(x)$ is graphed in Figure 18; on the same axes, sketch a
graph of $f^{-1}(x)$.

(b) Complete the following table with five entries in

each column:

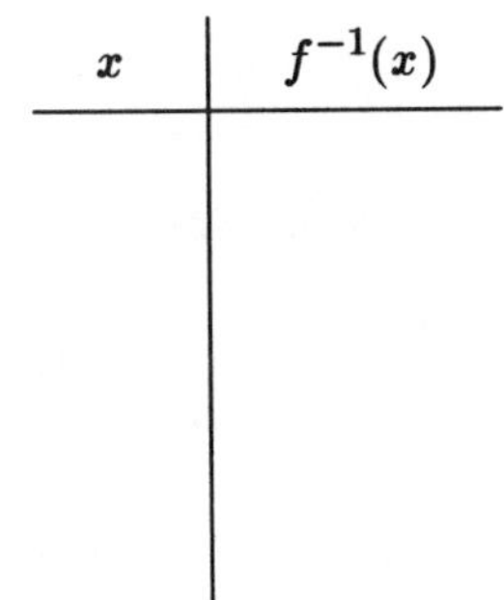

x	$f^{-1}(x)$

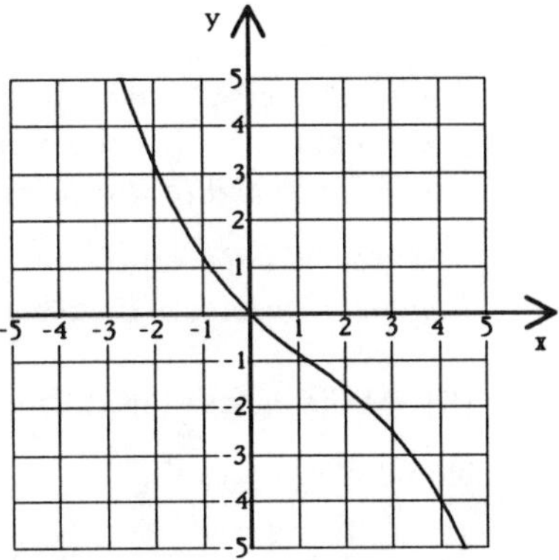

Figure 18. A function $f(x)$.

18. (a) A function $f(x)$ is graphed in Figure 19; decide whether or not this
function has an inverse, and give a reason for your answer.

(b) The function f has what kind of symmetry?

　　Circle one: Odd   Even   Neither

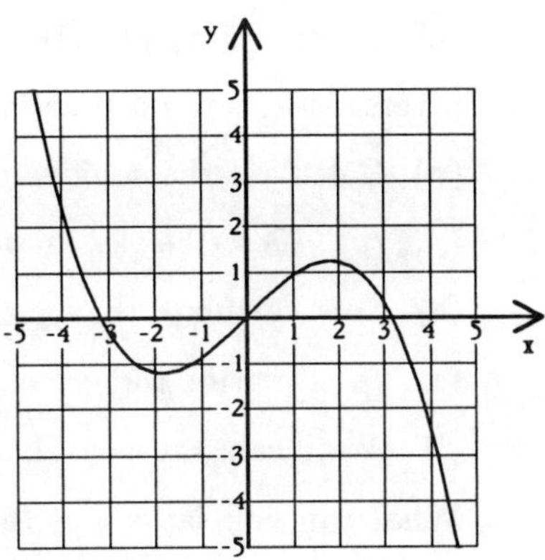

Figure 19. A function $f(x)$.

Conceptual Exercises

Exercises in this category use and/or further develop the concepts introduced in this chapter. The level of difficulty is similar to that of the exercises embedded in the text and at the ends of sections. While many of these exercises are presented in "realistic" contexts, they are not representative of real problems. Rather, these exercises are to help you get ready for tackling real problems.

You should expect to see exercises like these as a major portion of your homework assignments, quizzes, and tests.

1. (a) Make a scatter plot of the data in the following table. Describe the shape of the graph.

x	0.8	1.5	3.2	2.6	1.9	2.4	3.5	0.6	2.1
y	0.7	2.1	10.5	6.8	3.5	5.9	12.4	0.3	4.3

(b) Explain why the table defines a function. What is the domain of the function?

(c) Construct a table that defines the inverse function. Explain why the inverse is actually a function. What is the domain of the inverse function?

(d) Calculate the square roots of the y-values, and graph points of the form $(x, \sqrt{y})$. Describe the shape of this graph.

(e) Now calculate the squares of the x-values, and graph points of the form (x^2, y). Describe the graph.

(f) What can you say about the relationship between the last two graphs? What can you say about the relationship between x and y?

2. (a) Sketch a graph of speed as a function of time for a typical trip to get a pizza (one way or round trip, but be sure to explain your graph).

(b) Sketch a graph of distance traveled for the same run, also as a function of time.

3. What function of time is measured by the odometer in a car? Is the measurement discrete or continuous? Is the function itself discrete or continuous? Explain your answers in each case.

4. (a) Explain why the data in the table below defines a function. What is a function of what?

(b) Choose appropriate axes and scales for plotting the data on the 50×50 grid below. Plot the ten points.

Time (seconds)	Distance (meters)
1.0	5.0
2.0	19.4
3.0	44.1
4.0	78.0
5.0	122.8
6.0	175.8
7.0	240.0
8.0	312.8
9.0	396.1
10.0	489.0

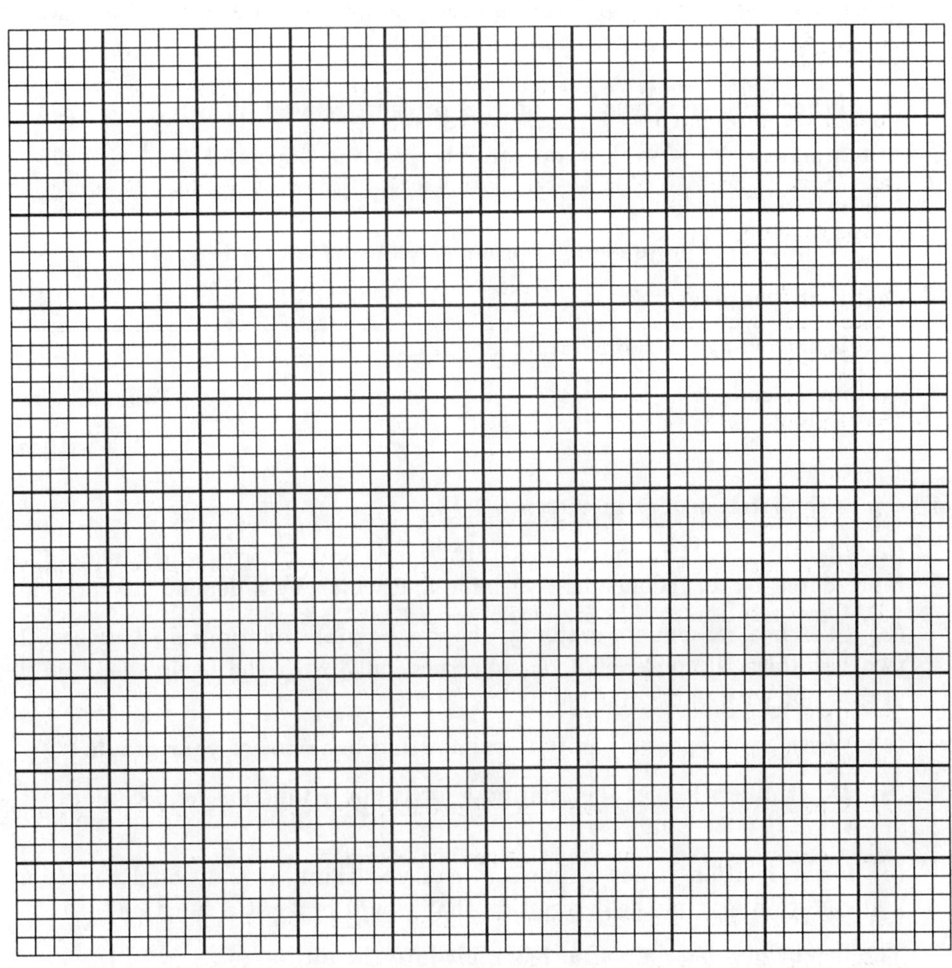

5.[43] Consider the situation described in this sentence: *The child's temperature has been rising for the last two hours, but not as rapidly since we gave her the antibiotic an hour ago.* Sketch a graph of temperature as a function of time, consistent with the situation described.

---

[43]Adapted from *Calculus Problems for a New Century*, edited by Robert Fraga, MAA Notes Vol. 28, 1993.

6.   The data shown in Table 9 were collected by chemistry students, who varied the temperature of a gas in a closed container and recorded the pressure exerted by the gas.

(a) Make a scatter plot of the data, with temperature graphed on the horizontal axis and pressure graphed on the vertical axis.

(b) When volume is held constant, is pressure related linearly to temperature? If so, how do you interpret the slope and y-intercept of the line?

(c) What change in pressure is brought about by a one-degree change in temperature?

(d) What change in temperature would cause a one-millimeter change in pressure?

Temperature (degrees Kelvin)	Pressure (mm mercury)
263	752.2
268	755.3
278	776.5
293	811.3
298	834.4
303	839.7
308	853.6
318	891.5
323	906.1

Table 9. Pressure vs. Temperature with constant volume

For each of the relationships in exercises 7-14:

(a) Identify the two quantities that vary, and decide which should be represented by an independent variable and which by a dependent variable.

(b) Sketch a reasonable graph to show a possible relationship between the two variables. (You should be concerned only with the basic shape of the graph, not with particular points.)

(c) Write a sentence to justify the shape and behavior of your graph.

(d) Determine whether the relation is or is not a function, and explain why you think it is or is not.

(e) If the relation is not a function, can you think of additional conditions you could impose on the description that would make it a function?

7.   The height of a baseball after being hit into the air.

8.   The speed of an egg dropped from the tenth floor of a building.

9.   The height of an individual as he or she ages.

10.   The amount of time it takes a person to run a mile as he or she ages.

11.   The amount of money in a savings account over an extended period of time.

12.   The size of a person's vocabulary from birth onwards.

13.   The number of bacteria in a laboratory culture over a period of time.

14.   The daily high temperature at your local airport over a calendar year.

15. (a) Sketch a graph of the relationship presented in Table 10.

    (b) Identify dependent and independent variables.

    (c) Discuss any conclusions you can draw from the shape of your graph.

Weight	Rate		Weight	Rate
not more than 1 oz.	$0.25		greater than 6 and not more than 8 ozs.	$1.10
greater than 1 and not more than 2 ozs.	0.45		greater than 8 and not more than 10 ozs.	1.20
greater than 2 and not more than 3 ozs.	0.65		greater than 10 and not more than 12 ozs.	1.30
greater than 3 and not more than 4 ozs.	0.85		greater than 12 and not more than 14 ozs.	1.40
greater than 4 and not more than 6 ozs.	1.00		greater than 14 and not more than 16 ozs.	1.50

Table 10. Third Class Postage Rates

16. (a) Sketch a graph of the relationship presented in Table 11.

    (b) Identify dependent and independent variables.

    (c) Discuss any conclusions you can draw from the shape of your graph.

Taxable Income	Income Tax
$0 but less than $10,000	$0
$10,000 but less than $50,000	15% of income over $10,000
$50,000 or more	$6,000 + 27% of income over $50,000

Table 11. Federal Income Tax

17. Some thermometers measure temperatures on both Fahrenheit and Celsius scales. These side-by-side scales suggest a functional relationship.

    (a) Define Fahrenheit temperature as a function of Celsius temperature by a simple formula.

    (b) Now express Celsius temperature as a function of Fahrenheit temperature.

    (c) Graph both of these functions. Would you describe either of these functional relationships as "linear"? Why or why not?

    (d) Is there a temperature that is assigned the same number on both the Fahrenheit and Celsius scales? If so, what is it, and how do you find it? Is it a high temperature, a low temperature, or an in-between temperature?

18.   On the grid below, make a scatter plot to analyze the relationship between rebounds and assists for the basketball data provided in Table 3. Start by labeling and scaling your axes. Use the graph to comment on the important characteristics of the relationship.

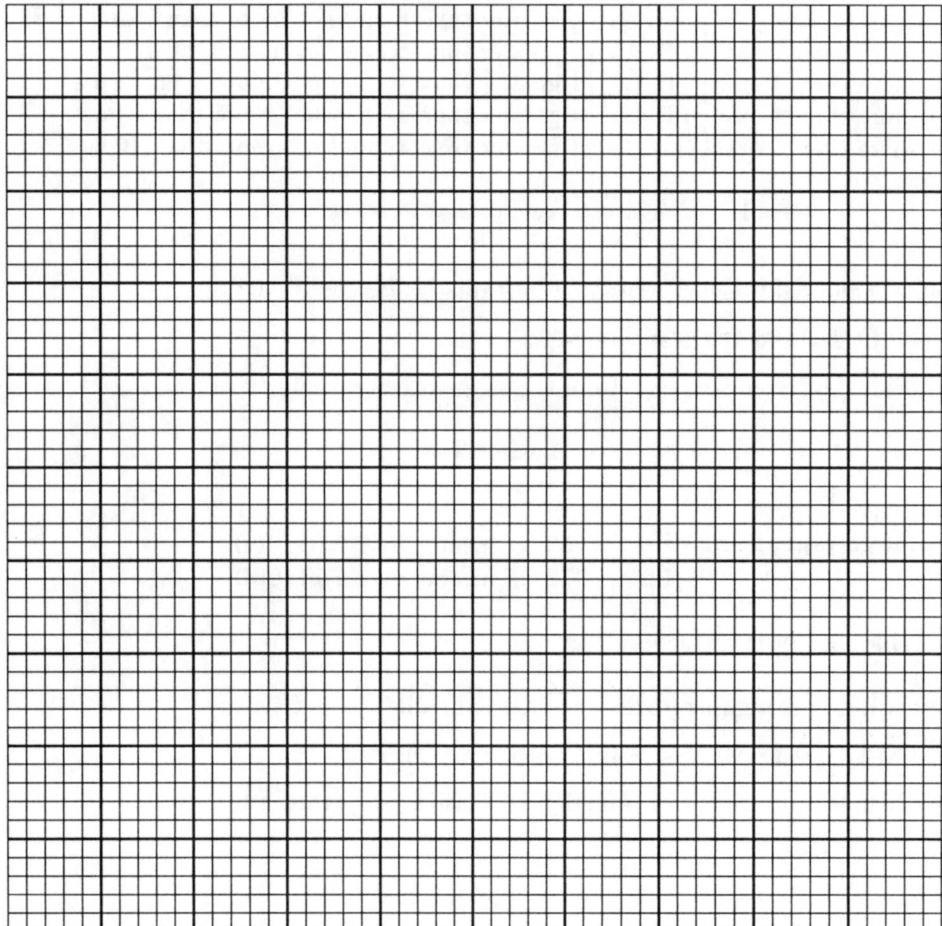

19.   In Section 1.5, you explored the concepts of "additive" and "multiplicative" functions.

(a) Define what it would mean for a function to be "subtractive." Can you find any functions that satisfy your definition?

(b) Define what it would mean for a function to be "divisive." Can you find any functions that satisfy your definition?

20. (a) Experiment with your calculator to determine the smallest positive number it can recognize and display.

(b) Similarly, determine the largest number it can recognize and display.

(c) Estimate the total number of different numbers your calculator can recognize and display. Is this number finite and moderately large, finite and enormously large, or infinite? Why?

21. (a) Your calculator has a lot of keys called "function keys." Discuss the extent to which each such key satisfies our description of a function.

(b) What is the domain of the $\boxed{\sin}$ function key?

(c) What is the domain of the $\boxed{\sin^{-1}}$ function key?

(Your answers to these questions may depend on what calculator you have. Don't try to "look up" the answers; experiment with the calculator itself.)

22. (a) Use the graphical information in Figure 6 (repeated below) to sketch your own idea of the "real" U.S. population function over the continuous time interval from 1789 to the present.

(b) Do you see your graph as being continuous or discrete? Why?

(c) Now imagine a view through a high-powered microscope of a very small portion of your graph, say, between 2:30 P.M. and 3:17 P.M. last Wednesday. Sketch what you think this microscopic view should look like. Is this graph continuous or discrete? Why?

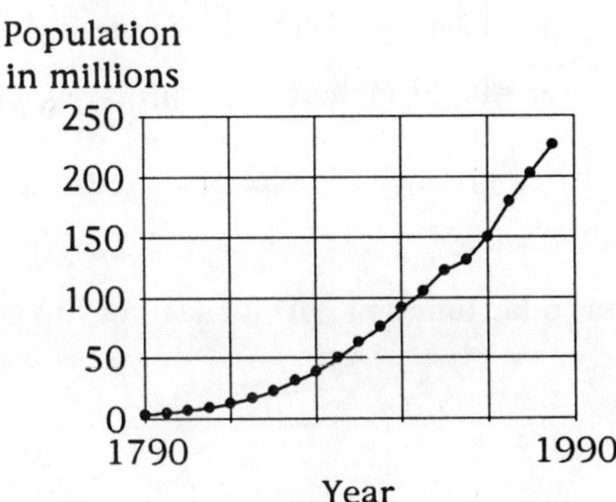

Figure 6R. U.S. population, 1790–1980.

23.[44] Figure 20 shows the number of deer in a forest at time t years after the beginning of a conservation study.

(a) During which of the following time periods did the deer population decline at the rate of 50 deer per year?

    (i) 1 to 2   (ii) 1 to 3   (iii) 1 to 4   (iv) 2 to 3   (v) 5 to 6

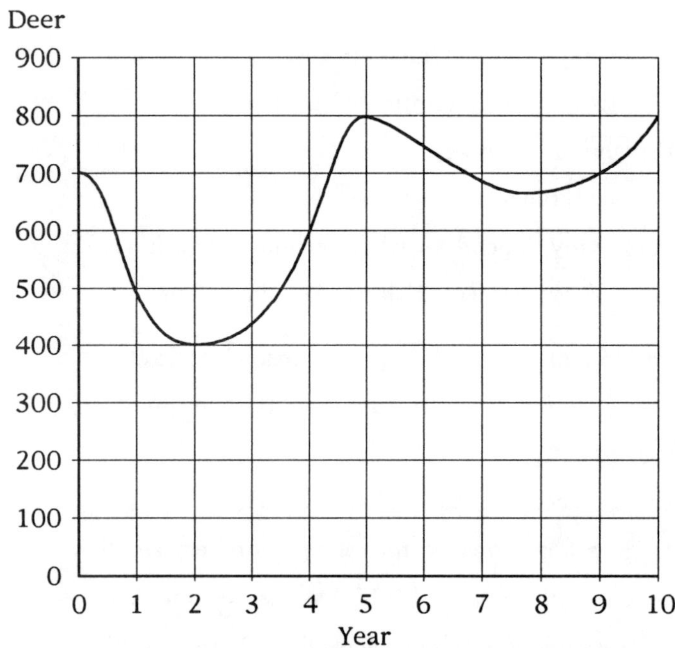

Figure 20. Data from a conservation study.

(b) When was the deer population increasing most rapidly?

(c) Approximately how fast was the deer population increasing or decreasing at time $t = 1.5$? Your answer should include appropriate units.

24. (a) Explain why an even function (if its graph has more than one point) cannot have an inverse function.

(b) Explain why the inverse of an odd function (if it has an inverse function) is odd.

---

[44]Adapted from *Calculus Problems for a New Century*, edited by Robert Fraga, MAA Notes Vol. 28, 1993.

25. Table 12 shows the total number of people in the armed forces and the percentage of women, over the period 1984-1992.

Year	Total number (millions)	Women (%)
1984	2.14	9.5
1985	2.15	9.8
1986	2.16	10.1
1987	2.17	10.2
1988	2.13	10.4
1989	2.12	10.7
1990	2.04	11.0
1991	2.03	10.9
1992	1.76	11.6

Table 12. Total number in the armed forces and percentage of women.[45]

(a) Let $T(t)$ be the total number function and $P(t)$ the percentage of women function. Plot each of these functions of time on separate coordinate axes.

(b) Let $W(t)$ be the number of women in the armed forces at time t, and let $M(t)$ be the number of men in the armed forces. How are the functions W and M related to the functions T and P?

(c) Add columns to Table 12 for the functions W and M, and fill in the numbers.

(d) Plot the functions W and M on separate coordinate axes.

(e) In what year between 1984 and 1992 were the largest number of women in the armed forces? In what year were the largest number of men?

(f) Make up another question that can be answered from this data. Then answer your question.

---

[45]Sources: Department of Defense, Defense Advisory Committee on Women in the Services, cited in "Up in Arms," by Rick R. Smith, *The News and Observer* (Raleigh, NC), April 11, 1993.

26. Table 13 shows data on total applications and applications by blacks to the Duke University Graduate School in the years 1985 and 1992.

	Black student applications	Total applications	Black students enrolled	Total enrolled
1985	67	3190	26	1715
1992	205	6654	71	2297

Table 13. Applications and enrollments, Duke University Graduate School.[46]

(a) How did the percentage of black applicants admitted compare with the percentage of all applicants admitted in 1985? In 1992?

(b) In 1985, what percentage of applicants were black? In 1992?

(c) In 1985, what percentage of students enrolled were black? In 1992?

(d) Suppose the numbers of applicants and of black applicants grew linearly from 1985 to 1992. How many applications were there in 1990, and how many of these came from black students?

(e) Make up another question that can be answered from this data. Then answer your question.

---

[46]Source: "Diversifying Duke's ivory tower," by Craig Whitlock, *The News and Observer* (Raleigh, NC), Dec. 20, 1992.

27.[47] A "flow restrictor" is a low-cost device (essentially a beveled washer) that can be inserted in a shower head to slow the rate of water flow. In this problem, you will use some facts and some assumptions to determine the significance of using such a device.

Facts: The flow restrictor reduces the water flow rate through the shower head by 42%. A gallon of water weighs 8.36 pounds. Heating a pound of water $1°F$ requires one British thermal unit (Btu) of heat energy. Electric energy is measured in kilowatt-hours (kwh); one kwh is equivalent to $3,412$ Btu's. An electric power plant emits 1.5 pounds of carbon dioxide for every kwh of energy produced. At 33% efficiency, a power plant produces twice as much waste heat as useful energy.

Assumptions: A family averages four showers a day, each of five minutes duration. The shower flow rate without the restrictor is six gallons per minute. The family's electric water heater heats water from $60°F$ to $120°F$. Their cost for electricity is 10¢ per kwh.

(a) How much water can this family save in a year by using the flow restrictor?

(b) How much energy will they save in a year?

(c) How much money will they save in a year?

(d) What will be the yearly reduction in carbon dioxide emissions?

(e) What will be the yearly reduction in thermal pollution from the power plant?

(f) If this family is your family, how many years of flow-restricted showers would it take to pay for one year of your college tuition?

---

[47] Adapted from "Quantitative Aspects of an Energy Conservation Device," by J. R. Shanebrook, *NLA News*, May, 1991.

Problems and Projects

The problems presented here are not intended for individual homework assignments or for tests. These problems should be attempted by groups of two to four students sharing ideas, whether in the classroom or elsewhere. Some of the more extensive problems are suitable for projects with time frames ranging from a class period to a week.

1. A student who was helping at a yard sale watched as a customer looked through the pile of jeans for sale. The man would bend his hand back at the wrist, bend his arm at the elbow, and then wrap the waist of the pants around his forearm. The man explained that his waist was the same size as his forearm, so he never needed to try slacks on for size. The student decided to gather some data and model the relationship between forearm circumference (along the arm) and waist size. The scatter plot in Figure 21, with forearm circumference (measured in inches) on the horizontal axis, is a graph of his data set.

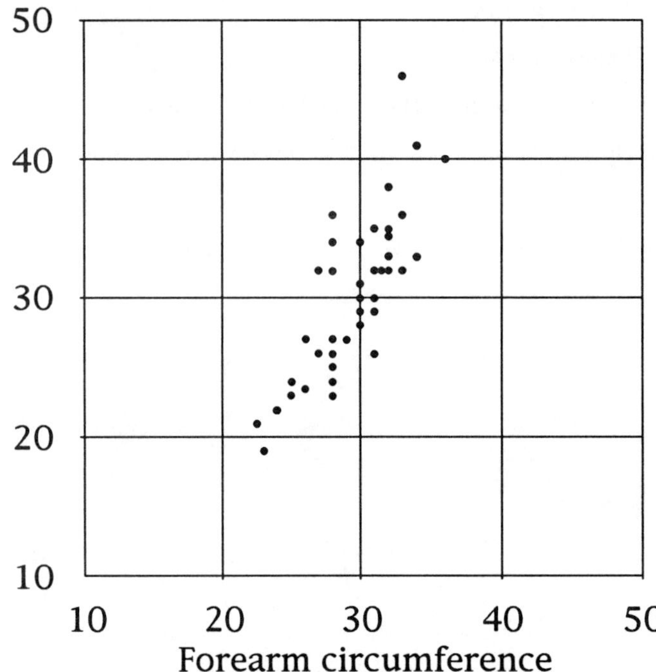

Figure 21. Waist measurement (inches) versus forearm circumference (inches).

(a) Find an equation of the form $y = mx + b$ to "fit" the points. (You might want to experiment by moving a clear ruler or dark thread through the scatter plot until you find the line that seems to fit the points best.) After you find an equation of the line you feel is best, comment on the criteria you used to determine this line. Compare your line with those found by other students.

(b) What information is provided by the slope of the line you found in Part (a)? What information is provided by the y-intercept? Use the model to predict your waist size. How accurate is the prediction? What message or advice would you give to a person whose data point lies above the line? What message or advice would you give to a person whose data point lies below the line?

(c) In Parts (a) and (b), you considered two different relations: that represented by the data plotted in Figure 22, and that represented by your fitted linear equation. For one of these relations, but not for the other, each value of the independent variable determines a unique value of the dependent variable. Which of the two relations has this property? Why? Which of the two relations (if either) is a function?

2.[48] (a) It appears from the data in Table 14 that average income in the U. S. depends on the level of education a person has achieved. In particular, the table suggests three distinct functional relationships with average income as a function of years of education, one each for males, females, and the combined average. State the domain of each of these functions. Make a separate table to represent each of the three functions. (Assume that a high school dropout has completed 10 years of education.)

Education	Men	Women	Combined
High school dropout	$13,655	$7,004	$10,326
High school graduate	$21,583	$11,143	$15,886
College degree	$37,002	$19,215	$28,406
Graduate degree	NA	NA	$38,604

Table 14. Average income by educational level.

---

[48]This problem is based on the article "Education Does Pay!" in the February 1992 issue of *TI-81 Newsletter*. The editors credit Karen Longhart of the Montana Project as the source of the ideas. The data is from a survey of 58,000 American households by the Census Bureau of the Department of Commerce.

(b) Make a scatter plot for each of your three functions. (If you put all three plots on a single coordinate grid, use different point symbols to distinguish them. You may do the plotting on graph paper or use a graphing calculator or a computer.)

(c) For each of the three average income functions, find a linear function that approximates the data reasonably well. What do the slopes of these linear functions represent?

(d) What do you think a person with two years of college could expect to earn on average? An 11th grade dropout? A person with five years of post-secondary education? (How do your answers depend on gender?)

(e) Estimate the missing entries in the table.

3. The Hanford, Washington, Atomic Energy Plant has been a plutonium production facility since World War II, and some of the wastes have been stored in pits in the same area. Radioactive waste has been seeping into the Columbia River since that time, and eight Oregon counties and the city of Portland have been exposed to radioactive contamination. Table 15 lists the number of cancer deaths per 100,000 residents for Portland and these counties.[49] It also lists an index of exposure that measures the proximity of the residents to the contamination. The index is formulated on the assumption that county or city exposure is directly proportional to river frontage and inversely proportional both to the distance from the Hanford site and to the square of the county's (or city's) average depth away from the river.

The scatter plot in Figure 22 has the exposure index on the horizontal axis. Find an equation of a line that seems to best fit the data in the scatter plot. What information does the algebraic equation provide? In particular, what is the significance of the y-intercept in the equation? Explain why it should give the cancer death rate when the index of exposure is zero, that is, when there is no radioactive contamination. Does it seem to give a reasonable value for the number of cancer deaths per 100,000 residents in an area without exposure to radioactive contamination?

---

[49] The data for this problem is taken from an article by Robert Fadeley in the *Journal of Environmental Health,* May-June 1965, Volume 27, Number 6, pages 883-897.

What values can each of the variables assume? What real numbers are not reasonable values for each of the variables? What is the slope of your fitted line? What is its significance in terms of the situation being modeled?

County/City	Index	Deaths
Umatilla	2.5	147
Morrow	2.6	130
Gilliam	3.4	130
Sherman	1.3	114
Wasco	1.6	138
Hood River	3.8	162
Portland	11.6	208
Columbia	6.4	178
Clatsop	8.3	210

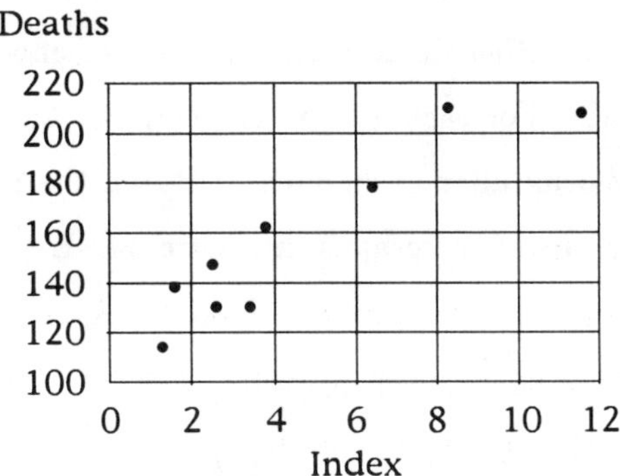

Table 15. Cancer deaths in Oregon.   **Figure 22. Scatter plot of data in Table 15.**

4. The graph shown in Figure 23 represents the "math pipeline." It shows the number of ninth grade students studying mathematics in 1972 and the numbers of the same students at various stages of preparation for and achievement of a Ph.D. in the mathematical sciences. The graph appears to show a nearly linear relationship. What quantity do you think is linear as a function of time?

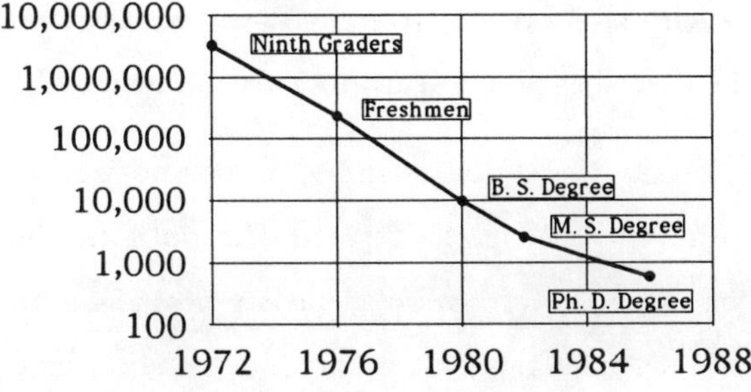

Figure 23. The Math Pipeline.[50]

---

[50]From *Everybody Counts: A Report to the Nation on the Future of Mathematics Education*, National Academy Press, 1989, page 6.

5.[51] (a) A sociological study examined the process by which doctors decide to adopt a new drug. The doctors were divided into two groups. The doctors in Group A had little interaction with other doctors and so received most of their information via mass media. The doctors in Group B had extensive interaction with other doctors and so received most of their information via word of mouth. For each group, let $f(t)$ be the number who have learned about a new drug after t months. Match the graph of $f(t)$ for each of the groups to one of the graphs in Figure 24. Explain your choice.

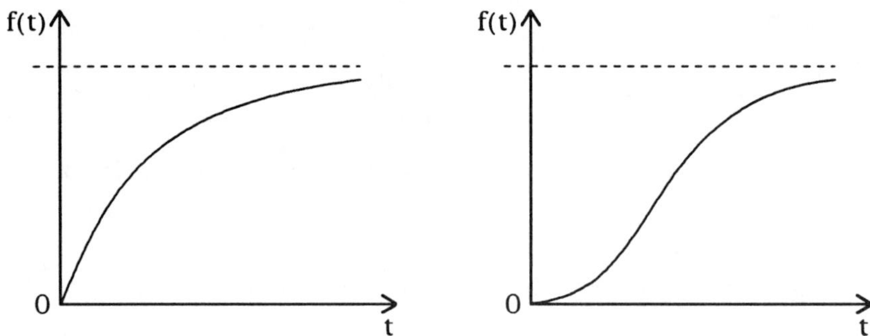

Figure 24. Spread of new information.

(b) The graphs in Figure 24 also describe the learning curves of two different types of jobs. If t is the time on the job, $f(t)$ describes how much of the required job skills the individual possesses. One type of job is skilled or semi-skilled, such as "operator of a word processor." The other type is primarily unskilled, such as "French fry chef" at a fast-food restaurant. Which of the graphs describes which learning curve, and why?

---

[51]Adapted from *Calculus Problems for a New Century*, edited by Robert Fraga, MAA Notes Vol. 28, 1993.

6. According to Hooke's "law," the force (weight) required to stretch a spring beyond its "natural" length is proportional to the distance stretched (see Figure 25).

(a) Express by a simple formula the force as a function of distance stretched.

(b) Sketch the graph of this formula.

(c) What happens to the spring when its "elastic limit" is exceeded? Hooke's law approximates the force required for stretching only up to the spring's elastic limit; add to your sketch what you think the force function looks like beyond that elastic limit.

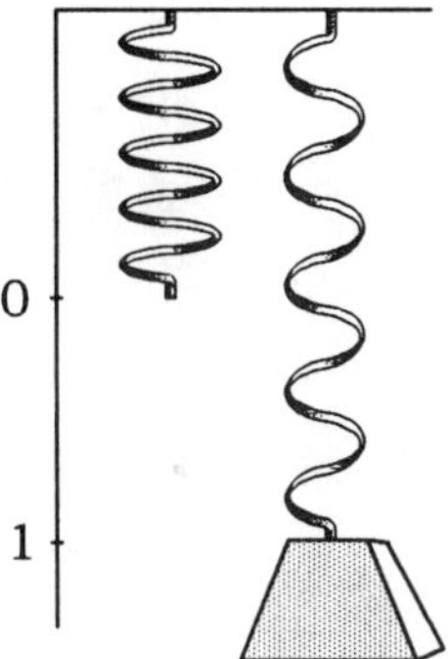

Figure 25. Stretching a spring.

CHAPTER 2

Rates of Change: Models of Growth

Writing in 1798, the British economist Thomas Malthus made some dire predictions about the human population of the Earth. His concept of the problem is shown in Figure 1: He observed that the food supply was growing only linearly and the population at a much faster rate. Thus, even if the food supply were more than adequate at the time, population growth would soon outstrip the ability of people to feed themselves. Unless some man-made or natural disaster wiped out large portions of the population first, widespread famine would be the inevitable result. Malthus may well have been wrong about both the food supply and the population. Nevertheless, we have been hearing, at least since the 1960's, even more dire predictions about overpopulation early in the 21st century, so it is clearly important to know what growth rates tell us about population size and overuse of essential resources.

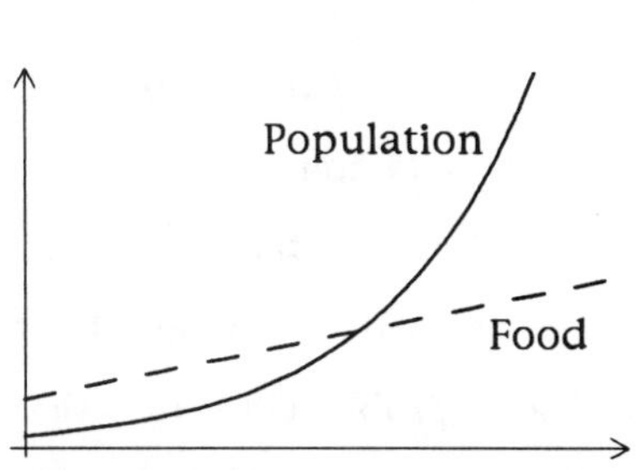

Figure 1. Malthus's view of population and food supply.

To analyze the population problem as Malthus saw it, we must understand and compare the *growth rate* of the world's human population and the *growth rate* of the food supply. Growth rates are examples of *rates of change* of functions — not all functions of time *grow* all the time — and the study of *rates of change of functions* is the subject of the first course in calculus.

We have stated the population problem because it is a real problem for our generation and because it serves nicely as a prototype for problems that can be understood, analyzed, and solved[1] by methods of calculus — but not by methods of algebra or other mathematics typically studied before calculus.

---

[1]We don't mean to say that mathematics has a solution for what to do about the world's burgeoning population — only that we can solve a mathematical formulation of the growth rate problem to find a function that models population growth.

Many other important phenomena can be described and analyzed by studying rates of change. For example, if an object falls from a height, both its distanced traveled and its speed are growing functions of time. Our first two examples happen to come from the fields of population biology and physics of motion; as the course progresses, we will encounter examples from engineering, chemistry, economics, psychology, and other areas. The common language of the sciences and social sciences is mathematics, and a key part of that mathematics is calculus.

We will discover very quickly that the problem of analyzing population growth is *hard* — or at least it *looks* hard, because the mathematics is relatively unfamiliar. We will actually start our study of growth rates on much more familiar ground: *linear* functions, such as the one Malthus envisioned as a model for the food supply. Next, we will study the falling body problem, because the functions involved and their rates of change are still relatively familiar. Later in the chapter, when we are ready to apply what we have learned about rates of change to population growth, we will put to work (in several different ways) the "inverse pairs" of exponential and logarithmic functions that we used as examples in Chapter 1. In particular, we will see that simple but reasonable assumptions about biological growth lead to conclusions about populations growing "exponentially." We will also see that logarithms provide us powerful tools for determining whether *data* on the growth of a population does or does not represent exponential growth.

2.1 Rates of Change

Sarah stretched luxuriously in the familiar bed, enjoying the fact that she didn't really have to get up now. It was great to be home from college, her first year successfully behind her, a week of laziness before she started her summer job. Right now her mother was probably fixing her something special for breakfast — toast, eggs, bacon. She watched the dust motes dancing in the shaft of June sunlight. Something was wrong. She didn't smell any bacon, or anything else cooking for that matter.

As Sarah came into the kitchen she confirmed that, indeed, nothing was cooking. Her mother was sitting at the breakfast table studying a large book.

"Pour yourself some cereal, Dear. The water on the stove is hot if you want a cup of coffee."

Sarah poured bran flakes in a bowl, added raisins and milk. Without much enthusiasm, she made a cup of instant coffee. What was her mother up to this time? As long as Sarah could remember, her mother had been learning about something or other. She had never finished college; as if to compensate, she had spent the rest of her life studying one thing after another — Latin, Elizabethan drama, micro-economics. What was it now?

*As she folded up the newspaper, she glanced at the book: **Calculus with Analytic Geometry**. Why in the world did she want to study calculus?*

"I'm so glad you're up. There is so much in this book that I don't know where to start."

"Why do you want to learn calculus? You are not thinking of going into engineering are you?"

Her mother looked at her sharply. Sarah's father was an engineer, and her mother was always asking questions — usually good questions, often difficult ones. Occasionally her father became exasperated and said that, if she wanted to know, she should become an engineer.

"No," her mother responded, "I'll leave that to your father. I want to understand what Malthus was talking about. How do they make these predictions about population growth? You know about all this I guess." She shoved a magazine article in Sarah's direction.

Writing in 1798, the British economist Thomas Malthus made some dire predictions about the human population of the Earth. His concept of the problem is shown in Figure 1: He observed that the food supply was growing only linearly and the population at a much faster rate. Thus, even if the food supply were more than adequate at the time, population growth would soon outstrip the ability of people to feed themselves. Unless some man-made or natural disaster wiped out large portions of the population first, widespread famine would be the inevitable result. Malthus may well have been wrong about both the food supply and the population. Nevertheless, we have been hearing, at least since the 1960's, even more dire predictions about overpopulation early in the 21st century, so it is clearly important to know what growth rates tell us about population size and overuse of essential resources.

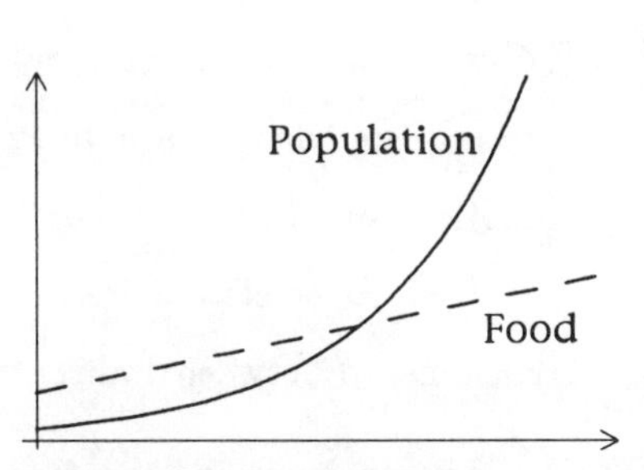

Figure 1. Malthus's view of population and food supply.

Her mother continued. "I have been sifting through these articles on population growth models; they all use differential equations, initial value problems. So I need to know what differential equations are and how they are connected to population growth."

Sarah sighed; her cereal was becoming soggy, her coffee getting cold. Her mother wanted to talk about calculus! She had just finished a year of calculus. It had been OK; the labs were sometimes fun, but writing those reports was hard work. She was ready for some relaxation — nothing more strenuous than reading Jane Austen. Still, her parents were paying for her education. And she had agreed to help in the calculus lab next year. That meant she was going to have to explain things to the new students; she might as well try first at home.

Her mother turned the pages of the text. "There are so many little sections. What do I really need to know?"

"Well, you need to start by understanding what a derivative is and how to calculate a derivative from a formula."

"Then what is a derivative? How is it related to the way things change?"

*"Derivatives are rates of change, instantaneous rates of change. I'll go get my **Calculus Reader**; I know where things are in that book."*

Sarah went up to her room, dragged a somewhat battered book out of a stack dumped on the floor, and went back downstairs. "Here, start by reading this section."

Average Rates of Change

We start with a remembrance of things past. The magazine article mentioned Malthus's belief in *linear* growth of the world's food supply. In Chapter 1 (including its end-of-chapter exercises and problems), we observed several linear or near-linear functional relationships: sales of prescription drugs as a function of time, the speed of a falling object as a function of time, car payment as a function of sticker price, the force required to stretch a spring as a function of displacement, and the *exact* relationship between Fahrenheit and Celsius temperature measurements.

Linear functions are familiar from your previous study of mathematics; you associate them with *linear equations*, such as

$$y = mx + b, \tag{1}$$

where x is the "input" (also called the "independent variable"), y is the "output" (also called the "dependent variable"), m is the "slope" (about which more very soon), and b is the "y-intercept" (i.e., the value of the function when $x = 0$). You further recall that

$$\text{slope} \ = \ \frac{\text{rise}}{\text{run}}. \tag{2}$$

That is, the slope of the line connecting two points (x_1, y_1) and (x_2, y_2) is the ratio described by *each* of the following expressions (all of which say the same thing):

- rise over run,
- change in y over change in x,
- $y_2 - y_1$ over $x_2 - x_1$.

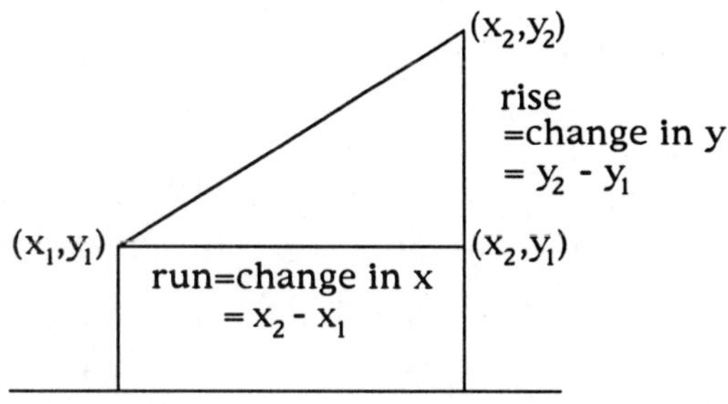

Figure 2. Slope equals rise over run.

Exercise 1. In the space below, explain why the slope of a line does not depend on which two points are selected for the computation. In other words, why is the ratio of rise to run the same no matter where you start and no matter how long the run is or in which direction? (This property is the essence of linearity.)

Exercise 2. On the first set of coordinate axes below, draw a line with slope $\frac{1}{4}$. On the second, draw a line with slope $-\frac{3}{2}$. On the third, draw a line with slope $-\frac{1}{3}$; on the fourth, a line with slope -3. (Each block is one unit square.)

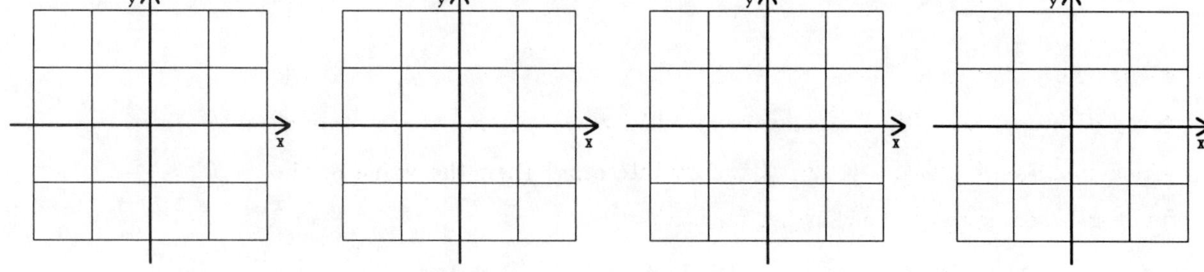

The geometric interpretation of slope is the *rate* at which y changes relative to a change in x. Indeed, your strategy in Exercise 2 may have been (but did not have to be!) to choose your x's exactly one unit apart (i.e., make the run $= 1$) and make the corresponding change in y match the desired slope. We want to discuss rates of change in general, not just in geometry. Thus, we rewrite equation (2) in the equivalent form:

$$\text{The (average) rate of change} \;=\; \frac{\text{the change in the dependent variable}}{\text{the change in the independent variable}}. \tag{3}$$

We put "average" in parentheses because a linear function has *only one* rate of change: the slope of its graph. Thus, in this case, we may talk about *the* rate of change. However, equation (3) makes perfectly good sense for *any* function. Change the input from one value to another, and there is a corresponding change in output values. The ratio of the change in output to the change in input is the average rate of change of the function over the interval defined by the input values. Let's look at a familiar example.

Exercise 3. *A common way to check the accuracy of your speedometer is to drive one or more "measured miles," holding the speedometer at a constant 60 MPH and checking your watch at the start and end of the measured distance.*

(a) Explain the method; how does your watch tell you whether the speedometer is accurate or not?

(b) In this experiment, why shouldn't you use your odometer to measure the mile(s)?

In the next exercise, we ask you to compute some average speeds in a familiar setting in which the rate (speed) is clearly not constant.

Exercise 4. An object is dropped from a height; Table 1 records at each second the distance it has fallen.

(a) What is the average speed of the object for the first second? For the first three seconds? For the first five seconds?

(b) Estimate the average speed for the first 4.5 seconds.

(c) Suppose the height from which the object is dropped is 489 meters; what is the average speed for the last two seconds of its fall?

Time (seconds)	Distance (meters)
1.0	5.0
2.0	19.4
3.0	44.1
4.0	78.0
5.0	122.8
6.0	175.8
7.0	240.0
8.0	312.8
9.0	396.1
10.0	489.0

Table 1. Falling body data.

Mathematical Notation: The Difference Quotient

The phrases you see over and over — "rate of change," for example — are clearly important. In fact, we are repeating them frequently to stress that importance. However, once we have made our point, it becomes equally important that we reveal the standard notation for the important concepts. There is an obvious efficiency in having a concise symbol for each concept. More important than mere efficiency of expression, we find great power in being able to manipulate our symbols to carry out computations, many of which would be much more difficult to do entirely in words. Nevertheless, *every* symbol is in fact a shorthand for one or more words; whenever you are not sure of the exact meaning of a symbolic expression, translate it into the equivalent words. As we observed in Chapter 1, the entire sentence, "The position of a falling body is a function of the time it has been falling," may be abbreviated to "$s = f(t)$." But this abbreviation is not useful —

indeed, it may cause confusion — unless the brief symbolic representation triggers recollection of the words for which it stands.

The next concept we shorten to a symbol is "change," specifically, the change in a variable from one value to another. If x_1 and x_2 are values of the variable x, the change from x_1 to x_2 is the difference, $x_2 - x_1$. A near-universal symbol for "difference" is the Greek capital delta, Δ, so we write "$\Delta x = x_2 - x_1$" to abbreviate "The change in the variable x is the difference between the second value and the first."

Finally, we combine these symbols to obtain a notation for "average rate of change of one variable with respect to another." If $s = f(t)$, then the average rate of change of s as t changes from t_1 to t_2 is the ratio of the change in s to the change in t, i.e.,

$$\frac{\Delta s}{\Delta t}.$$

This quotient of differences is called, naturally enough, a **"difference quotient."**

Exercises

5. A line with slope 1.5 passes through the point $(-1, 2.3)$ in the xy-plane.

  (a) Find an equation of the line.

  (b) Choose two x values, x_1 and x_2, and calculate $\frac{\Delta y}{\Delta x}$.

  (c) Choose another pair of x values, and calculate $\frac{\Delta y}{\Delta x}$.

6. A line passes through the points $(2.1, 1.7)$ and $(-1.5, 4.2)$ in the xy-plane.

  (a) Find an equation of the line.

  (b) Choose two x values, x_1 and x_2, and calculate $\frac{\Delta y}{\Delta x}$.

  (c) Choose another pair of x values, and calculate $\frac{\Delta y}{\Delta x}$.

7. Put your calculator in radian mode. Find the average rate of change of the sine function (use the sin button) over each of the intervals $[0, \pi/4]$, $[0, \pi/2]$, $[0, \pi]$.

8. Put your calculator in degree mode. Find the average rate of change of the sine function (use the sin button) over each of the intervals $[0°, 45°]$, $[0°, 90°]$, $[0°, 180°]$.

2.2. The Derivative: Instantaneous Rate of Change

Sarah came into the kitchen, followed by the family dog, a black lab named Coal. The screen door banged three times behind her. Both the dog and the human looked exhausted. Sweat dripped off Sarah's reddened face; Coal panted noisily. "It sure is great to be able to get outside and work off the tension of those exams. Did you know that the Carters have a new dog, some sort of shepherd? She and Coal spent 10 minutes sniffing. Thought I never would get started on my run."

No answer came from her mother who was back at the breakfast table studying the calculus book. Her mother was an elementary school teacher, but school was out, and she had plunged into her newest study full time. "How's it going?" Sarah inquired.

"I think I understand rates of change. But what is a derivative?"

"That's an instantaneous rate of change."

Her mother looked uncertain.

"Explain it to me in terms of the graph. I have your father's graphing calculator here. How can I use it to understand instantaneous rates of change?"

Sarah grabbed a dish towel to wipe her face and slumped into a chair. Her mother said absently, "Don't use the dish towels, Dear." But it was just a reflex, not a real concern. Sarah ignored it and explained to her mother how to use the zooming feature on the calculator.

Zooming In: Local Linearity

We now resume our study of the falling body problem, for which we have a theoretical model (i.e., a formula) for distance fallen, s, as a function of time t. Recall that, if the gravitational acceleration is constant, and air resistance and wind currents are negligible, elementary physics tells us (for reasons we will examine later) that $s = ct^2$, where c is a constant that depends on the gravitational force and on the units of measurement. For definiteness, we will measure time in seconds and distance in meters.[2] The question we address is this: How can we use the formula for distance

---

[2] You may know already (from some previous course in mathematics or physics) that the value of c (when time is in seconds and distance is in meters) is approximately 4.90; if not, don't worry about it — we will examine later where that number comes from. If it bothers you to have an extra letter floating around in your calculations, you can replace our c's by 4.90. However, we warn you that c is going to turn up often, and it's much easier to write one symbol rather than four.

as a function of time to determine the *speed* of the falling object at any instant of its fall?

Figure 3 shows the graph of our assumed relationship between time t and distance s. (Why does s *rise when the object falls*?) We choose a time, say $t_1 = 5$ seconds, at which we want to know the instantaneous speed. At this time, the object has fallen $s_1 = c(5)^2 = 25c$ meters. How fast would you *guess* the object is falling at that instant? [There is a hint in the picture, if you look at it in the right way.] Write your answer below for reference later, when you will *know* the answer.

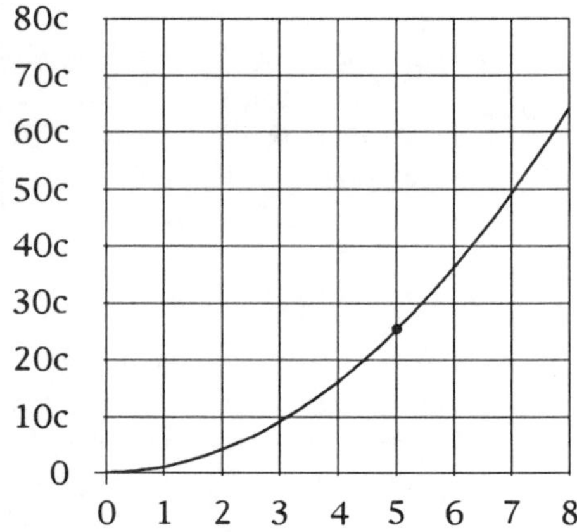

Figure 3. Graph of $s = ct^2$.

My guess is that the speed at 5 seconds is _________________ meters per second.

We repeat Figure 3 below with an added "zoom box," and we show in the result of "zooming in" by magnifying the portion of the graph in the zoom box.

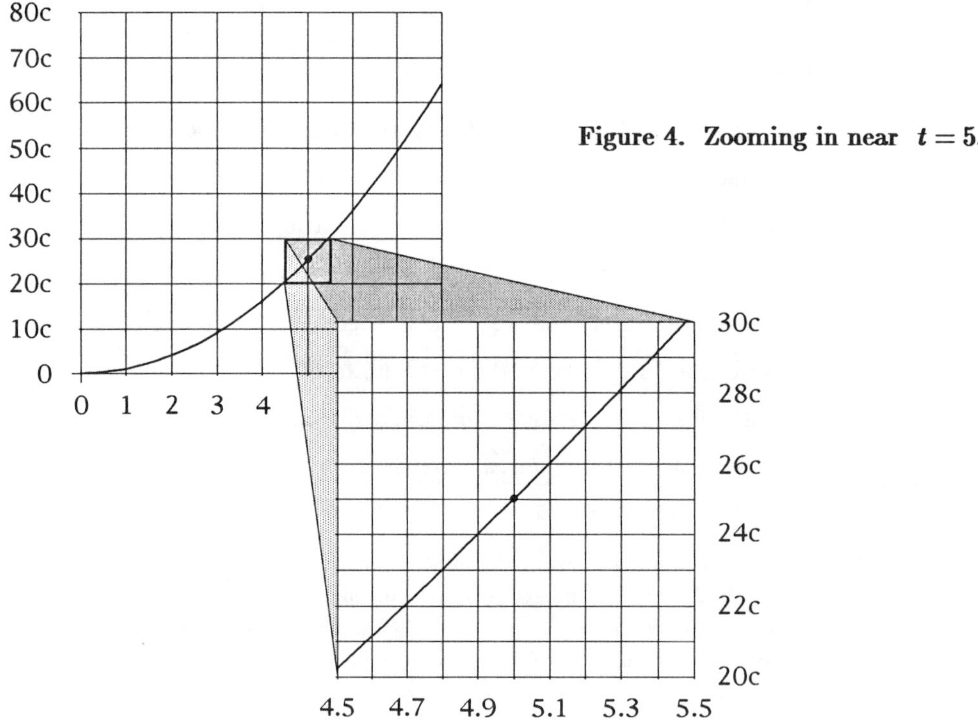

Figure 4. Zooming in near $t = 5$.

We continue the zooming process in Figure 5 by magnifying the portion of the graph in a smaller zoom box centered at $(5, 25c)$.

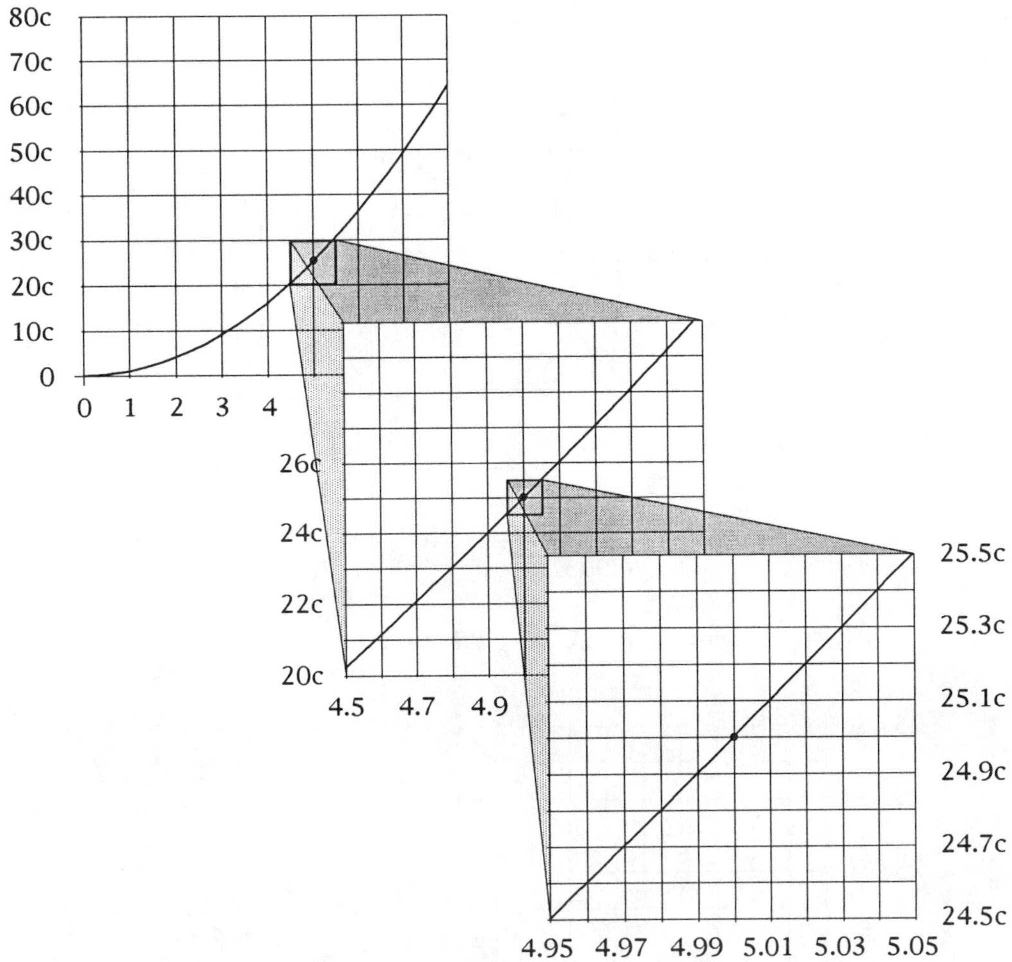

Figure 5. Further zooming in near $t = 5$: From $\Delta t = 1$ to $\Delta t = 0.1$.

It will get rather unwieldy to repeat the whole sequence each time we zoom in further; it should be enough to identify the zoom box and show its magnification for each zoom step. In the following figures, we start from the second step in Figure 5 and examine still smaller portions of the graph near $t = 5$ at still higher magnifications. Study the horizontal and vertical scales in all of the figures to make sure you understand exactly what is being graphed in each case. (In particular, note that the vertical scale in each picture depends on c — because the values of s depend on c.)

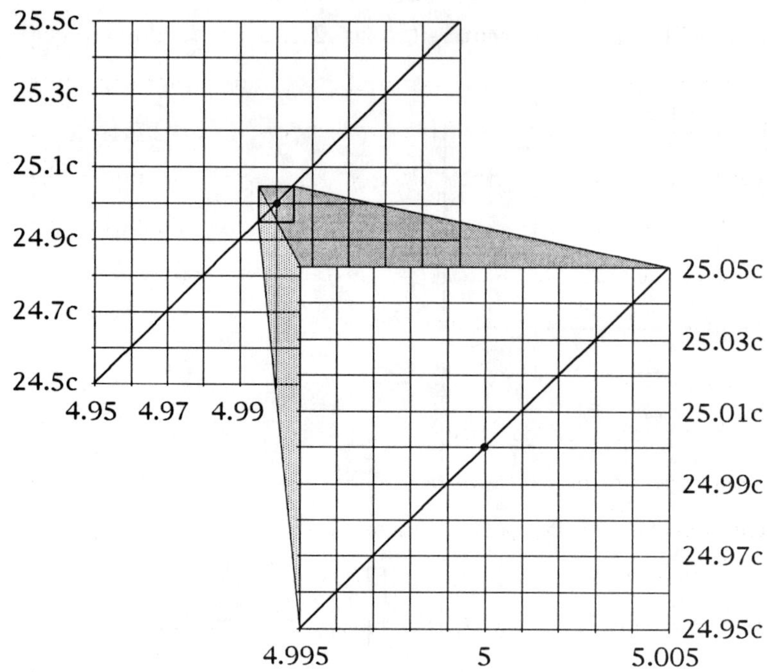

Figure 6. From $\Delta t = 0.1$ to $\Delta t = 0.01$.

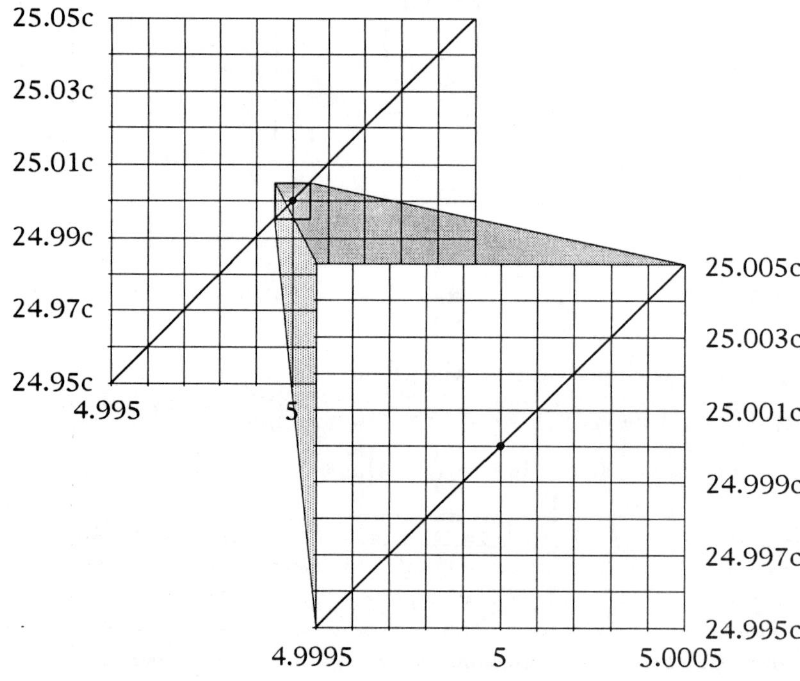

Figure 7. From $\Delta t = 0.01$ to $\Delta t = 0.001$.

What leaps out at you from the sequence of right-hand pictures in Figures 4, 5, 6, and 7? Lay a straightedge along each curve to check. The emerging "straightness" you see as you look at shorter and shorter segments of the curve is a property of *most* well-behaved curves. This property is called **"local linearity."** That is, in the "locality" of a particular point, the curve "looks like" a straight line! Now there is no problem calculating the instantaneous speed, the *rate* at which s is increasing relative to t, because, on a straight line, that rate is just the slope, i.e., "rise over run."

Exercise 1. (a) *We repeat in Figure 8 the final zoom box in Figure 7. Use this figure to finish the calculation of the instantaneous speed at the instant* $t = 5$. *(Your answer should contain the constant* c.)

(b) *Compare your answer with the guess you wrote under Figure 3.*

(c) *Recall from Note 2 that* c *is approximately 4.90. Write the numerical value of the speed (in meters per second) of an object that has been falling for 5 seconds.*[3]

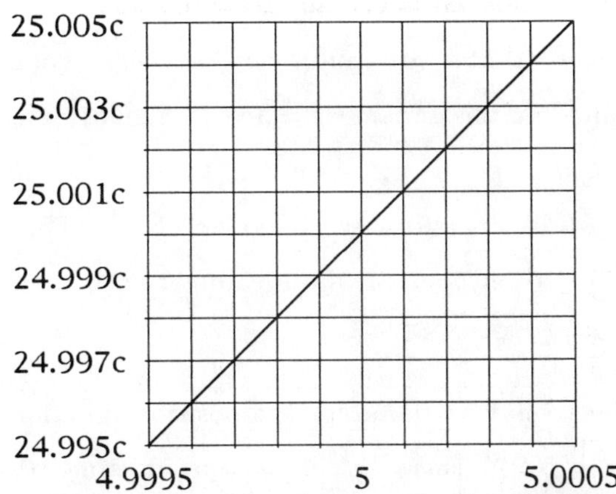

Figure 8. A zoom-in view of the graph of $f(t) = ct^2$ **with** $\Delta t = 0.001$.

---

[3]Note carefully the distinction between parts (a) and (c) of this exercise. The point of (a) is that we *don't need to know* the value of c in order to answer questions about instantaneous speed — if we are willing to accept answers that involve the "unknown" c. And often we have no choice. This important "parameter" of the falling body model happens to be *known* — we used that knowledge in part (c) — but many important models have parameters that are not known. Thus, you should not consider it strange to see formulas in which some or all of the constants are letters rather than numbers, and you should not see this as a barrier to getting useful information from the formulas.

Our "solution" of the speed problem relied heavily on a technological tool, the graphing computer that drew Figures 3 through 8. Any graphing calculator or computer could do as well, but, to be candid, only if we give it a value of c. Furthermore, the graphical calculation told us almost nothing about rates of change in general, not even the speed of the falling object at other times during its fall. The primary value of this graphical calculation was to reveal the local linearity of the curve, which turns out to be a very general phenomenon. How can we do the speed calculation in such a way as to learn something about speeds in general? In the next two sections, we give two answers to that question, one based on using graphics in a different way, the other based on a more ancient tool: algebra.

Graphical Calculation of Rates of Change

We have just used the graphical viewpoint to determine instantaneous speed (at a particular instant) from the position function for a falling object, but we also learned something that will enable us to use our graphical tool more effectively.   Specifically,   we   saw   that   we   could   get   a   very   good approximation to the rate of change at the instant $t = 5$ by calculating a difference quotient $\dfrac{\Delta s}{\Delta t}$ (that is, a slope, a "rise-over-run") with a *small* value of Δt. Why not do that *all at once* for "every" value of t? That should give us a good approximation to a *function* whose value at each t is the instantaneous rate of change at t.

We write $f(t) = ct^2$ in order to have a name for the position function for the falling object [that is, $s = f(t)$, where s is position at time t], and we write $g(t)$ for the difference quotient of f with a suitably small Δt, say, $\Delta t = 0.001$:

$$ g(t) \; = \; \frac{\Delta f}{\Delta t} \; = \; \frac{f(t + 0.001) - f(t)}{0.001} \, . \qquad (4) $$

This new function $g(t)$ is *not* the same function as the instantaneous-rate-of-change function at t, but its value at each t should be very close to the instantaneous rate of change. Thus, a graph of $g(t)$ should closely approximate a graph of the unknown "speed function."

For the time being, we ignore the fact that we actually know a value for the constant c; instead of focusing exclusively on the falling-body position function, we study the whole family of functions of the form $f(t) = kt^2$. In Figures 9-16, we show the graphs of $f(t)$ and $g(t)$ for four different values of k. As you study these graphs, notice the shape(s) of the approximate speed functions $g(t)$, and answer the questions we pose as exercises.

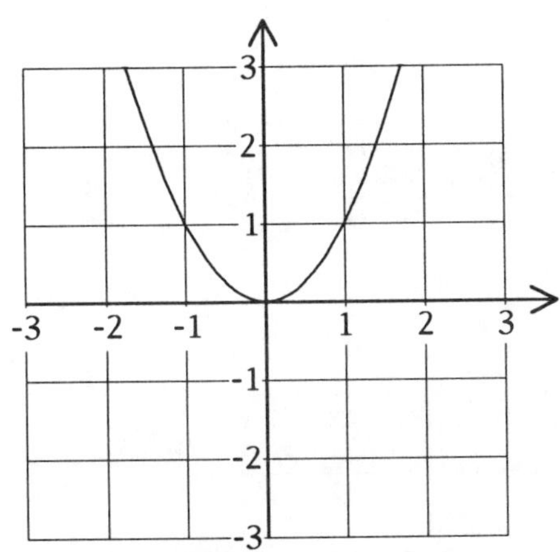

Figure 9. $f(t) = kt^2$ for $k = 1$.   Figure 10. $g(t) = \dfrac{f(t+0.001) - f(t)}{0.001}$ for $k = 1$.

Exercise 2. (a) The approximate speed function in Figure 10 has a familiar shape; what is it?

(b) What is the slope of the line you see in the figure?

(c) Write a simple formula, without using f, for the linear function whose graph appears to match the graph of g.

Exercise 3. Figures 12 and 14 also show linear graphs for the approximate speed functions.

(a) What is the slope of the graph in Figure 12? Write a simple linear formula for a function that approximates $g(t)$ when $k = \frac{1}{2}$.

(b) What is the slope of the graph in Figure 14? Write a simple linear formula for a function that approximates $g(t)$ when $k = -\frac{1}{2}$.

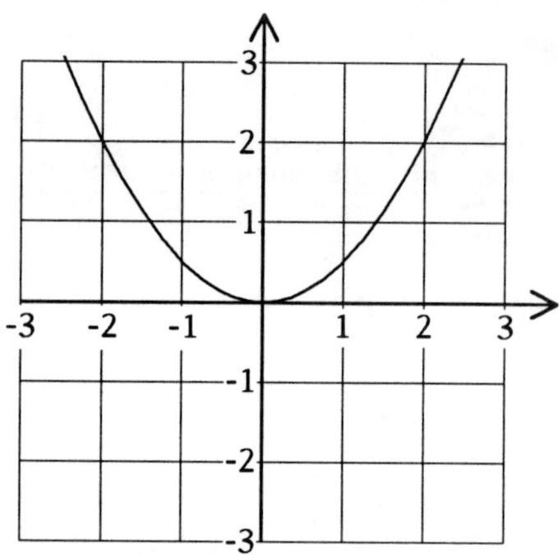

Figure 11. $f(t) = kt^2$ for $k = \frac{1}{2}$.

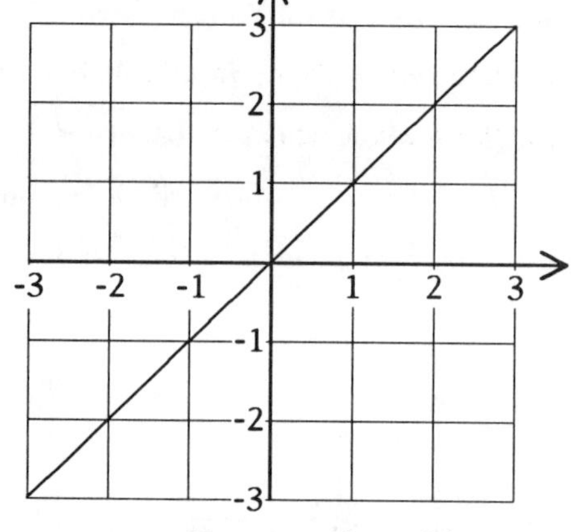

Figure 12. $g(t) = \dfrac{f(t+0.001) - f(t)}{0.001}$ for $k = \frac{1}{2}$.

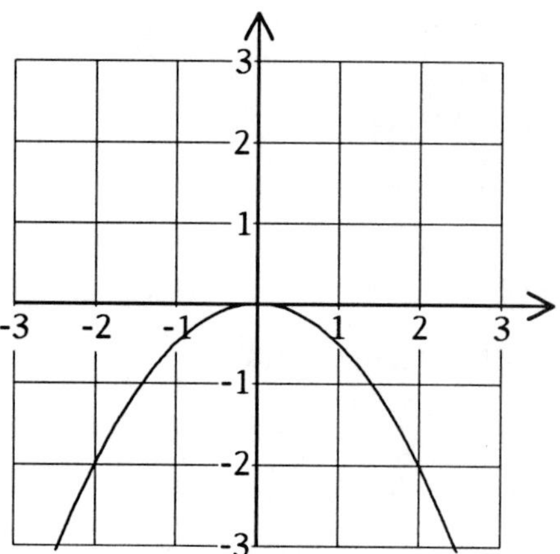

Figure 13. $f(t) = kt^2$ for $k = -\frac{1}{2}$.

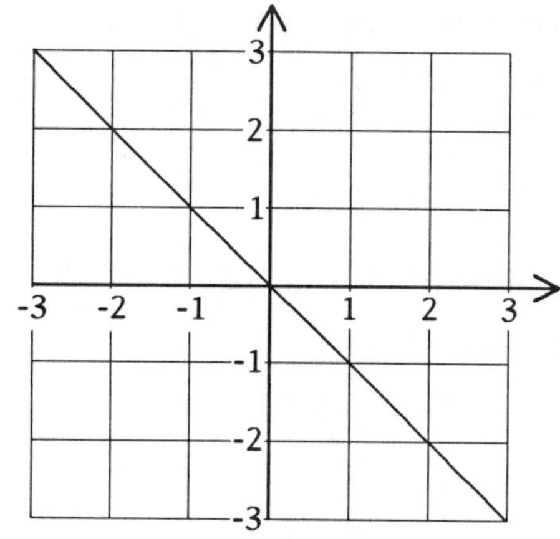

Figure 14. $g(t) = \dfrac{f(t+0.001) - f(t)}{0.001}$ for $k = -\frac{1}{2}$.

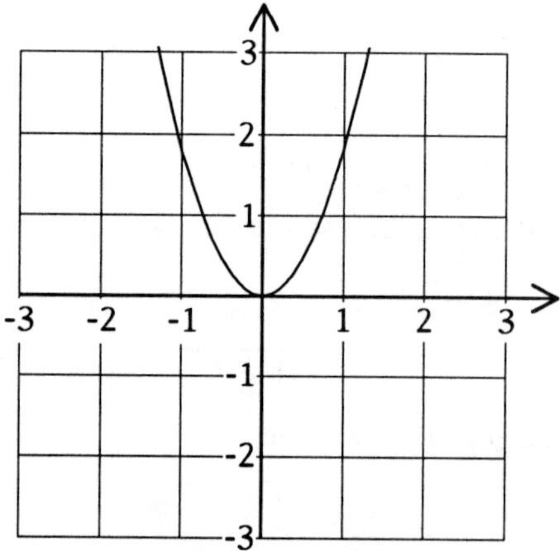

Figure 15. $f(t) = kt^2$ for $k = 1.8$.

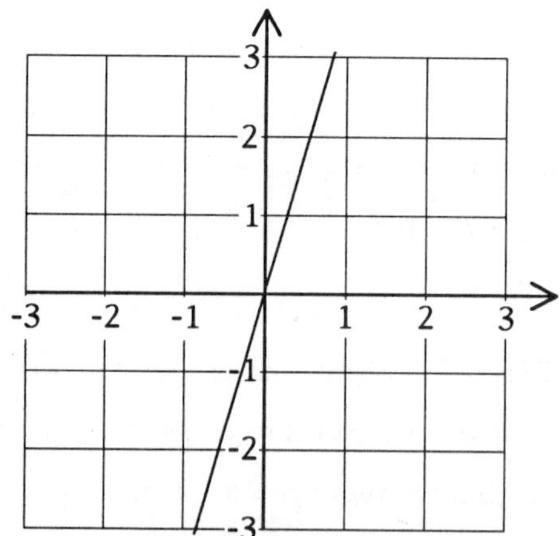

Figure 16. $g(t) = \dfrac{f(t+0.001) - f(t)}{0.001}$ for $k = 1.8$.

Exercise 4. (a) *Our fourth case is a little harder; estimate the slope of the line in Figure 16.*

(b) *Fill in the table at the right with an appropriate formula for the speed function in each case.*

$f(t)$	speed at time t
t^2	
$\frac{1}{2}t^2$	
$-\frac{1}{2}t^2$	
$1.8\,t^2$	

(c) *If* $f(t) = kt^2$ *gives position at time* t *for an object moving in a straight line, what is your best guess at a formula for the speed of the object at time* t? *(Check your formula against the result of Exercise 1; does it give the right answer when* $t = 5$ *and* $k = c$?)

Algebraic Calculation of Rates of Change

At the end of the "Zooming In: Local Linearity" section, we asked, how can we find a formula for the speed at every time t? Our first answer (just discussed in the "Graphical Calculation" section) is to use a "high-tech" graphing tool. Our second answer is to use a "low-tech" tool: algebra. That's harder than looking at computer-drawn pictures but also, as we shall see, more satisfying, because there will not be any *guessing* at formulas.

We first do the calculations that exactly parallel Figures 4 through 8, which require no more than arithmetic, except for keeping track of c. Then we will "abstract" our arithmetic calculations to an algebraic formulation that will (a) tell us a formula for the speed of the falling body at *every* instant and (b) help us develop some tools for dealing with rates of change in general.

In Figures 4 through 8 we used time intervals *centered* on $t = 5$, the time "of interest." For purposes of our algebraic calculation, we will use intervals that have $t_1 = 5$ and t_2 some other time. This has the advantage that only one of the two time values changes when we change Δt. For the time being, think of using only the right-hand half of each of the final zoom boxes in Figures 4 through 7, which we repeat here; your calculation of the slope *at 5* would clearly give the same result (see Exercise 1 again).

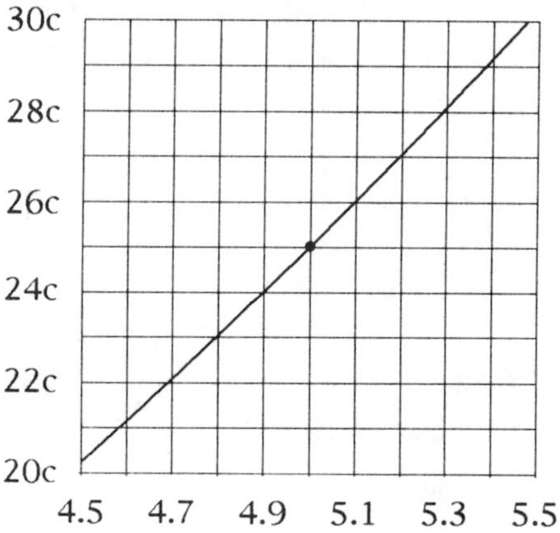

Figure 17. The graph of $f(t) = ct^2$ from $t = 4.5$ to $t = 5.5$.

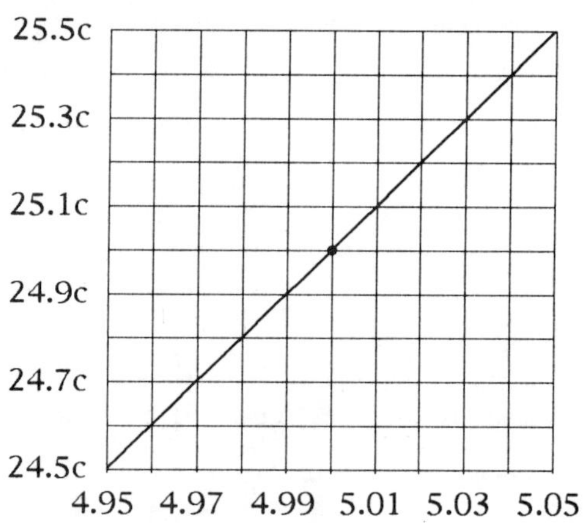

Figure 18. The graph of $f(t) = ct^2$ from $t = 4.95$ to $t = 5.05$.

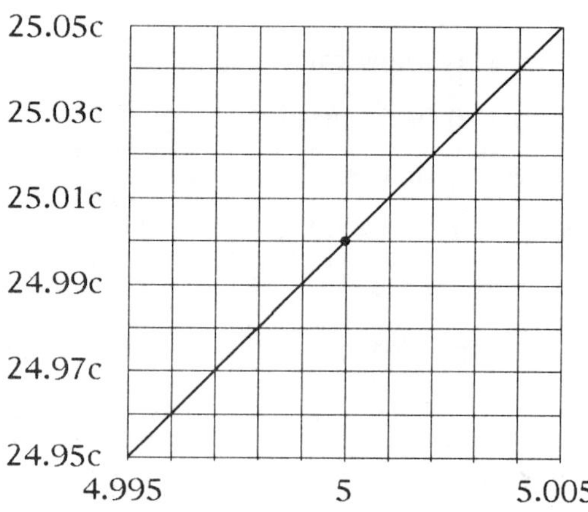

Figure 19. The graph of $f(t) = ct^2$ from $t = 4.995$ to $t = 5.005$.

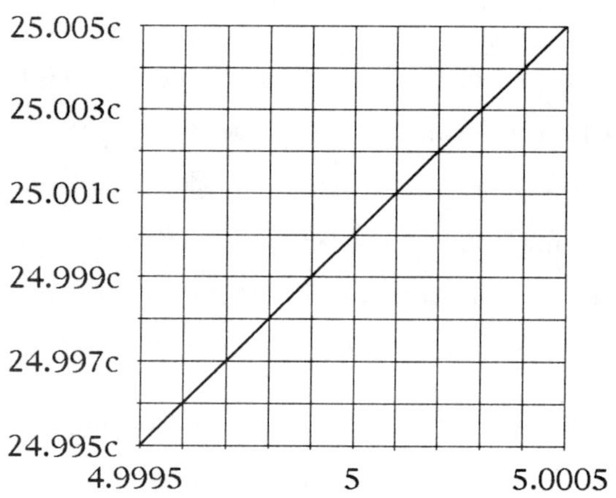

Figure 20. The graph of $f(t) = ct^2$ from $t = 4.9995$ to $t = 5.0005$.

With $t_1 = 5$, we consider first the small change in time, $\Delta t = 0.5$, that leads to $t_2 = 5.5$ seconds, the right-hand edge of Figure 17. The distance fallen at time t_2 is $s_2 = (5.5)^2 c = 30.25c$ meters. Thus, the average speed (rate of change of distance) from t_1 to t_2 is

$$\frac{\Delta s}{\Delta t} = \frac{s_2 - s_1}{t_2 - t_1} = \frac{30.25\,c \, - \, 25\,c}{0.5} = 10.5\,c \text{ meters/second.}$$

This only approximates the speed *at* t_1, but we get a closer approximation by choosing a smaller Δt, as in Figure 18, namely, $\Delta t = 0.05$. Then

$$\frac{\Delta s}{\Delta t} = \frac{s_2 - s_1}{t_2 - t_1} = \frac{25.5025\,c\ -\ 25\,c}{0.05} = 10.05\,c \ \text{meters/second}.$$

We still have only an approximation to instantaneous speed at $t = 5$, but the approximation is better, because the time interval is smaller. What do you think the average speed will be if we choose $\Delta t = 0.005$, as in Figure 19? Would it surprise you if the answer is $10.005\,c$ m/sec? Check it! Then do the calculation one more time, as in Figure 20, with $\Delta t = 0.0005$.

These calculations illustrate both (a) our strategy for finding instantaneous speed and (b) the result of pursuing this strategy. What we know already about rates of change requires an actual *change* in the independent variable, i.e., two time values, not a single instant. However, we can choose those two values as close together as we like, thereby approximating instantaneous rate of change by average rate of change over a very small interval. That's the strategy; what's the result?

Exercise 5. Would we get the same result if we chose negative values of Δt, such as -0.5, -0.05, -0.005, -0.0005? [These numbers correspond to using the left-hand halves of Figures 17, 18, 19, and 20.] Do the calculations to find out.

Exercise 6. When you did the graphical calculation of speed at time $t = 5$ seconds, you may not have taken $t_1 = 5$ as we have been doing in our algebraic calculations; rather, you may have computed the slope all the way across Figure 20, which amounts to taking t_1 at the left edge of the graph. Do this for all four of the "zoom" figures: Calculate the average slope (speed) for the entire interval shown. How do these slopes compare with the instantaneous speed at $t = 5$ and with the approximations you and we computed taking t_1 at 5?

Exercise 6 illustrates a phenomenon that will play an important role later: At a point of local linearity on a curve, a very good estimate of the instantaneous rate of change can often be computed by taking a ratio of differences ("rise over run") at points that are equally spaced on opposite sides of the point in question. Problem 1 at the end of this chapter shows that this procedure can lead to a strange result if the point happens *not* to be one of local linearity.

Now we observe that our speed calculation for the falling object does not depend on the specific instant $t_1 = 5$. Indeed, the whole process often turns out to be *easier* if we do it algebraically (for an arbitrary but unspecified t_1) rather than arithmetically. Compare the algebraic computation below with the work we just did with the calculator:

<u>Algebraic step</u>   <u>Reason</u>

$$\frac{\Delta s}{\Delta t} = \frac{s_2 - s_1}{t_2 - t_1}$$

$$= \frac{ct_2^2 - ct_1^2}{t_2 - t_1}$$

$$= \frac{c(t_2 - t_1)(t_2 + t_1)}{t_2 - t_1} \qquad \text{We factored the numerator.}$$

$$= c(t_2 + t_1)$$

$$= c(t_1 + \Delta t + t_1)$$

$$= c(2t_1 + \Delta t).$$

Exercise 7. Write a sentence in the "Reason" column to explain each step of the calculation. [One such sentence is already provided as an example.]

Now we can see clearly what happens as Δt becomes smaller and smaller; the Δt term in our computed average speed disappears, and the average speed approaches $2ct_1$. That agrees with our result when t_1 is 5 (Exercise 1), and it should also confirm your conjectured formula in Exercise 4, since it tells us the instantaneous speed at *every* instant t_1. Furthermore, the conclusion is clearly the same for both positive and negative values of Δt; we don't need a separate calculation for the two cases. That should confirm your answer to Exercise 5.

Exercise 8. If the position s of a falling body at time t is given by $s = ct^2$, where $c = 4.90$, then the instantaneous speed at $t = 1$ is ______; at $t = 2.5$ is ______; at $t = 17$ is ______; at $t = \sqrt{34}$ is ______ .

What just happened here is important, because we will go through this process many times. Our "defining" concept of instantaneous rate of change, as a limiting value of average rates of change, is too cumbersome for "routine" calculation. We will have to use it for each new *family* of functions that comes along, and occasionally for deriving "general rules" for finding rates of change. But once we have the general rules and the formulas for specific types of functions, we will use them to make the calculation of a rate of change a routine algebraic manipulation. That will free us up to concentrate our energies on things that are *not* routine.

The Derivative

To calculate instantaneous rate of change from average rate of change, we have to let the change in the independent variable (Δt in the example above) *approach* zero without actually letting it *equal* zero — because we need *two* values of the independent variable to compute an average rate of change. We abbreviate the process of letting Δt approach zero by "$\Delta t \to 0$."

Of course, when the change in the independent variable approaches 0, the change in the dependent variable usually does also. Thus, we have $\Delta t \to 0$ and $\Delta s \to 0$ simultaneously, and we can't necessarily tell from that information alone what $\frac{\Delta s}{\Delta t}$ is approaching. Whatever that limiting value of difference quotients is, the standard notation for it is

$$\frac{ds}{dt}.$$

Each upper case delta, standing for "difference," is replaced by a lower case d, which stands for "differential."[4] Thus, $\frac{ds}{dt}$ is what $\frac{\Delta s}{\Delta t}$ approaches when Δt and Δs both approach zero. Of course, what the average rate of change approaches when the time interval shrinks to zero is the *instantaneous rate of change*, so now we have a notation for instantaneous rate of change. Think back to the "zooming" process in Figures 4 through 8; when we look at a segment of the graph of s (as a function of t) that is so small it appears straight, then $\frac{ds}{dt}$ is the slope of that straight line.

---

[4]And that explains why we didn't use "d" for "distance" — also see Footnote 21 in Chapter 1. We will take up differentials in Chapter 4.

Exercise 9. Fill in the blank: If $s = kt^2$, *then* $\dfrac{ds}{dt} =$ _______.

The concept of instantaneous rate of change has a name: the **"derivative."** Using that name spares us from having to use the phrase "instantaneous rate of change," but at the cost of having to learn a new word, one whose English meaning has no obvious connection to what we just did.

There *is* a connection — it's just not obvious. In our example, we started with the functional relationship $s = kt^2$, and we *derived* the functional relationship $\dfrac{ds}{dt} = 2kt$. Lots of other things in mathematics are derived, but this special "derived function" is called the ***derivative.*** On the other hand, the *process* of calculating derivatives of functions is *not* called "derivation," but rather ***differentiation.*** Bear with us — this is not as arbitrary as it sounds. First, "derivation" is too easily confused with other important mathematical processes, many of which involve derivatives, but not necessarily calculation of derivatives. Second, as we shall see later, the process of calculating a derivative is virtually identical to the process of calculating a differential.

However, one aspect of this naming business *is* arbitrary: The standard naming conventions for the most important mathematical and scientific concepts have been in place for a long time, and we would do you no service by attempting to change them. We will do our best to explain them as we go along, including the logic or illogic behind them. Your job is to learn to use them correctly.

Alas, there is still more notation and terminology, even without a single additional concept.[5]   First, the instantaneous rate of change (think "derivative") of position or distance is what we have been calling *speed* — at least for an object that moves in only one direction along a straight line. For motion in general, this rate is called ***"velocity."***[6]   Good thing, too,

---

[5]This is not a sinister plot to make your studies more difficult — it's just what happens to concepts that have been around for 300 years.

[6]The technical distinction is this: Speed is the absolute value of velocity. When the motion is along a line and in only one direction (no backing up), we can take that direction to be the positive direction, and the two concepts coincide.

because we can't use "s" for "speed"; it's already in use for something else. We abbreviate the sentence, "Velocity is the instantaneous rate of change of position," to "$v = \frac{ds}{dt}$."

Finally, when we want to focus on functional notation, we use a notation for "the derived function": If $s = f(t)$, then $v = f'(t)$ (which is read, "f prime of t"). The notations f' and $\frac{ds}{dt}$ are interchangeable; each will be used when it is convenient for the task at hand.

Exercise 10. *Suppose $s = 2t^2 + 3$. Use the zooming feature on your calculator or computer program to determine the values of $\frac{ds}{dt}$ at $t = 1$, $t = 2$, and $t = 3$.*

Exercise 11. *Suppose $s = 2t^2 + 3$.*

(a) Use algebra to calculate $\frac{\Delta s}{\Delta t}$, where $\Delta t = t_2 - t_1$.

(b) From your calculation in part (a), write down a formula for $\frac{ds}{dt}$ at t_1.

(c) Write down a formula for $\frac{ds}{dt}$ as a function of t. [This requires only a simple modification of your formula in part (b).]

(d) Evaluate your formula for $\frac{ds}{dt}$ at $t = 1, 2$, and 3. Compare the results with your answers to the previous exercise.

Section Summary

Use the spaces provided to summarize the mathematical content of Section 2.2. Your summary should include a discussion of zooming in, local linearity, difference quotients, average and instantaneous rates of change, and derivatives.

Zooming In and Local Linearity

Difference Quotients and Average Rates of Change

Derivatives and Instantaneous Rates of Change

Exercises

In all of the following exercises, $f(t)$ is the reciprocal function: $f(t) = \frac{1}{t}$.

12. *Sketch the graph of $f(t)$. [Check your graph with your graphing calculator.]*

13. *Let $g(t)$ be the difference quotient function for a difference of $\Delta t = 0.001$:*

$$g(t) = \frac{f(t + 0.001) - f(t)}{0.001} .$$

(a) Carefully select 8 values of t, and calculate the corresponding values of $g(t)$. [Your "careful" selection should enable you to sketch the graph of g.]

(b) Sketch the graph of $g(t)$. [Check your graph with your graphing calculator.]

(c) Describe the graph of $g(t)$ in words. [For example, you might say how it compares to the graph of $f(t)$.]

14. *(a) Calculate the difference quotient $\frac{\Delta f}{\Delta t}$ algebraically for an arbitrary Δt.*

(b) Simplify your algebraic expression for $\frac{\Delta f}{\Delta t}$ until the Δt in the denominator cancels. What does the difference quotient approach as $\Delta t \to 0$?

(c) What is the derivative of the reciprocal function?

(d) Graph the derivative function you just obtained algebraically. How does it compare with the function g in the previous exercise?

2.3 Symbolic Calculation of Derivatives: Polynomial Functions

The next evening Sarah was washing dishes while her mother worked at the table.   Of all the household chores, Sarah disliked this one the least. Particularly in the summer, when the windows were open and you could hear snatches of conversation, radio and TV sounds, and, if you were lucky, an occasional bird calling.

*"Well, Sarah, I have read all that stuff in the **Reader**; I guess I know what an instantaneous rate of change is.   And I like to think about the derivative as the slope of the line when you zoom in on the graph. I suppose I can deal with the different notations, though I am not quite sure what a differential is.   But I only know how to calculate the derivative of linear functions and things like kt^2. ... You know, I've been working with Anthony on his precalculus ... "*

Anthony is Sarah's brother.   He's into basketball, rap music, and impressing girls; precalculus is not high on his list of priorities. Sarah's mother had been working with him almost every night — in Sarah's view, with amazing patience.   Sarah had heard them when she was home for spring break.   Her mother would try to explain the concept behind the particular problem; Anthony would break in, saying that he just wanted to write down solutions and quit.

Sarah's mother continued, "How do you calculate the derivatives of some of the other functions you see in precalculus — say, a third-degree polynomial like $5t^3 + 7t^2 - 3t + 6$?"

Sarah replied, "That's not too hard.   You need to learn the rule for differentiating powers and then just use the rules for derivatives of sums and constant multiples."

Derivatives of Power Functions

What is the derivative of a polynomial function?   Let's start with the simpler question of finding derivatives of *power* functions. We know some special cases:

$$\frac{d}{dt}\, k \; = \; 0. \qquad \text{(If there is \textit{no} change, the \textit{rate} of change is zero.)}$$

$$\frac{d}{dt}\, kt \; = \; k. \qquad \text{(The rate of change of a linear function is the \textit{slope} of its graph.)}$$

$$\frac{d}{dt}\, kt^2 \; = \; 2kt. \qquad \text{(Recall Exercise 9 in section 2.2.)}$$

What about the general power function t^n, where n may be any positive integer? We repeat the cases we know (with k set equal to 1); note that $t^0 = 1$, and $t^1 = t$.

$$\frac{d}{dt}\, 1 \;=\; 0.$$

$$\frac{d}{dt}\, t \;=\; 1.$$

$$\frac{d}{dt}\, t^2 \;=\; 2t.$$

Let's do one more case by direct calculation of $\frac{\Delta s}{\Delta t}$, where $s = t^3$. Just to display the use of alternate tools, we will do this one by writing $\Delta s = s(t+\Delta t) - s(t)$. As before, we ask you to provide the missing reasons.

<u>Algebraic step</u>   <u>Reason</u>

$$\frac{\Delta s}{\Delta t} \;=\; \frac{(t+\Delta t)^3 - t^3}{\Delta t}$$

$$= \; \frac{\left[t^3 + 3t^2(\Delta t) + 3t(\Delta t)^2 + (\Delta t)^3\right] - t^3}{\Delta t}$$

$$= \; \frac{3t^2(\Delta t) + 3t(\Delta t)^2 + (\Delta t)^3}{\Delta t}$$

We removed the brackets; $t^3 - t^3 = 0$.

$$= \; 3t^2 + 3t(\Delta t) + (\Delta t)^2 .$$

Now, as $\Delta t \to 0$, the second and third terms on the right also approach zero, so

$$\frac{d}{dt}\, t^3 \;=\; 3t^2.$$

Exercise 1. Carry out a similar calculation to find $\frac{d}{dt}\, t^4$. You may use either the factoring technique or the binomial expansion technique above. Which do you prefer?

Exercise 2. Formulate a general "**power rule**" for differentiating power functions of the form t^n. Explain how this rule follows for arbitrary n from either the factoring technique or the expansion technique. Write your rule in the space below to complete formula (5):

$$\frac{d}{dt}\, t^n \;= \hspace{5cm} (5)$$

What About General Polynomials?

Sarah's mother looked up. "OK, I know that the derivative of t^n is nt^{n-1}. What about general polynomials?"

Sarah placed a stack of dishes on the shelf. "That's where the constant multiple rule and the sum rule come in."

The Constant Multiple Rule

If $f(t)$ is any function that has a derivative, and c is any constant, we will show that

$$\frac{d}{dt}\, cf(t) \;=\; c\,\frac{df}{dt}. \tag{6}$$

That is, constant factors can be factored out of derivative calculations. So, for example,

$$\frac{d}{dt}\, 6t^4 \;=\; 6\frac{d}{dt}\, t^4 \;=\; 6\!\left(4t^3\right) \;=\; 24t^3.$$

The Constant Multiple Rule follows from some simple algebra with difference quotients. If $s = cf(t)$, then

$$\frac{\Delta s}{\Delta t} \;=\; \frac{cf(t+\Delta t) - cf(t)}{\Delta t}$$

$$\;=\; c\,\frac{f(t+\Delta t) - f(t)}{\Delta t}.$$

As $\Delta t \to 0$, we obtain

$$\frac{d}{dt}\, cf(t) \;=\; c\,\frac{df}{dt}.$$

The Sum Rule

In words, the Sum Rule says that the derivative of the sum of two functions is the sum of the derivatives of the two functions. This can be established in much the same way as the Constant Multiple Rule; we leave the calculation to you in the next exercise.

Exercise 3. (Sum Rule) Suppose $f(t)$ and $g(t)$ are any functions that have derivatives. Show that

$$\frac{d}{dt}\, [f(t) + g(t)] \;=\; \frac{df}{dt} + \frac{dg}{dt}. \tag{7}$$

All polynomials are just sums of terms of the form kt^n for various values of k and n. We now know how to differentiate each such term, and we also know how to differentiate sums. For example, the derivative of $3t^5 - 2t^3 + 7t^2 + 3$ may be calculated in the following way:

$$\frac{d}{dt}\left(3t^5 - 2t^3 + 7t^2 + 3\right) = \frac{d}{dt}\left(3t^5\right) + \frac{d}{dt}\left(-2t^3\right) + \frac{d}{dt}\left(7t^2\right) + \frac{d}{dt}(3)$$

$$= 3\frac{d}{dt}t^5 - 2\frac{d}{dt}t^3 + 7\frac{d}{dt}t^2 + 0$$

$$= 3 \cdot 5\, t^4 - 2 \cdot 3\, t^2 + 7 \cdot 2\, t$$

$$= 15\, t^4 - 6\, t^2 + 14\, t.$$

Exercise 4. *Find the derivatives of*

(a) $5t^3 + 7t^2 - 3t + 6;$

(b) $t^4 - 18t^2 - 10t + 39.$

Section Summary

Use the space below to summarize the mathematical content of Section 2.3. In particular, generalize your result in Exercise 4:   Write one or more sentences that describe a rule for calculating the derivative of any polynomial.

Exercises

Find the derivative of each of the following functions:

5. $2t^5 - 7t^3 + 4t^2 - 6t + 1$ **6.** $t^6 + 2t^4 - 4t^2 - 6$

7. $3t^4 - t^3 + 7t^2 - t + 17$ **8.** $-t^3 + 5t^2 - 27t + 13$

2.4 Exponential Functions

Sarah's mother pushed back her chair and stretched out her arms. Sarah came in with coffee for both of them.

"So now I know how to differentiate polynomials. I still don't see any relation between all this and population growth."

Sarah sipped her coffee. Her mother always wanted the big picture, to know where she was going. Often Sarah was content just to do the work assigned and assume that at some point it would all make sense. However, her mother clearly wanted a response.

"Yeah, you need to learn how to find the derivatives of a lot more functions — square roots and sines and cosines and ... ."

Sarah's mother looked a little discouraged. Sarah hurried to add, "But to understand the simplest population models, you only need to look at functions whose derivatives are multiples of the original function."

"What do you mean, 'multiples of the original function'?"

"Like a function whose derivative is twice the original function."

"Hmm," her mother muttered. She paused with her cup in mid-air, frozen in thought. Sarah looked at her watch and realized that Stan would be by in a couple of minutes.

Her mother drank some coffee and set the cup down. "But we don't know any functions like that! Well, maybe the function that is always zero, but we can't use that to describe anything interesting. The linear functions have derivatives that are constants, and the derivative of kt^2 is just $2kt$. Maybe, higher powers ... . No, you always decrease the degree of a polynomial when you differentiate."

Sarah was impressed that her mother had picked up on the degree thing. She was pretty good at this. It was amazing how much smarter her parents appeared after she had been away at college for a year. But she needed to move this along or she was never going to be ready when Stan arrived.

"You need to look at the exponential functions."

"Isn't t^2 an exponential function? There's a 2 in the exponent."

"Well, yeah, but what they mean by an exponential function is something like 2^t or 3^t or 10^t."

"I've never looked at functions like that before. Why in the world would something like that turn up when we are talking about population growth?"

*Sarah tried to remember — what had they looked at just before exponential functions. She picked up the **Reader** and flipped through the pages. Oh yeah, that stuff about interest rates.*

"Here, look at this next section. A bank account at fixed interest grows at a rate that is proportional to how much money you have in it, just like a

population. Except it only grows each time they calculate interest. It's all in the book. I've got to get ready for Stan."

Sarah felt guilty as she hurried from the room before her mother could ask any more questions. Her mother still looked a little confused. However, in a few minutes Sarah had forgotten all about it.

Interest Rates and Exponential Growth

Your savings account grows (because interest is added) at a rate proportional to the amount present in your account. Suppose you start with \$300 in an account that pays interest once a year (i.e., interest is "compounded annually") at a rate of 5%, and you do not make any additional deposits or withdraw any of the money. At the end of one year, you will have the original 300 plus $300 \times .05$ (or 15) dollars. That is, your balance will grow by \$15 to become $300 + 15 = \$315$. At the end of two years, your interest will be $315 \times .05$ (or \$15.75), so your new balance will be $315 + 15.75 = \$330.75$. At the end of three years, your interest will be $330.75 \times .05 = \$16.54$, and your new balance will be $330.75 + 16.54 = \$347.29$. Notice that the *growth* (interest) each year is the same number as the *rate of growth* (because the time step Δt is 1 year), and this number is proportional to the present balance, with proportionality constant .05.

Now we repeat this "interesting" calculation in another way. At the end of the first year, your new balance will be the starting amount plus interest. That is, your balance will be $300 + 300 \times .05 = 300 \times 1.05 = 315$ dollars. At the end of the second year, you will have $315 + 315 \times .05$ $= 315 \times 1.05$ or $300 \times (1.05)^2$ dollars. (Use your calculator to check that this last expression gives \$330.75, as we got before by adding.) At the end of the third year, you will have $300 \times (1.05)^3$ dollars, or \$347.29. In general, after n years you will have $300 \times (1.05)^n$ dollars. Notice how the repeated *addition* of interest can also be calculated by repeated *multiplication* by $(1 + \text{the interest rate})$.

If you start with A dollars instead of 300, then the amount you would have after n years is $A(1.05)^n$ dollars. If the percentage interest rate is $r\%$ rather than 5%, then the decimal interest rate is $k = \frac{r}{100}$, and the amount you have after n years is $A(1+k)^n$ dollars.

Exercise 1. *(a) If you invest $300 at an annual interest rate of 7.35% compounded annually, how much would you have in 15 years?*

(b) How much would you have to invest at 7.35% compounded annually so that you would have $20,000 after 10 years?

Functions like 1.05^t or 2^t or Whatever

Sarah's Uncle Harold came to visit that evening and stayed for the next few days. There was no further discussion of exponential growth during this time, just interminable discussions of distant relatives. It was almost a week later that Sarah's mother returned to the breakfast table with the **Reader** *in hand. Sarah was eating breakfast, about to go off to work.*

"Well, I guess I am willing to look at functions like 1.05^t or 2^t or whatever. It still bothers me that bank accounts grow in jumps — once a year, or however often the interest is figured — but most biological populations grow all the time ... well, not animals that have litters once a year. I guess those populations jump at regular intervals, in the spring or whenever."

Graphical Determination of Derivatives of Exponential Functions

It appears that *exponential* functions — constant base and variable exponent — have something to do with growth for which the rate of growth is proportional to the amount present. Our goal now is to find the derivative of an exponential function, b^t. The base b might be 2 (if we want to talk about things doubling) or 10 (to invert the "common" logarithm function discussed in Chapter 1) or 1.05 (if we want to talk about bank balances) or any other constant that makes sense as a base for exponentiation. As we did in finding the derivative of kt^2, we will approach this task first from a *graphical* point of view (with the aid of computer-drawn graphics) and then from a *numerical* point of view (with the aid of your calculator). In Section 2.7, we will see what can be added by exploring the *algebraic* point of view.

In our study of functions of the form kt^2, we saw that we could get a very good approximation to the derivative function at every instant by calculating difference quotients $\dfrac{\Delta s}{\Delta t}$ with a *small* value of Δt. We use the same idea now to see what we can learn about the derivative of an exponential function.

We write $f(t) = b^t$ in order to have a name for the exponential function with base b, and we write $g(t)$ for the difference quotient of f with a suitably small Δt, say, $\Delta t = 0.001$:

$$g(t) \;=\; \frac{\Delta f}{\Delta t} \;=\; \frac{f(t + 0.001) - f(t)}{0.001} . \tag{8}$$

This new function $g(t)$ is *not* the same function as $f'(t)$, but its value at each t should be very close to $f'(t)$. Thus, a graph of $g(t)$ should closely approximate a graph of $f'(t)$.

We can determine whether $f'(t) = cf(t)$ for some constant c by asking whether the ratio $\dfrac{f'(t)}{f(t)}$ is a *constant function*. And, since $g(t)$ approximates $f'(t)$, we can examine that question graphically by determining whether the graph of $\dfrac{g(t)}{f(t)}$ looks "approximately constant."

In Figures 21 and 22, we graph $f(t)$, $g(t)$, and their ratio $\dfrac{g(t)}{f(t)}$ for six different choices of the base b. In each case, the graph of $f(t) = b^t$ is the solid curve, the graph of $g(t)$ is the dashed curve, and the graph of the ratio is the dash-dotted curve. The graphs of the ratios provide compelling evidence that the derivative of an exponential function, no matter what the base, is indeed a constant multiple of the function itself.

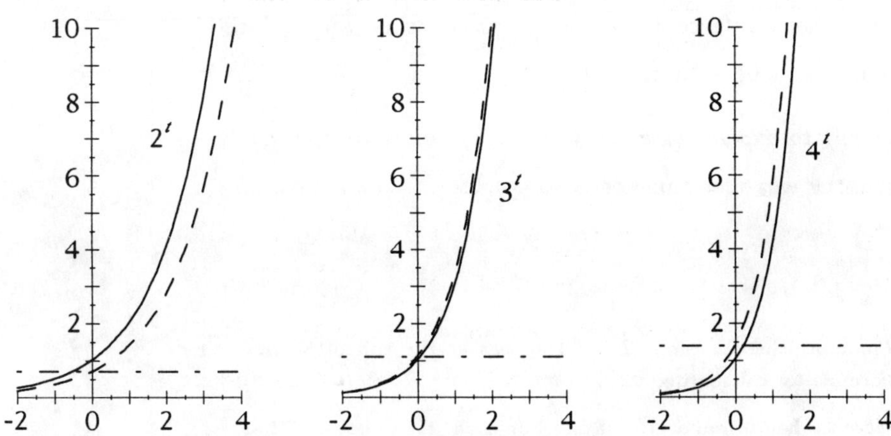

Figure 21. Graphs of $f(t) = b^t$ (solid), difference quotients $g(t)$ (dashed), and their ratios (dash-dotted) for $b = 2,\ 3,\ 4$.

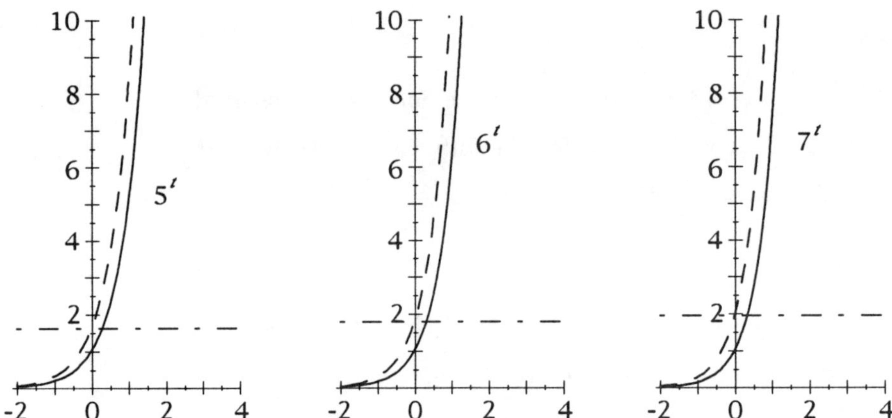

Figure 22. Graphs of $f(t) = b^t$ (solid), difference quotients $g(t)$
(dashed), and their ratios (dash-dotted) for $b = 5, 6, 7$.

Figures 21 and 22 reveal more: First, the constant c in the formula
$f'(t) = cf(t)$ is not the same for every choice of b;   larger values of b
produce larger values of c.   There appears to be some (non-constant)
functional relationship between b and c;   to indicate the dependence of c
on b,   we write $c = L(b)$,[7]   and we rewrite the formula $f'(t) = cf(t)$ in
the form[8]

$$\frac{db^t}{dt} \; = \; L(b)\, b^t. \tag{9}$$

Second, in each of the six cases graphed in Figure 22, the horizontal line
$y = L(b)$ has the same y-intercept as the graph of the (approximate)
derivative $g(t)$. That can't be an accident!

*Exercise 2. Use Equation (9) to explain carefully why $L(b)$ must be the value of
$\frac{db^t}{dt}$ at $t = 0$,   no matter what the value of b is.*

---

[7]The reason for naming this function L will appear at the end of Section 2.4,
where we will find a formula for calculating $L(b)$ from b.

[8]Let's be clear about what formula (10) says.   For each fixed choice of base b,
there is a *constant*, which we have just named $L(b)$,   that is the proportionality
constant relating the derivative to the exponential function. Because many different
numbers can be bases for exponential functions, the association of proportionality
constants to bases defines a *function*. But a single value of a function is just a number;
in spite of the notation for that number, the coefficient of b^t on the right-hand side of
(10) is still a constant.

This last observation gives us a way to actually calculate $L(b)$ for each value of b: Find the slope of the graph (or the instantaneous rate of change) of the function $f(t) = b^t$ at $t = 0$. With a little help from our graphing computer, you can do just that for the particular case of $b = 2$ in the next exercise.

Exercise 3. Figures 23, 24, and 25 show successive zooms in the vicinity of the point $(0, 1)$ on the graph of $f(t) = 2^t$; use these figures to calculate an approximate value for $L(2)$. Write your answer here: The slope of the graph of $f(t) = 2^t$ at $t = 0$ is approximately ________________.

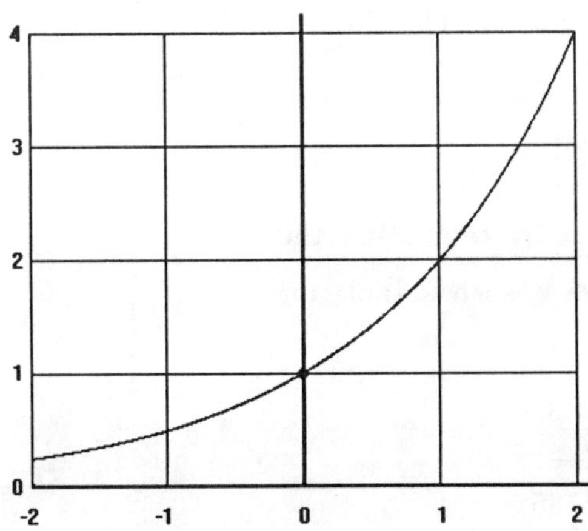

Figure 23. Graph of $f(t) = 2^t$.

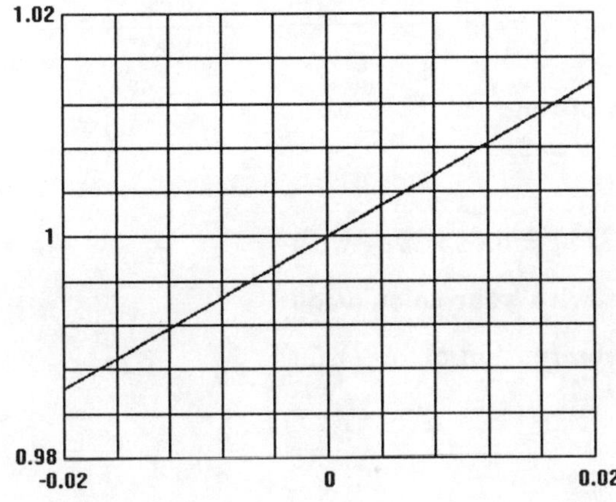

Figure 24. Zooming in by a factor of 0.01.

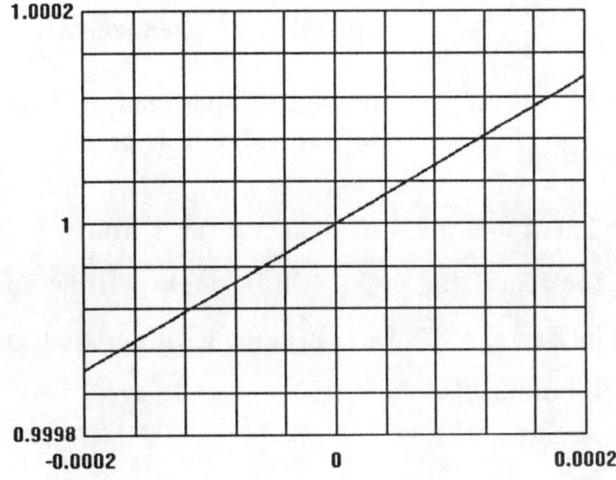

Figure 25. Zooming by another factor of 0.01.

Experimental Determination of Limiting Values

It's time to abbreviate "what something approaches as something else approaches yet another thing." A standard term for the "what" (usually a number) being approached is **"limiting value."**[9] Thus, the derivative of f at t is the *limiting value* of $\dfrac{\Delta f}{\Delta t}$ as $\Delta t \to 0$.

You can use your calculator to determine the slope at $t = 0$ on the graph of $f(t) = b^t$ (for any particular value of b) by a procedure that does not depend on our graphing computer. In the terminology just introduced in the preceding paragraph, this slope is the limiting value of the difference quotients $\dfrac{f(0 + \Delta t) - f(0)}{\Delta t}$ as Δt shrinks to zero. Therefore, if we replace $f(t)$ by b^t, we have

$$L(b) \text{ is the limiting value as } \Delta t \to 0 \text{ of } \frac{b^{\Delta t} - 1}{\Delta t}. \tag{10}$$

We can get a good estimate of such a limiting value by tabulating the difference quotients for decreasing values of Δt. Here is such a table for the case of $b = 2$:

Δt	$2^{\Delta t}$	$\dfrac{2^{\Delta t} - 1}{\Delta t}$
0.01	1.00695555	0.695555
0.001	1.000693387	0.693387
0.0001	1.0000693171	0.693171
0.00001	1.00000693150	0.693150

Table 2. Experimental determination of the limiting value in expression (10) when $b = 2$.

This table gives us evidence for a limiting value of $L(2) = 0.6931\ldots$, with the remaining digits still in doubt. [How does this fit with your calculation in Exercise 3? Does it agree with our first graphical observation of $L(2)$ in Figure 21?]

---

[9] Another standard name for this same concept is **"limit."** We choose not to use that name, for two reasons: (a) It is also the standard name for a very different concept that appears later in the course — and it's confusing to use the same name for two different things. (b) The *noun* "value" and the *adjective* "limiting" together convey the right idea, a *number* that is approached through some limiting process.

Exercise 4. (a) Check our calculations in Table 2 with your calculator — you may get slightly different answers, depending on your calculator.

(b) Why are we in doubt about digits of $L(2)$ beyond the first four?

(c) Make a corresponding table with $\Delta t = -.01, \ -.001, \ -.0001, \ -.00001.$ Do you get more evidence, less evidence, or about the same evidence about the value of $L(2)$? Do you get evidence for a different value of $L(2)$?

Exercise 5. (a) Why did we carry one additional digit in each entry of the second column in Table 2? What would happen if we didn't do that? (If your calculator carries fewer digits than ours, you already found out in the previous exercise. If you find these questions confusing, review Section 1.7.)

(b) What happens if you extend the table several lines further? Try it and see. Does this alter your opinion about the correct value for $L(2)$? Why or why not?

Exercise 6. You may have already associated the number "0.6931" with "2" in some other context. If so, you may have a conjecture about the relationship that gives $L(b)$ as a function of b — in particular, why we called this function L. Write down your conjecture. (If you can't do this yet, don't worry; we will return to this question later.)

Exercise 7. (a) Make similar tables (in the spaces below) for $b = 3$ and $b = 4$.

(b) What do you conclude about $L(3)$ and $L(4)$?

(c) If you made a conjecture in response to the previous exercise, test your conjecture against these results.

Δt	$3^{\Delta t}$	$\dfrac{3^{\Delta t} - 1}{\Delta t}$	Δt	$4^{\Delta t}$	$\dfrac{4^{\Delta t} - 1}{\Delta t}$
0.01			0.01		
0.001			0.001		
0.0001			0.0001		
0.00001			0.00001		

Touching Base

Sarah and her mother were driving to the mall; Sarah was driving, her mother was offering advice. "Look out for that road coming in on your right. I was almost hit by a pizza delivery car there last week. Just came whizzing out of the parking lot. You need to move over into the left lane, Dear."

Sarah was becoming irritated. She decided to shift her mother's attention. "How are you coming with the exponential functions?"

"Well, I believe that the derivatives are multiples of the given functions. The graphs are pretty convincing. But I don't know what those multiples are! This mysterious $L(b)$. I guess I know approximately what these numbers are when b is 2 or 3 or 4. But that's all. I asked your father last night; he just said to use the exponential key on the calculator. But that is for some function e^x. What is e, and why does the calculator have that key and not a key for 2^x or 3^x?"

Sarah tried to decide how to reply. Why was e so important, after all? Her mother brought her back to her driving. "I think you should have turned left, Dear."

Sarah pulled into a fast food restaurant, cruised through the drive-in window without stopping to explain, and headed back the other direction. Her mother was frowning. Quick...get her back onto exponential functions. "I think e is the one value of the base b for which $L(b)$ is exactly one. So that means that the derivative of e^t is again the function e^t."

Sarah made the right turn to put her back on the road to the mall. Then she glanced at her mother. She had pulled the calculator out and was graphing something. "What are you doing mother?"

"Just graphing the approximate derivative of e^t together with the function. They do seem to overlap."

*"You know mother, they discuss all that in the **Reader**," Sarah said in an exasperated tone.*

"Yes, but they sound just like your father. Have you ever had him give you directions? He always leaves something out or adds something else in. Besides, this is what the book says you should do: Try things out for yourself."

By this time they had reached the mall. Sarah's mother put away the calculator, and no more was said about exponential functions.

The Natural Base

One lesson we can draw from the section on "Experimental Determination of Limiting Values" is that the connection between the base b and the constant factor $L(b)$ in equation (9) is rather intricate. Here is the equation again, for your convenience:

$$\frac{db^t}{dt} \;=\; L(b)\, b^t, \tag{9R}$$

where

$$L(b) \text{ is the limiting value as } \Delta t \to 0 \text{ of } \frac{b^{\Delta t}-1}{\Delta t}. \tag{10R}$$

This description of $L(b)$ defines it as a function of b but does not tell us how to find values of that function. We have determined three such values approximately, those for $b=2$ (in Table 2) and for $b=3$ and 4 (in Exercise 7); here are those values:

b	$L(b)$
2	0.6931...
3	(Fill in from Exercise 7.)
4	(Fill in from Exercise 7.)

The same computer that drew the graphs in Figure 21 is happy to tell us the y-coordinates at which it drew the dotted lines in that figure, so we can extend our table of values from three to six entries. In Figure 26, we plot this scanty information about $L(b)$ as a function of b.

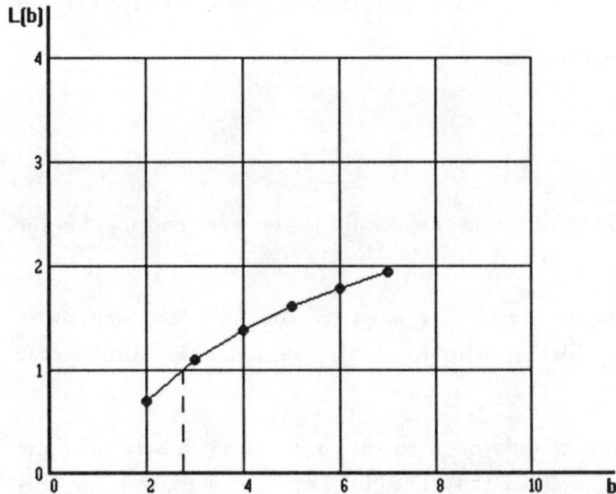

Figure 26. Known values of $L(b)$ as a function of b.

Figure 26 suggests the possibility of a smooth curve that passes through the points $(b, L(b))$ for $b = 2, 3, 4, 5, 6, 7$; to enhance that suggestion, we have "connected the dots" with straight line segments that should *approximate* such a curve. Pencil in the curve being approximated. Have you seen a curve with this shape somewhere before? [Think back — or look back — to our work with logarithms in Chapter 1.]

The figure now suggests a way to finish our derivative calculation, at least for one base b, *without* having to calculate $L(b)$: Your curve in Figure 26 crosses the horizontal line $L(b) = 1$ for some b; we can *choose* b cleverly so that $L(b)$ turns out to be 1.

Exercise 8. *Estimate from your graph about how big b has to be to make $L(b) = 1$. [You can use linear interpolation if you wish; you have the coordinates of the points in the table preceding Figure 26.]*

In class, you will do a numerical experiment to determine the first few digits of this "natural" base, the number b such that $L(b) = 1$.[10] The entire world of science and technology knows the natural base for exponentials as "e."[11]   Thus, "**the natural exponential function**" is $f(t) = e^t$. The usual name for this function is "**exp**," so we may write $\exp(t)$ in place of $f(t)$; the value of $\exp(t)$ is e^t.

Your calculator knows the natural base and its exponential function; the key for the function is either $\boxed{e^x}$ or $\boxed{\exp}$.[12] Enter 1 in your calculator and press the exponential key;[13] the result will be the calculator's best estimate of $\exp(1)$, which equals e.

---

[10]You may have already done this experiment, especially if you are reading this for the second or third time.

[11]The symbol "e" for the natural base was first used by the 18th century Swiss genius, Léonhard Euler (pronounced "oiler"), who modestly named this number for himself.

[12]If you don't find a key on your calculator with either of these labels, use the "inverse function" key (probably $\boxed{\text{Inv}}$) followed by the $\boxed{\ln}$ key. The reason for this combination will become clear in Section 2.6.

[13]On some calculators, the correct procedure is the other way around: Press the exponential key, then 1 and Enter.

Now we can see what is "natural" about the natural exponential function: Let's write down equation (9) with b replaced by e and $L(b)$ by 1:

$$\frac{de^t}{dt} = e^t.\tag{11}$$

Thus, the derivative of exp is *itself*, not just proportional to itself. In particular, we don't need to find the limiting value in expression (10) to complete this derivative calculation.

Exercise 9. *(a) Expression (10),*

$$L(b) \text{ is the limiting value as } \Delta t \to 0 \text{ of } \frac{b^{\Delta t}-1}{\Delta t}.\tag{10R}$$

with b replaced by e and the limiting value $L(b)$ replaced by 1, gives us a definition of e: "e is the number such that ...". Finish the definition.

(b) Interpret this definition (as you did in Exercise 3) as giving us the slope of the graph of exp at a particular value of t. What value of t? What is the slope there?

(c) Fill in the blanks in the following characterization of exp: "Among all exponential functions, the natural one is the one that has slope __________ at $t =$ __________."

Exponential Functions and Population Growth

Several days later Sarah's mother came into Sarah's room. Sarah was lying on the bed, reading a murder mystery. **Pride and Prejudice** *lay unopened on the table beside her.*

"I hope you don't mind my asking these questions, Dear. It's hard when there isn't anyone to talk to about these ideas. It all begins to mush together in my mind."

Sarah hadn't thought much about learning from other people in her class. But they had talked a lot in the lab and when they were working on the group projects. She guessed that the talks had helped her — except when she was working with her third lab partner. What a jerk! He had to tell everyone the proper way to work every problem. It wouldn't have been so bad if he hadn't been wrong more than he was right.

"Sure, Mom, I don't mind. How's it going?"

"Well, I know that the derivative of any function of the form b^t is a multiple of the original function. But I don't know what that multiple is except when b is e. For that function, I know the multiple is one. It seems like a lot of work to only know how to differentiate one new function.

Besides that, you never said why I want to work with these functions that have derivatives that are multiples of themselves."

"That's coming. Basically, if there aren't any natural forces to limit the growth of a population, like disease or a limited food supply, then the rate of growth of a population is proportional to the population itself. So if, say, $P(t)$ is the population at time t, then the rate of change of P is a multiple of P."

Sarah had written

$$\frac{dP}{dt} = kP$$

on the back of an envelope. She felt very pleased with herself. Maybe she was going to be good at this. Maybe she should consider being a teacher ... if only teachers made more money.

Her mother pointed to the envelope. "What's the difference between $P(t)$ and P?"

"There isn't any difference — they mean the same thing."

"Then why do the authors sometimes use one and sometimes another?"

That was a hard question. She had worried about that too. Then eventually she had got used to it, and it didn't bother her any more. "I guess it depends on whether you want to stress the t or not. Let me see; they say something about that in here."

She searched around and found the section she was looking for.

Suppose you want to study, over a given period of time, a population of animals in some region. That population might consist of bacteria in a Petri dish, insects in an agricultural area, bear in a national forest, or crabs in the Chesapeake Bay. Or you might be interested in studying higher life forms, such as the human population in a city, a nation, or the entire world. Whatever the population of interest, let's denote its number at time t by P; then P is a function of time t whose domain is the time period of interest.   When we need a functional notation, we will write $P = P(t)$. This is an ***abuse*** of notation, as it deliberately confuses the function with the dependent (or output) variable.   However, this is also a common convention in the scientific literature, and usually no real confusion results. In fact, there is a definite advantage in not using any more symbols than are absolutely necessary to make ourselves clear.

"Well, I suppose I'll get used to it," her mother said, without too much conviction.

"Anyway," Sarah went on, "once you have the derivative of one exponential function, it's easy to describe the others; that's what happens in the next section."

Symbolic Differentiation of Exponential Functions: The Natural Base

Have we really accomplished something by introducing e and concentrating on the one function $f(t) = e^t$? Suppose we consider the function $g(t) = e^{3t}$. We know from a law of exponents that this can be rewritten as $g(t) = \left(e^3\right)^t$. This gives $g(t)$ the form b^t, where b is e^3, or, what amounts to the same thing,[14] $3 = \log_e b$. Suppose we turn this around; in general, $b = e^{\log_e b}$, so $b^t = \left(e^{\log_e b}\right)^t = e^{t\,\log_e b}$. Thus, any function of the form b^t is just e^{kt}, where $k = \log_e b$.

So now we need to know how to differentiate functions of the form e^{kt}, given that $\frac{de^t}{dt} = e^t$. As often happens, it is both *simpler* and *more efficient* to study a more general problem. Here is the more general problem: Suppose k is a constant and we know the derivative of $f(t)$. What is the derivative of $f(kt)$?

We claim that the following statement[15] is true — and we will soon show why:

If $f(u)$ is any function that has a derivative, and $u = kt$ for some constant k, then

$$\frac{d}{dt}\, f(kt) \;=\; k\, \frac{d}{du}\, f(u). \tag{12}$$

Before we do a little algebra to show that (12) is true, let's check it out in two cases for which we know how to calculate both sides of the proposed equation.

Exercise 10. (a) *Let $f(u) = u^3$ and $k = 2$, i.e., $u = 2t$. Show that $f(kt) = 8t^3$ and that the derivative of this function is $24t^2$. Now show that $k\, \frac{d}{du}\, f(u)$ is also $24t^2$.*

(b) *Verify that the two expressions in (12) are equal if $f(u) = u^3$ and $u = 7t$.*

---

[14]If this equivalence is not clear, see Chapter 1 for a review of logarithms.

[15]The differentiation rule in (12) is a different "factoring-out-constants" rule from the one we saw in the Constant Multiple Rule. Here the constant k multiplies the *input* to f, not the output. Rule (12) is a special case of the "Chain Rule," which we will take up in Chapter 4.

We consider now a familiar instance of (12). Suppose u represents time measured in minutes, and t represents the same time measured in hours. Then $t = \frac{u}{60}$ or $u = 60\,t$. Now suppose position s is given by a function of time *in minutes*, say $s = f(u)$. Then the corresponding expression for s as a function of time *in hours* is $s = f(60\,t) = g(t)$. Note that f and g are *different* functions; g is computed by multiplying the input by 60 and then applying f.

How do you think $\frac{ds}{du}$ and $\frac{ds}{dt}$ are related? Is the rate of change in units per minute *larger* or *smaller* than the rate of change in units per hour? How much larger or smaller? If you are not sure of your answers, think instead of a given speed on your speedometer, say 45 miles *per hour*. How many miles *per minute* is that? Which number is larger, and by how much?

The calculation of the exact relationship is straightforward and should confirm your intuition. Observe that

$$\Delta u \;=\; u_2 - u_1 \;=\; 60\,t_2 - 60\,t_1 \;=\; 60\,(t_2 - t_1) \;=\; 60\,\Delta t.$$

That is, a time interval measured in minutes is (numerically) 60 times as long as the same time interval measured in hours. Thus,

$$\frac{\Delta s}{\Delta u} \;=\; \frac{\Delta s}{60\,\Delta t} \;=\; \frac{1}{60}\frac{\Delta s}{\Delta t}\,;$$

the average rate of change per minute is one-sixtieth the average rate of change per hour. Since $\Delta t \to 0$ when $\Delta u \to 0$, the same is true for instantaneous rates of change:

$$\frac{ds}{du} \;=\; \frac{1}{60}\frac{ds}{dt}.$$

In particular, when your speedometer reads 45 MPH, your speed can also be expressed as three-fourths of a mile per minute.

Another way to write this result is $\frac{ds}{dt} = 60\,\frac{ds}{du}$, or

$$\frac{d}{dt}\,f(60\,t) \;=\; 60\,\frac{df}{du}.$$

In words, when the unit of time is changed from minutes to hours, the rate per hour is 60 times the rate per minute. A mile per minute is 60 miles per hour.

But the implications of this simple observation go far beyond what you've always known (at least since Driver's Ed.) about your speedometer. First, the "60" in our calculation obviously could have been *any* constant. Second, s could have been *any* function. In particular, it could have been an exponential function, $s = f(u) = e^u$, where $u = kt$.

Exercise 11. (a) *Let's try this out. Let* $k = 3$, *and let* f *be the exponential function. Use (12) to show that* $\frac{d}{dt} e^{3t} = 3e^{3t}$.
(b) *Show that, for any constants* k *and* A,

$$\frac{d}{dt} A e^{kt} = k A e^{kt}. \tag{13}$$

Symbolic Differentiation of Exponential Functions: Other Bases (Optional)

Equation (13) is written in a form that we will apply to population growth problems in Section 2.5. It is also of interest just as a differentiation formula in the particular case in which $A = 1$:

$$\frac{d}{dt} e^{kt} = k e^{kt}. \tag{14}$$

At the start of the section "Symbolic Differentiation of Exponential Functions: The Natural Base," we saw that we could calculate the derivative of b^t for an arbitrary base b by rewriting b^t in a form that uses the natural base: $b = e^{\log_e b}$, so $b^t = \left(e^{\log_e b}\right)^t = e^{t \log_e b}$, so b^t is just e^{kt}, where $k = \log_e b$. Hence, equation (14) can be rewritten in the form

$$\frac{d}{dt} b^t = (\log_e b)\, b^t. \tag{15}$$

Notice that we have now solved a problem we left dangling in the section "Graphical Determination of Derivatives of Exponential Functions": how to calculate $L(b)$ in the formula

$$\frac{d}{dt} b^t = L(b)\, b^t. \tag{9R}$$

What's the answer? In particular, what is $L(2)$? (Use your calculator.[16]) Look back at Exercises 6 and 7 in the section "Experimental Determination

---

[16]For reasons that will be explained in Section 2.6, the button on your calculator for "logarithm-base-e" is labeled "ln."

of Limiting Values," and use what you know now to check both your calculations and your conjecture.

Exercise 12. You have now found the limiting value in display (10):

$$L(b) \text{ is the limiting value as } \Delta t \to 0 \text{ of } \frac{b^{\Delta t} - 1}{\Delta t}. \qquad (10R)$$

What is it? What did the "L" stand for when we wrote the proportionality constant in (9) as $L(b)$? Fill in the blank in this updated version of (10):

$$\text{The limiting value as } \Delta t \to 0 \text{ of } \frac{b^{\Delta t} - 1}{\Delta t} \text{ is } \underline{\hspace{2cm}}. \qquad (16)$$

Exercise 13. As an added bonus, equation (15) also reveals, for an arbitrary constant k, the base b such that

$$\frac{d}{dt}\, b^t \;=\; k\, b^t. \qquad (17)$$

How is b determined from k? For example, what is b if $k = 0.7$?

We could use the result of Exercise 13 to find functions that model "natural" population growth, i.e., functions whose rates of change are proportional to the functions themselves, with a given proportionality constant. However, we already have in equation (13) a sufficient family of such functions — the *same* functions, really — expressed in terms of the "natural" base e. In Section 2.5, we will pursue the population modeling problem using only base-e exponential functions.

Section Summary

Use the following spaces to summarize the mathematical content of Section 2.4. Your summary should include a discussion of growth rates that are proportional to the amount present (for example, compound interest on a bank account), using a calculator to find limiting values, the natural base for logarithms and exponentials, and derivative formulas for exponential functions.

Growth Rates Proportional to Amount Present (Compound Interest)

How to Find Limiting Values with a Calculator

The Natural Base for Logarithms and Exponentials — What Makes It Natural?

Derivative Formulas for Exponential Functions

Exercises

14. *Rewrite each of the following equations in exponential form:*

(a) $\log_2 4 = 2$ 　　　　(b) $\log_{10} 100 = 2$ 　　　　(c) $\log_{10} 0.01 = -2$

15. *Solve each of the following equations for t:*

(a) $t^6 = 10$ 　　　　(b) $e^t = 10$ 　　　　(c) $10^t = 6$

16. *Calculate each of the following derivatives:*

(a) $\dfrac{d}{dt}\, e^{-2t}$ 　　　　(b) $\dfrac{d}{dt}\, e^{0.07t}$ 　　　　(c) $\dfrac{d}{dt}\, 2^t$

(d) $\dfrac{d}{dt}\,(2t - 5e^t)$ 　　　　(e) $\dfrac{d}{dt}\, 4\,t^3$ 　　　　(f) $\dfrac{d}{dt}\, e$

17. (a) *Express $15\, e^{0.15t}$ in the form cb^t.*

(b) *Express $3 \cdot 10^t$ in the form ce^{kt}.*

18. *Find the derivative of each of the following functions:*

(a) $t^4 - 2t^3 + 2t^2 - t - 1$ 　　　　(b) $t^5 + t^3 - t^2 - 9 + e^{-t/3}$

(c) $4e^4 - 3e^3 + 2e^2 - e + 7$ 　　　　(d) $13 - 26t + 6t^2 + e^t$

2.5 Modeling Population Growth

After supper, Sarah was out for a walk with her mother. Her father and brother were washing the dishes — it was about time her brother had to do his share. Sarah was uneasy. Her mother usually wanted to talk to her about something when she suggested a walk. She knew it was going to be about Stan. Why did Stan always bring up politics when he was talking to her parents?

"Sarah, I've done all this work and now I'm just discouraged."

"Mother, I can explain about Stan —"

"Stan! I'm not talking about Stan. I've done all this work, learning about derivatives and exponential functions. But I still haven't seen how this can apply to population growth. I don't want to give up, but it seems I am spending all my time just getting ready. Like preparing for a trip — packing suitcases, getting someone to pick up the mail, someone to look after the dog — but never going anywhere."

Sarah took a while to respond. She had been deep in her defense of Stan; it took time to switch gears. Finally, she replied: "You should be ready to start now. You have everything you need to look at exponential growth. That should be coming up in the next section."

"Oh, good. I'll keep at it."

They walked on, listening to the traffic noises and children's shouts. Sarah's mother looked deep in thought. "I'm glad you did mention Stan though. I do wish he wouldn't contradict your father on everything."

A Difference Equation and a Differential Equation

We return to our investigation of models of population growth. In Sections 2.1 and 2.2, we digressed into the realm of physical motion as a metaphor for rates of change because the odometer-speedometer connection and falling objects are familiar to almost everyone. Along the way we established the conceptual and notational context in which we can discuss problems of population growth.

Our first biological assumption is that populations tend to grow at a rate proportional to their numbers. In particular, this says that the more individuals there are in a population, the faster the population grows. However, "proportional" says something more specific about the rate of growth: For a given time step Δt, the mathematical interpretation of this biological principle is

$$\frac{\Delta P}{\Delta t} = kP, \qquad (18)$$

where k is a constant, and P is the population at the start of a time interval of length Δt. Because of the *differences* (ΔP and Δt) and the assertion of *equality* appearing in formula (18), it is called a **"difference equation."**[17]

For some populations (bears, for example), Δt is essentially constant; that is, there are fixed times at which reproduction is possible. If we assume that conditions are so favorable for our population that few or no deaths occur in the time frame of interest, and that there are no migrations in or out of the population, then reproduction accounts for *all* the change in the population, and we may reasonably assume that Δt *is* constant, that it measures the smallest interval between reproduction times. On the other hand, there are biological populations (bacteria and humans, for example) in which reproduction can and does occur at *any* time; that is, there is no fixed or smallest value that Δt can assume.[18]

Now let's reexamine the "natural" biological growth equation,

$$\frac{\Delta P}{\Delta t} = kP, \qquad (18R)$$

under conditions of *continuous* change. This equation says that, no matter how small Δt is, the *average* rate of change of P is proportional to P. We need to state that a little more carefully: The proportionality constant

---

[17]We encountered a concrete example of such a difference equation in the section on "Interest Rates and Exponential Growth." There, our P represented the bank balance (think "principal"), Δt was the time between interest compoundings (one year in our example), and k was the interest rate for the compounding period (0.05 in our example). The upshot of our calculations was that the bank balance at time t could be represented by an exponential function with base $b = 1 + k$. At the end of the chapter, we will see that *all* solutions of difference equations of the form (18) are exponential functions, no matter what the time step or the proportionality constant.

[18]News item in *The News and Observer* (Raleigh, NC), June 10, 1989: The official time and date of birth of a boy born in Eldersberg, Maryland, was 1:23:45 P.M. on 6/7/89. The mother was "convinced the stars promise a bright future for her 8-pound, 6-ounce baby." The same paper reported on July 14, 1990, that a Massachusetts couple and their obstetrician made a deliberate — and successful — effort to have a baby born at 12:34:56 on 7/8/90. How remarkable do you think either of these events is?

k actually depends on the time step (but not on the population); for each *fixed* Δt, no matter how small Δt is, the average rate of change of P is proportional to P. If we let $\Delta t \to 0$, we have already seen that the left-hand side of (18) approaches the *instantaneous* rate of change of P. Of the two factors on the right, only the k can change as Δt changes; P stands for the population at the start of a time interval, and that is independent of the length of the interval. Thus, the instantaneous rate of change of P must also be proportional to P, i.e.,

$$\frac{dP}{dt} = KP, \tag{19}$$

where K is whatever k approaches as Δt approaches zero. For example, an instantaneous growth rate of slightly less than 2% is necessary if the average annual rate is to be 2%, because the instantaneous rate is being applied later in the year to a bigger population than was present at the start of the year. Thus, if $k = .02$ when $\Delta t = 1$, then k shrinks a little as Δt gets smaller, which means K will be a little smaller than $.02$ for the same population. Later in the chapter we will work out how to find K.

Since the *differences* in equation (18) have become *differentials* in equation (19), we call (19) a **"differential equation."**[19] To summarize, our model of "natural" biological growth is equation (18) if the reproductive times are discrete, equation (19) if reproduction is occurring continuously. As we consider influences that hinder or enhance natural growth rates, we will modify the models accordingly. Our immediate goal is to determine the implications of the differential equation (19), i.e., to *solve* the equation in order to find the functions that can represent natural populations.

In Chapter 1 we raised some questions for which the answers had to be expressed in terms of *functions* rather than in terms of numbers or variables — for example, "What functions are additive?" or "What is the inverse of base-10 exponentiation?" Now we have posed another kind of problem that

---

[19]In the section on "The Derivative," we introduced the concept of differential, but we have yet to say what a differential is. The presence of a *derivative* (the apparent "ratio" of differentials) in (19) suggests the name "derivational equation," but that term is *never* used.

calls for an answer of the same type: Given an equation that describes an *instantaneous* rate of change, what **functions** "satisfy" that equation?

Direction Fields

We have just seen that "natural" biological growth of populations that reproduce continuously can be modeled reasonably by equation (19), i.e., it is reasonable to assume that the *instantaneous* rate of change is proportional[20] to the population:

$$\frac{dP}{dt} = kP. \tag{19R}$$

In this section we will learn how to draw a picture of the differential equation that will help us understand what it means for a function to be a solution.

In order to have a specific example in mind, we set $k = 0.25$, so our specific objective is to draw a picture of the differential equation

$$\frac{dP}{dt} = 0.25P. \tag{20}$$

We already know some solutions of equation (20), such as $f(t) = 3e^{0.25\,t}$. This function is a solution, because, according to formula (13), $f'(t) = 0.25 \cdot 3e^{0.25\,t} = 0.25 f(t)$.

Exercise 1. Write down three more functions that are solutions of (20).

In Figure 27 we have drawn the graphs of a number of solutions of equation (20). In the following figure, we plot the same solutions, but only short segments of each solution curve. The short segments still give us a good idea of what the solution curves look like, and we could draw in the rest of the curves if we wanted to.

---

[20]We are changing the name of our proportionality constant in equation (20) from K to k, because the lower case letter is more commonly used for this purpose. In the previous section, we needed both symbols in order to make a distinction between continuous and discrete models. We will not take up the discrete model again until Section 2.8.

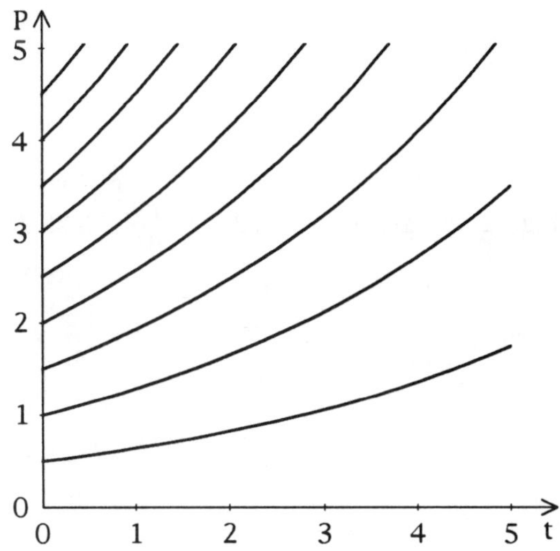

Figure 27. Solution curves for the differential equation $\dfrac{dP}{dt} = 0.25P$.

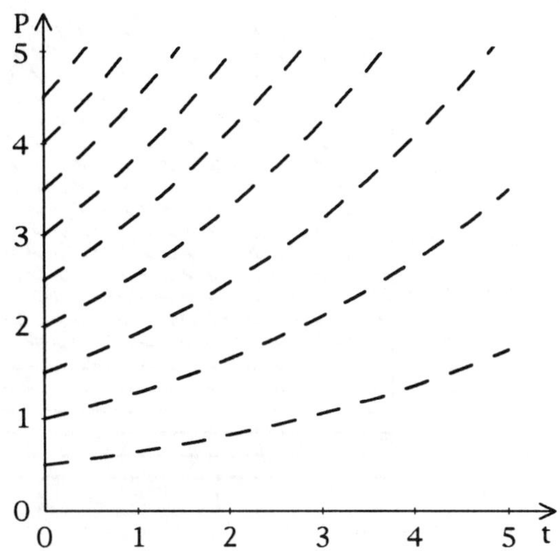

Figure 28. Segments of solution curves for the differential equation $\dfrac{dP}{dt} = 0.25P$.

Notice that the curve segments are essentially straight; i.e., they are essentially line segments. Here is the important point: **We know what the *slopes* of these line segments must be from the differential equation; we do not need to know formulas for the solutions in order to determine directions.** For this particular differential equation, we know that the slope of a very short piece of solution curve (a line segment) at any point (t,P) is $0.25P$. So, for example, the slope at $(1,2)$ is $0.25 \times 2 = 0.5$, the slope at $(2,6)$ is 1.5, and so on.

This gives us a way to represent the differential equation (20) graphically without reference to its solutions: At lots of points in the t,P-plane, we draw short line segments with slopes that match the equation. Figure 29 is a such picture of equation (20): At each point (t,P) on a closely spaced grid, we have drawn a small line segment whose slope is $0.25\,P$. The resulting picture is called a **"direction field"** or a **"slope field."**[21]

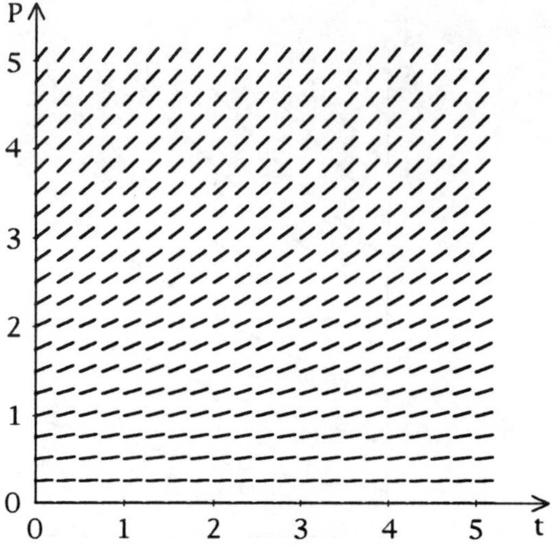

Figure 29. Direction field for $\frac{dP}{dt} = 0.25\,P$.

The selected points are only "representative" of the infinity of points in the t, P-plane, but they show us a clear picture of the direction field described by the differential equation.   Any solution of the differential equation is a function, $P = P(t)$, whose graph passes through the direction field in the directions of the line segments it meets along the way.   In Figure 30, we show the same field with several solutions superimposed.

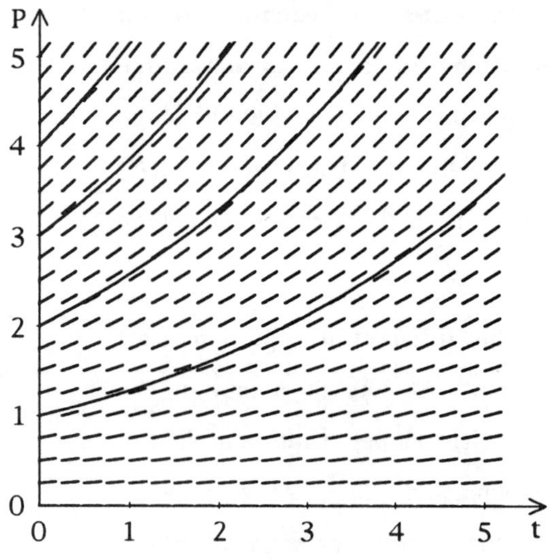

Figure 30. Same direction field with several solution curves.

---

[21]This conversation actually took place in a classroom:

Student: But what *is* a direction field?

Teacher: Let's think of something analogous. What's a sunflower field?

Student: A field full of sunflowers.

Teacher: Right. Now, what's a direction field?

Student: A field full of directions!

Exercise 2. Sketch two more solution curves on the direction field in Figure 28.

Tracing Out Solution Curves

*Sarah was on the phone talking to Charlene, who was having trouble with her boy friend. This was nothing new — Charlene was always having trouble with her boy friend. Sarah saw her mother come in the room with her finger in the **Reader**. Her mother frowned and sighed heavily. Sarah listened to Charlene's complaint a little longer, then broke in: "Well Charlene, Richard may have the sensitivity of a pumpkin at times. Still, he did come through last month on your birthday. Look, can I call you back? I need to discuss something with Mom."*

Sarah's mother sat down on the sofa. Sarah joined her. "Looks like the calculus is getting you down."

"Well, Sarah, things were going fine until I got to these direction fields. What are all these little lines?"

"Oh, those. They're directions," Sarah replied.

"What do you mean, they're directions?"

"That bothered me too. I finally figured out that you should think of them as arrows pointing from left to right."

"All right. So they are arrows. What does that tell me?" Sarah's mother still did not seem satisfied.

"One of the TA's told me to think of it like this. Suppose you are a little person riding around in a little car on the t,P-plane. Where you start is determined by the initial condition. Once you start, where you go is completely determined by the differential equation. Whenever you come to a point in the plane that is the start of one of the arrows, you will be headed in the direction of the arrow. If you put your finger down here, then you can pretty much see what the graph of the solution has to be."

Sarah traced out a curve with the tip of her finger. "Oh, like this." Sarah's mother traced out another curve. "I think I see. So that is why the differential equation together with the starting value determines a unique function graph. ... Now tell me who a TA is."

"A teaching assistant. Often they run the labs."

Sarah's mother looked up from the book. "Oh, isn't that what you said Stan was?"

"No, Mother. Stan was a lab monitor. He is just a year older than I am; the TA's are either advanced undergraduates or graduate students."

"Yes, he does look rather young."

Differential-Equation-with-Initial-Value Problems

In general, there are many solutions of the differential equation such as (20) — *infinitely* many, in fact. However, it is certainly plausible that, if we choose an "initial point" (t_0, P_0), and if we know the direction of the solution at every point (t, P), then there must be only *one* solution that passes through (t_0, P_0). Thus, in hope of finding a uniquely determined solution (and because it's reasonable to assume we know a starting population), we turn our attention to a **"differential-equation-with-initial-value problem"**:

$$\frac{dP}{dt} = kP \quad and \quad P = P_0 \ \text{ when } \ t = t_0. \tag{21}$$

In spite of the plausibility of there being a unique solution $P(t)$ to such a problem, the process of actually *solving* a differential-equation-with-initial-value problem is challenging.   In particular, strictly algebraic operations are not adequate tools, because the derivative is not a strictly algebraic object. In fact, we will sometimes be forced to treat the existence of a unique solution of an initial value problem as "evident" and to speak of "*the* function defined by" a differential-equation-with-initial-value problem, whether we know a formula for the function or not.

Fortunately, problem (21) is not difficult. We have already done most of the work. Indeed, if we let $P = Ae^{kt}$, then we know from Exercise 11 in Section 2.4 that

$$\frac{dP}{dt} = \frac{d}{dt} Ae^{kt} = kAe^{kt} = kP. \tag{13R}$$

Thus, for every constant A, the function $P = Ae^{kt}$ is a solution of the differential equation

$$\frac{dP}{dt} = kP.$$

For our differential-equation-with-initial-value problem, we need some way of picking out the particular A that we want. We have to find a function that will match a known population P_0 at a given time t_0.

Suppose we wanted to match a population of, say, 1000 at time $t = 0$.   Then we could simply declare the unit for population to be "thousands"; that would mean our desired solution would have value 1 (thousand) at time 0. Then, since 1 *is* the value of e^{kt} at $t = 0$, the problem has the solution $P = e^{kt}$. Sneaky! But how do we solve the problem if we really wanted the units to be "millions"?

We "guess" (but not blindly) that the solution to (21) is a function of the form $P = Ae^{kt}$ for some constant A (still to be determined). Now we choose A so that $P = P_0$ when $t = t_0$; we want $P_0 = Ae^{kt_0}$, so we solve this last equation for A: $A = P_0 e^{-kt_0}$. Then our solution to

$$\frac{dP}{dt} = kP \quad and \quad P = P_0 \quad \text{when} \quad t = t_0 \tag{21R}$$

is

$$P = Ae^{kt} = P_0 e^{-kt_0} e^{kt}$$

or

$$P = P_0 e^{k(t-t_0)}. \tag{22}$$

In particular, if the starting time is $t = 0$ (and we can usually arrange this by choosing our own time scale), then the solution function is the initial population times the exponential function of kt, where k is the growth rate constant.

Exercise 3. Write the last assertion of the previous paragraph in symbols — that is, write down the problem (21) and the solution (22) with the simplifying assumption that $t_0 = 0$:

Problem:

Solution:

At the start of this chapter, Sarah's mother said, *"I have been sifting through these articles on population growth models; they all use differential equations, initial value problems. So I need to know what differential equations are and how they are connected to population growth."* She has followed us through a long, difficult development, but we (and she) have now reached a point of major accomplishment — summarized in displays (21) and (22) — one marking the beginning of our quest, and the other the end.   Along the way, we formulated the central concept for the first semester of calculus, the concept of *derivative*.   We discovered the importance of exponential functions for modeling growth in which the growth *rate* is proportional to the amount present.   And, among all possible bases for logarithms and exponentials, we discovered one (a very unlikely one!) that is "natural" from the point of view of calculus, in that it leads to the simplest possible derivative formula: The derivative of the natural exponential function is itself.

Section Summary

Use the spaces provided to summarize the mathematical content of Section 2.5. Your summary should include a discussion of "natural" population growth rates, representation of these rates by differential equations, direction fields for differential equations, and the significance of initial values for singling out unique solutions of differential equations.

Natural Population Growth Rates

Differential Equations that Represent Natural Growth Rates

Direction Fields as Pictures of Differential Equations and their Solutions

Differential-Equation-with-Initial-Value Problems

Exercises

4. (a) *Write down three functions that are solutions of the differential equation*

$$\frac{dP}{dt} = 0.02P.$$

(b) *Describe the infinite family of functions that satisfy the differential equation in part (a).*

(c) *Solve the differential-equation-with-initial-value problem*

$$\frac{dP}{dt} = 0.02P \quad with \quad P = 5 \quad at \quad t = 0.$$

(d) *Solve the differential-equation-with-initial-value problem*

$$\frac{dP}{dt} = 0.02P \quad with \quad P = 5 \quad at \quad t = 2.$$

5. (a) *Write down three functions that are solutions of the differential equation*

$$\frac{dP}{dt} = 0.5P.$$

(b) *Describe the infinite family of functions that satisfy the differential equation in part (a).*

(c) *Solve the differential-equation-with-initial-value problem*

$$\frac{dP}{dt} = 0.5P \quad with \quad P = 2 \quad at \quad t = 0.$$

(d) *Solve the differential-equation-with-initial-value problem*

$$\frac{dP}{dt} = 0.5P \quad with \quad P = 2 \quad at \quad t = 5.$$

2.6. Logarithms and Representation of Data

We turn our attention now to an aspect of growth problems that is not addressed directly by the formal manipulations of calculus: Now that we have a model for growth of populations, do biological populations actually grow that way? We will see that the same exponential and logarithmic functions help us answer this and related questions. More important, they provide a set of tools that we will continue to use throughout the course.

Semilogarithmic Plots of Data

Early in this chapter we considered Thomas Malthus's concept of human population growth, which we subsequently identified as "exponential." Indeed, we have seen that the "natural" assumption about growth of biological populations (growth rate proportional to the population) leads to exponential functions representing population sizes. How can we tell whether populations really grow that way?

We consider two case studies. The first uses data on fruit flies in an ideal laboratory setting,[22] the other, historical data on the human population of the Earth. In each case we want to examine whether the data could have come from a population that is represented *approximately* by a function of the form $P = P_0 e^{kt}$. Here are the data:

---

[22]We use this same data already in Problem 15 at the end of the chapter, where we consider the same question from a different point of view. Candor compels us to admit that we made up this data — it did not come from a "real" lab.

Case 1. Table 3 lists numbers of fruit flies in a laboratory colony, under ideal conditions, for the first ten days of an experiment.

Day Number	Number of Flies
0	111
1	122
2	134
3	147
4	161
5	177
6	195
7	214
8	235
9	258
10	283

Table 3. Fruit fly data.

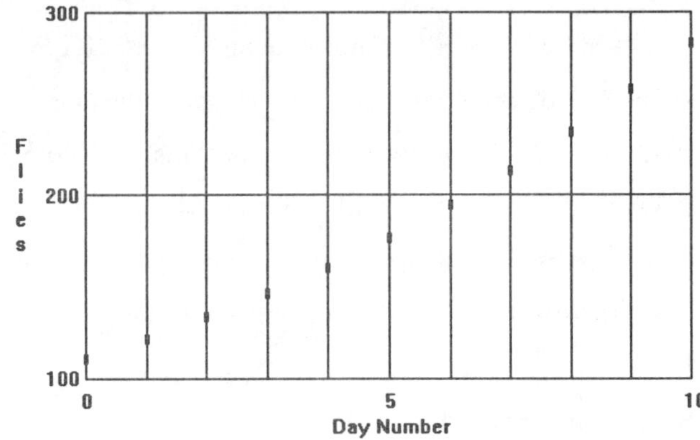

Figure 31. Graph of fruit fly data.

Case 2. Table 4 lists authoritative estimates of world population in the indicated years.

Date (A.D.)	Population (billions)
1650	0.545
1750	0.728
1800	0.906
1850	1.171
1900	1.608
1950	2.517

Table 4. World population.

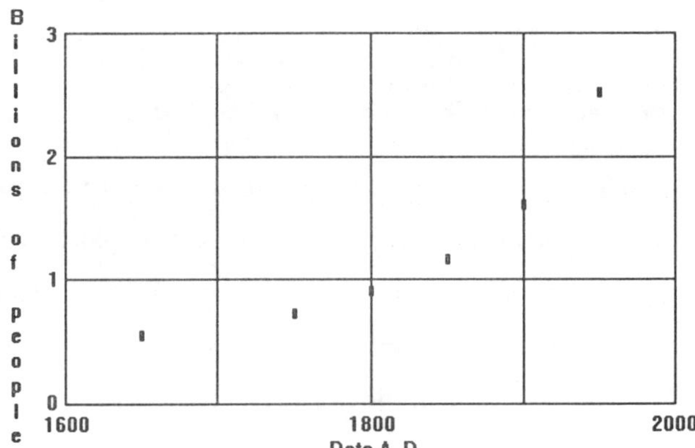

Figure 32. Graph of world population.

In Figures 31 and 32, you will have to imagine smooth curves connecting the data points. Could those curves be exponential, that is, like segments of the curves shown in Figure 27 or 30? Write your tentative answers in the margin next to each figure. It's hard to be certain at this point; that's a question we have to investigate. When we have done it, you can look back here to see if our analysis confirms or refutes your intuition.

In your previous study of algebra, as well as in Chapter 1, you learned that "taking logs" is a good thing to do when confronted with problems involving exponents, and that's the kind of problem we have here. Recall the reason for this: "logarithm" is essentially defined by inversion (undoing) of exponentiation.

In Chapter 1, we reviewed logarithms with an arbitrary base b; in this chapter, we have already determined a special interest in the "natural" base e. The corresponding logarithm is also called "natural"; its abbreviated name is "**ln**," which stands for "**l**ogarithm, **n**atural" (but which is read "**natural logarithm**"). Thus, $\ln x = \log_e x$. At the risk of belaboring the obvious, we repeat the definition of logarithm for this important case:

$$y = \ln x \quad \text{if and only if} \quad x = e^y. \tag{23}$$

Note, in particular, that if $y = 1$, then x must be e, and if x is e, then y must be 1. That is, $\ln e = 1$.

What happens if we "take logs" on both sides of the equation $P = P_0 e^{kt}$? The answer depends on properties of logarithmic functions that are the most important reasons for studying these functions: Logarithms turn products into sums and exponents into multipliers (see Problem 2 at the end of the chapter). Use those properties to fill in reasons for each step in the following calculation. Because our exponential has base e, we will use base e for the logarithms, as well. Write a complete sentence for each reason.

<u>Algebraic step</u>   <u>Reason</u>

$\ln P = \ln P_0 e^{kt}$

$\quad\quad = \ln P_0 + \ln e^{kt}$ Log of a product is the sum of the logs of the factors.

$\quad\quad = \ln P_0 + (kt)\ln e$

$\quad\quad = \ln P_0 + kt$

The result of this calculation is that $\ln P$ ("log of the population") turns out to be a constant ($\ln P_0$) plus another constant (k) times t. That is, (natural) log of the population is a *linear* function of time! Now, that's important, because it is easy to tell whether data points come from a linear function: Just plot them, and see if they lie on a straight line. Of course, we need some numbers to work with, since we were given values of P, not of $\ln P$. Turn back now to Tables 3 and 4 for Case 1 and Case 2. Make a third column for each table, headed "$\ln P$." Then use the $\boxed{\ln}$ key on your calculator to fill in the third column for each case. Those are the numbers we want to plot as functions of time.

You will get your chance to do the plotting in the next exercise. We show the plots of $\ln P$ for Case 1 (fruit flies) and Case 2 (humans) in Figures 33 and 34. (Check the numbers in your tables to see if they match our graphs.) Now you should be able to answer definitively which of these populations grows exponentially. For the one that does not, describe in your own words the *way* in which it fails to be exponential — for example, by growing too fast or too slow, or perhaps something else. (Write your description next to the figure.)

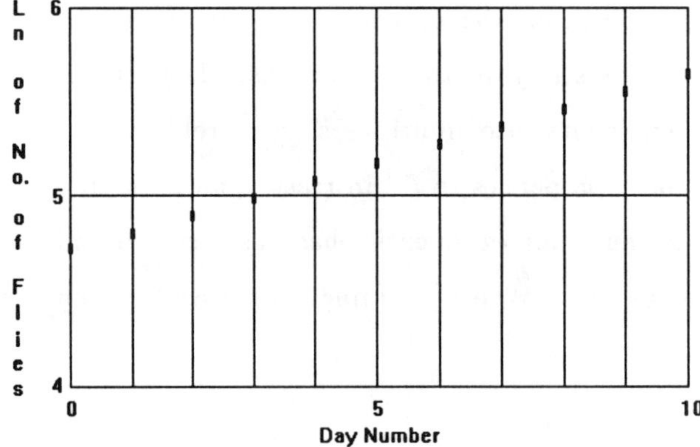

Figure 33. Graph of logarithms of the fruit fly data (Case 1).

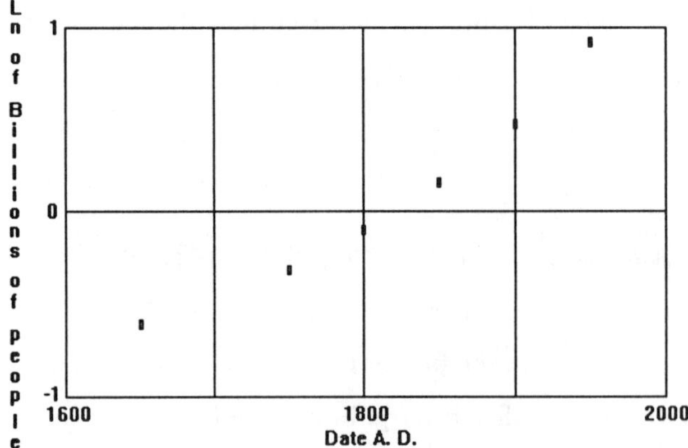

Figure 34. Graph of logarithms of the world population data (Case 2).

Exercise 1. *(a) The data in Table 5 come from the early U. S. censuses. Fill in the third column of the table, and decide whether the U. S. population grew exponentially during its first century. If so, explain why you think so. If not, explain the way in which the data deviates from being exponential.*

(b) Answer the same question for the time from the first census to the start of the Civil War.

[If you have access to a computer or graphing calculator, you can use it for both the computation of the $\ln P$ column and the graphing. Otherwise, use your (non-graphing) calculator to fill in the table, and use the graph paper on the next page for the graphing.]

Date (A.D.)	Population (millions)	ln of Population
1790	3.929	
1800	5.308	
1810	7.240	
1820	9.638	
1830	12.866	
1840	17.069	
1850	23.192	
1860	31.443	
1870	38.558	
1880	50.156	
1890	62.948	

Table 5. U. S. population for the first 100 years.

Exercise 2. *In our calculation of the linear relationship between $\ln P$ and t, we chose "ln" because the base of our exponential function was e. Show that this choice was unnecessary. That is, given $P = P_0 e^{kt}$, show that we could use logarithms with any base b and still find that "log of P is a linear function of t."*

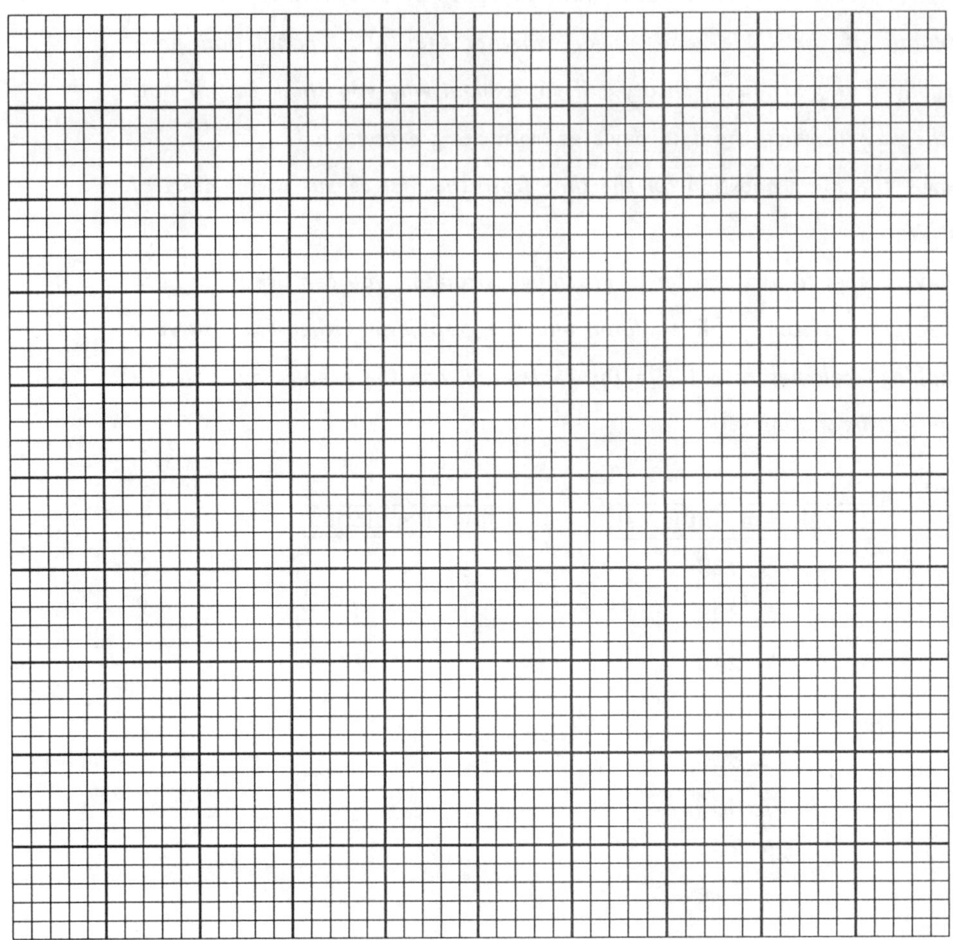

The graphs we presented in Figures 33 and 34, and the one you constructed for Exercise 1, even though they serve their intended purpose, have a distinct disadvantage as "pictures" of our data: You can't look at these pictures and read off approximate populations at particular dates. Contrast this with Figure 35 (repeated from Problem 7 at the end of Chapter 1): That's also a picture of approximately exponential data (as we shall see), and the population figures can at least be estimated from the graph.

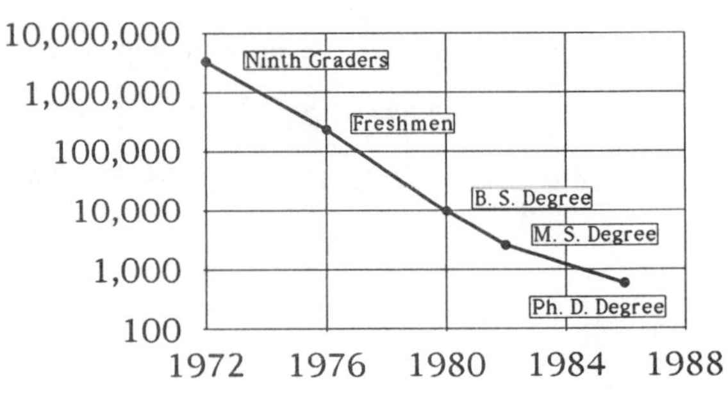

Figure 35. The Math Pipeline.

The secret of the graph in Figure 35 is its vertical scale. The numbers on that scale increase *exponentially* rather than linearly; each is ten times the one below it. But those numbers, 10^2, 10^3, 10^4, 10^5, 10^6, and 10^7, are arranged on a *linear* scale, so the effect of graphing against that scale is to "undo exponentiation," i.e., to graphically take a logarithm (what base?). This kind of graphing is called **"semilog"** plotting.[23] You can buy semilog graph paper, but of course it is much easier if your computer will do semilog plotting for you.

In Figures 36 and 37, we show semilog plots of the data from Case 1 and Case 2. Compare these figures with Figures 33 and 34, and also with Tables 3 and 4. The data in both Case 1 and Case 2 vary by less than a factor of 10, so we set the top level of the vertical scale to be 10 times the bottom level in each case. The horizontal lines show the locations of 2 times the bottom level, 3 times the bottom level, and so on. Verify that the number of fruit flies passes 200 on the right day. Check that each of the data points in Figure 37 appears to be at the right multiple of 0.3.

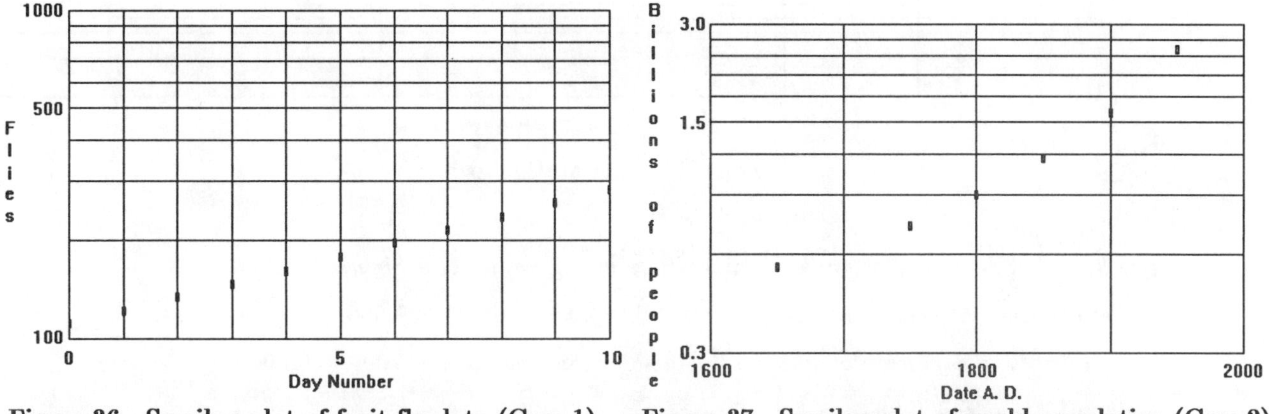

Figure 36. Semilog plot of fruit fly data (Case 1).   Figure 37. Semilog plot of world population (Case 2).

The vertical scale in Figure 35 ("The Math Pipeline") varies by many factors of 10 (how many?), or **"orders of magnitude,"** as they are called. The only problem with reading values from the graph is that the horizontal lines for "2 times," "3 times," etc., have not been drawn in as we did in

---

[23] "Semi" means half. In this case, it means we are using a logarithmic scale in half the places possible, that is, on one out of two axes. Two out of two is also possible, as we shall see.

Figures 36 and 37. Figure 38 shows approximately the same "pipeline" data[24] plotted on a true semilog graph, with the appropriate number of orders of magnitude in the vertical direction, and the logarithmically scaled reference lines within each order of magnitude.

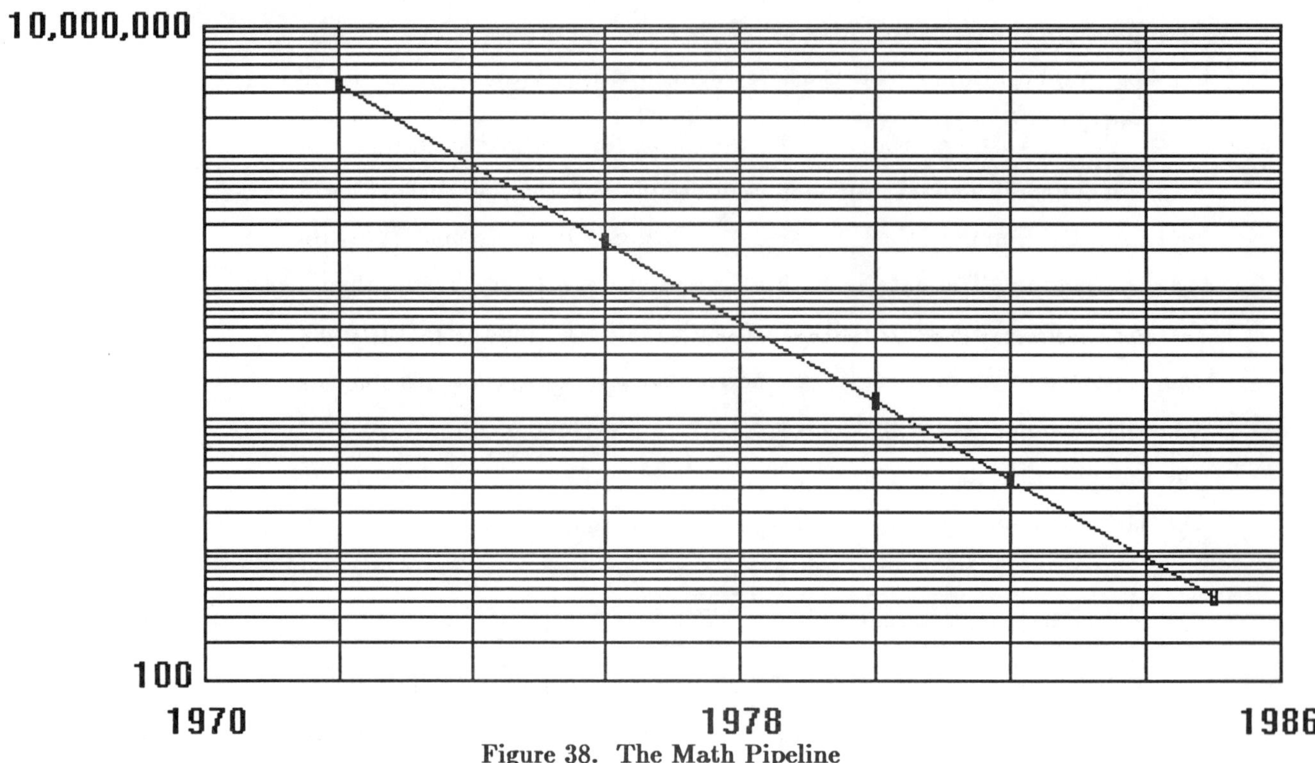

Figure 38. The Math Pipeline

Exercise 3. (a) *Compare each order of magnitude in Figure 38 with Figure 35, and label the lines that represent* 1000, 10,000, 100,000, *and* 1,000,000.
(b) *Identify the horizontal lines in Figure 38 that represent* 700, 20,000, 400,000, *and* 5,000,000.
(c) *For each of the five data points, estimate the number of students directly from the graph.*

---

[24]The numbers we used in Figure 38 are based on a discussion on pages 35 and 36 of *A Challenge of Numbers: People in the Mathematical Sciences*, by B. L. Madison and T. A. Hart, National Academy Press, 1990. These numbers are not the actual reported figures in the indicated years; for example, there were only 11,000 bachelor's degrees in the mathematical sciences in 1980, not 14,000. However, the yield that year happened to be lower than in previous and subsequent years. Thus, the near-linear relationship in Figure 38 is "typical," if not completely accurate for the given starting year.

Recall what the data points in Figure 38 represent: They track a single cohort of students, ninth graders in 1972, through the years those students would have been in college and graduate school, counting those who eventually got degrees in the mathematical sciences. The relationship on the semilog graph is nearly linear. Thus, if P represents the population in the "math pipeline," then $\log_{10}P = \log_{10}P_0 + kt$, where $t = 0$ in the starting year (1972), and P_0 is the population in that year.

Exercise 4. (a) Estimate the slope k of the linear function giving $\log_{10}P$ in Figure 38: Choose your own run, and calculate the corresponding rise.

(b) Solve the equation $\log_{10}P = \log_{10}P_0 + kt$ for P to express P as an exponential function (what base?).

(c) By what factor does P change in one year's time? Explain the following statement from Everybody Counts[25]: *"[O]n average, we lose half the students from mathematics each year."*

Power Functions and Log-log Plotting

We turn now to data of another sort, namely, (simulated) data giving the position of a falling body at half-second time intervals. Table 6 contains a refinement of the falling body data given in Table 1 earlier in this chapter, where we explored a theoretical model for this data from elementary physics: $s = ct^2$, where s is distance fallen at time t, and c is a constant. We call such a function — a constant multiple of a constant *power* of the independent variable — a **"power function."** (This is to distinguish such functions from "exponential" functions, which have constant base and variable exponent.) We can now address the question of whether the data actually fits a power function model.

---

[25] *Everybody Counts: A Report to the Nation on the Future of Mathematics Education*, National Academy Press, 1989. This report is the source of Figure 35; it is a companion to *A Challenge of Numbers*, the report from which we got the numbers for Figure 38.

Time (seconds)	Distance (meters)	Log of Time	Log of Distance
0.5	1.237		
1.0	4.962		
1.5	11.052		
2.0	19.398		
2.5	30.579		
3.0	44.113		
3.5	60.258		
4.0	78.024		
4.5	99.216		
5.0	122.757		
5.5	147.661		
6.0	175.769		
6.5	207.346		
7.0	239.973		
7.5	276.944		
8.0	312.798		
8.5	354.793		
9.0	396.130		
9.5	443.067		
10.0	489.007		

Table 6. Position of a falling object.

Exercise 5. (a) *Take logs of both sides of the proposed functional relationship,* $s = ct^2$, *and simplify as much as possible. (You decide what base to use for the logarithms.)*

(b) *The resulting equation says something is a linear function of something else. What is a linear function of what? What is the slope of that linear function?*

(c) *Propose a graphical test for deciding whether the data in Table 6 fits the theoretical model. (You may assume that a computer will do the necessary computations and graphing.)*

(d) *Fill in five rows of the "log of time" and "log of distance" columns in Table 6 (your choice of base), reasonably spaced over the whole table.*

(e) Use the graph paper below to carry out your proposed graphical test for the five points you selected. (If you have access to a computer or graphing calculator, you can let it fill in the third and fourth columns and do the graphing to carry out your test in full.) What do you conclude about whether the data fits the model?

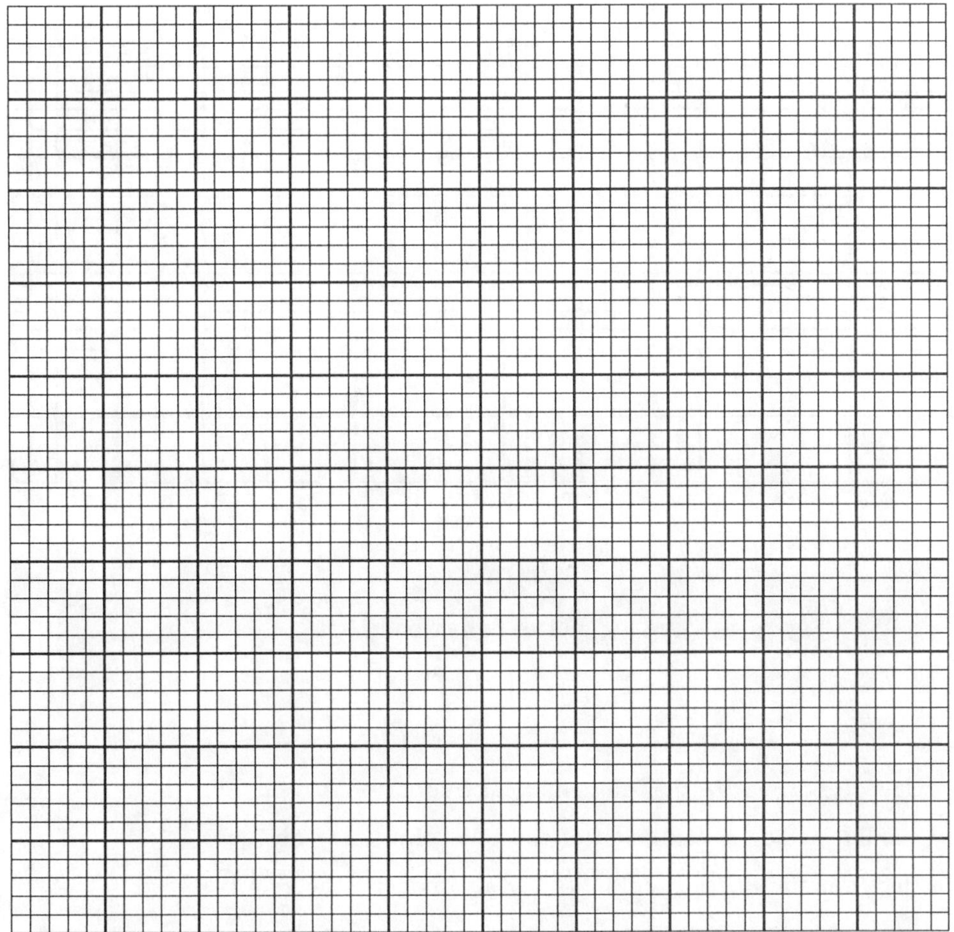

In Figure 40, we have plotted the Table 5 data with *both* scales treated logarithmically, as we did with only the vertical scale in the semilog plots. This is called a **"log-log"** plot or sometimes just a "log" plot.[26] How does this picture fit with the test you proposed in Exercise 5? Does Figure 40 confirm or refute your conclusion in Exercise 5? What do you think now about whether the data fits the theoretical model?

---

[26]You would think one "log" would be enough, since we have twice as much logarithmic scaling as in the "semilog" case. However, both "log-log" and "log" are considered standard terminology for logarithmic scaling of both axes.

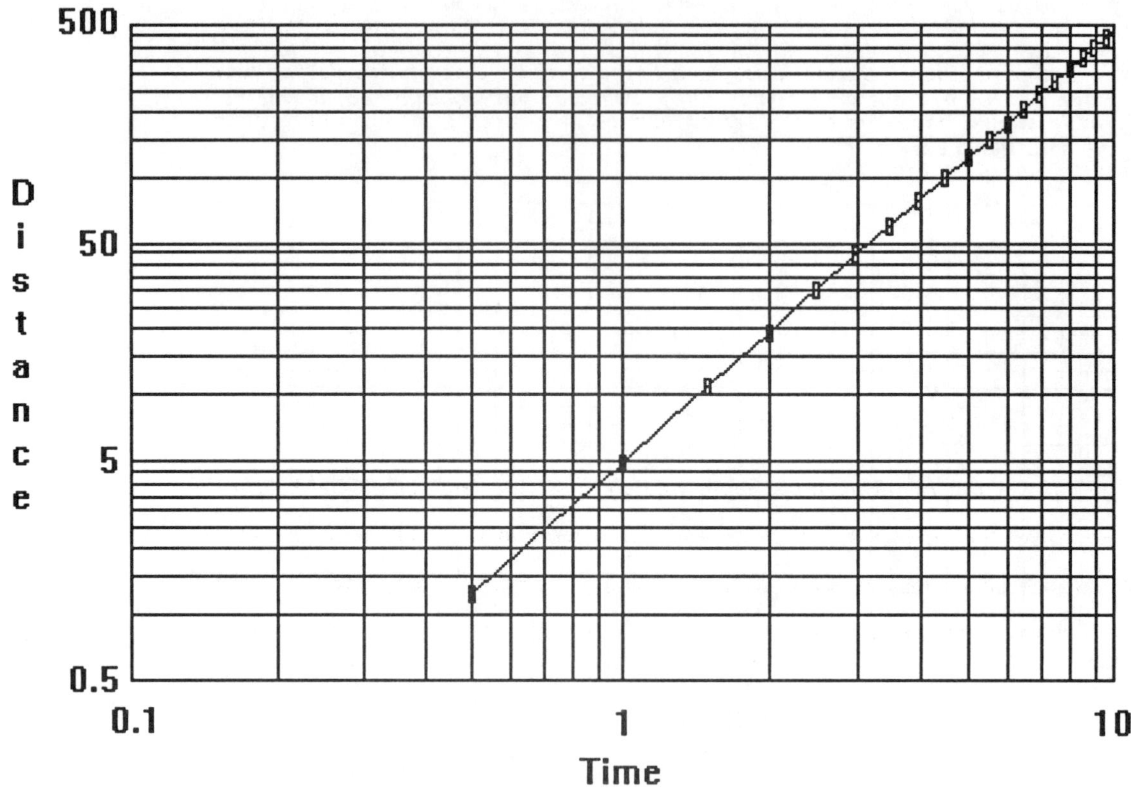

Figure 40. Log-log plot of falling body data.

Section Summary

Use the space below to summarize the mathematical content of Section 2.6. Your summary should include a discussion of the roles of semilog and log-log plotting in determining whether data fits certain models, the relative merits of plotting logarithms on Cartesian (regular) grids vs. using a computer to construct logarithmic scales, and the significance of a choice of base.

Exercises

6. *The data in the following table approximately fits either an exponential function or a power function. Decide which, and find a formula that fits the data. Your formula should have one of the forms $y = cb^t$, $y = ce^{kt}$, or $y = ct^k$, with explicit values for the constants.*

t	y
0	0.0
5	14.855
10	20.292
15	24.354
20	27.720
25	30.648

7. *The data in the following table approximately fits either an exponential function or a power function. Decide which, and find a formula that fits the data. Your formula should have one of the forms $y = cb^t$, $y = ce^{kt}$, or $y = ct^k$, with explicit values for the constants.*

t	y
0	1.0
5	4.452
10	7.339
15	12.101
20	19.950
25	32.893

8. *(a) We now know three ways to plot data: Cartesian (evenly spaced grid in both directions), semilog (logarithmic scale in one direction), log-log (logarithmic scale in both directions). In each case, seeing a straight line tells you something about the data. What does it tell you in each of these cases?*

   (b) What can you conclude about data if its plot is not straight in any of the three cases described in part (a)?

   (c) Consider the points $(0,0)$, $(1,3)$, $(2,6)$, $(3,9)$, and $(4,12)$, all of which have $y = 3x$. On how many of the three types of plots would this data appear as a straight line? Which ones, and why?

   (d) Consider the points $(0.1,-1)$, $(1,0)$, $(10,1)$, $(100,2)$, and $(1000,3)$, all of which satisfy $y = \log_{10} x$. What kind of plot of this data would appear to be a straight line? Why?

2.7 Differentiating Exponential Functions Algebraically (Optional)

In Section 2.4, we set out to find the derivative of an exponential function, b^t, a problem we attacked first from a geometric point of view. Here we revisit the problem from the perspective of algebra to see what else we might learn — about differentiation in general or about exponential functions in particular. Recall that our algebraic calculation in Section 2.2 — which showed that the derivative of ct^2 is $2ct$ — was spectacularly successful: With a few lines of relatively simple algebra, we were able to determine the instantaneous rate of change at *every* number t in the domain of the function. Now we attempt to do the same thing with b^t.

Let's lay out our tools and the job to be done on our metaphorical workbench. We have a function, $f(t) = b^t$. We want to calculate its derivative, $f'(t)$, which is the limiting value of $\dfrac{\Delta f}{\Delta t}$ as Δt approaches zero. So we have to calculate the quotient of differences, and then examine what happens as $\Delta t \to 0$. When t changes from a given value t_1 to another value $t_2 = t_1 + \Delta t$, $f(t)$ changes from $f(t_1)$ to $f(t_2)$, i.e., from b^{t_1} to b^{t_2}. In other notation,[27] with $t_1 = t$ and $t_2 = t + \Delta t$, f changes from $f(t)$, which means b^t, to $f(t + \Delta t)$, which means $b^{t + \Delta t}$. Thus, the *change* in f can be written as either $b^{t_2} - b^{t_1}$ or $b^{t + \Delta t} - b^t$.

Now we can write the quotient of differences in two ways:

$$\frac{\Delta f}{\Delta t} \; = \; \frac{b^{t_2} - b^{t_1}}{t_2 - t_1} \; = \; \frac{b^{t + \Delta t} - b^t}{\Delta t}.$$

Whichever form we use, our problem is to determine what happens to this expression as Δt approaches zero. We have two previous examples of such calculations; you wrote reasons for the steps in Exercise 7 in Section 2.2 and just before Exercise 1 in Section 2.3; you should review those calculations now.

---

[27]Just as we don't always know in advance whether to use a wrench or a pair of pliers, we won't know until we get into the calculation what notation works best for the task at hand. Thus it's a good idea to have all our tools handy, i.e., to write out each quantity in as many ways as we can.

In our present problem, there is no obvious factorization of $b^{t_2} - b^{t_1}$. There *is* an obvious factor in the other notation, however:

$$b^{t + \Delta t} - b^t = b^t b^{\Delta t} - b^t = b^t(b^{\Delta t} - 1).$$

It's not clear yet how much good this will do, but we can rewrite the difference quotient as

$$\frac{\Delta f}{\Delta t} = b^t \, \frac{b^{\Delta t} - 1}{\Delta t}. \tag{24}$$

Actually, this is more help than it may appear at first glance. We observed graphically (Figures 20 and 21) — and still need to confirm — that the exponential function b^t has a derivative that is proportional to itself. Equation (24) says exactly that, provided the second factor on the right actually approaches *something* as $\Delta t \to 0$. Why? Because the left side of (24) approaches $\dfrac{df}{dt}$, and the first factor on the right *does not involve* Δt.

In this terminology, the problem posed by equation (24) is to calculate a limiting value for

$$\frac{b^{\Delta t} - 1}{\Delta t}, \tag{25}$$

as $\Delta t \to 0$ — the same limiting value we called $L(b)$ in Section 2.4. Thus, the algebraic formulation of our problem has led us quickly to the same point we reached in a quite different way from the geometric formulation in 2.4 — but no further. And we can't go any further with algebra alone, because there isn't any algebraic way to find a factor of Δt in the numerator of (25).

What makes the calculation of $L(b)$ inherently difficult is this lack of a factorization. As $\Delta t \to 0$, so does the numerator in (25), because $b^{\Delta t} \to 1$. When numerator and denominator of a quotient are both approaching zero, we *can't tell* what the quotient is doing, at least not by casual inspection.[28] We call this situation **"indeterminate"** — which is just

---

[28]Your review of Exercise 7 in Section 2.2 should show that something very important happened between the third line and the fourth, and as a direct result of factorization: We canceled the common factor of Δt in numerator and denominator. The "answer," $2ct_1$, follows very easily from line four, but it could not have been guessed before that.

a name for the dilemma in which we find ourselves, not an aid to calculating our way out of that dilemma. Here is the fundamental problem of differential calculus: *Every derivative calculation requires the evaluation of an indeterminate limiting value, and only a handful of these can be evaluated* (as you did in Sections 2.2 and 2.3) *by algebraic manipulation.*

2.8. Difference-equation-with-initial-value problems (Optional)

In Section 2.5, we studied the implications for continuously growing populations of the biological assumption that population grows at a rate proportional to its size. We started with a mathematical formulation of this assumption in the form of a "difference equation" with a discrete time step Δt:

$$\frac{\Delta P}{\Delta t} \ = \ kP, \tag{18R}$$

where k is a constant, and P is the population at the start of a time interval of length Δt. Because our continuously growing population had no smallest time step Δt, we turned equation (18) into a differential equation, and we found that the process of solving that differential equation led in a natural way to exponential functions as the only possible solutions. In this section, we pursue the implications of equation (18) itself for populations that have a fixed inter-reproduction time Δt — for example, bears (and other wild and domestic animals), which breed once a year.

We have already studied an example of equation (18) in Section 2.4, where arithmetic calculations of compound interest led us to consider exponential functions in the first place. Indeed, compound interest and natural population growth are conceptually the *same idea*: Populations without constraints grow by adding "interest" (more individuals) to the population already present. Thus, in this section, we pursue algebraically (as opposed to arithmetically) the same kind of "compound interest" calculations we did in Section 2.4 in order to find all possible solutions of (18). Not surprisingly, the solutions turn out to be exponential functions.

We can rewrite equation (18) as

$$\Delta P \ = \ kP\Delta t. \tag{26}$$

In words, the change in population over a given time interval is *jointly* proportional to the population at the start of the interval and to the length of the interval. If, as we have assumed, this time interval is constant, then equation (26) becomes the simpler statement that the change in population in one reproductive cycle is proportional to the population at the start of that reproductive cycle.

Equation (26) can also tell us something about *continuous* population growth, as long as we limit our attention to *average* change over discrete time intervals. For example, suppose we are told that human population is growing at 2% per year, and, at the start of 1988, world population was 5 billion.[29] The information about the growth rate can now be written in the form of equation (26) as $\Delta P = 0.02\,P$, i.e., the change in one year is 2% of the population. Over the calendar year 1988, that would have been a change of 100 million, so the population at the start of 1989 would have been 5.1 billion. Then in 1989, the population would have grown another 2%, but this time it would be 2% of 5.1 billion, or 102 million. Thus, at the start of 1990, the population would have been 5.202 billion.

Exercise 1. (a) Suppose the 2% annual rate continued for the rest of the century. What would the population be at the start of 2001?[30]

(b) Do you notice any pattern in your computation in part (a)? Still assuming the continued 2% rate, can you write a formula for the population any number of years after 1988?

If you had any difficulty finding the formula in Exercise 1 (b), work through the following steps with us. If you know already what the formula is, skip to Exercise 4.

Suppose we set $t = 0$ at a time at which we know P, namely, at the start of 1988, when $P = P_0 = 5$ (billion). We calculate the *change* in P in 1988 (which is also the *rate* of change, since Δt is 1) as we did above: $0.02 \times 5 = 0.1$, so P_1 (i.e., the value of P at $t = 1$) is $5 + 0.1 \times 1 = 5.1$. In 1989, when $t = 1$ and P_1 is 5.1, the change is $0.02 \times 5.1 = 0.102$; that tells us what P must be at $t = 2$. As we saw above, this means that P_2 is 5.202. The next exercise asks you to continue this process up to $t = 7$.

---

[29]The population is approximately correct; the assumption of constant percentage growth is not.

[30]The year 2000 is the last year in the 20th century, not the first in the 21st.

Exercise 2. Fill in the rest of the following table for this example.

t	P	ΔP
0	5	0.1
1	5.1	0.102
2	5.202	
3		
4		
5		
6		
7		

Exercise 3. Make a table in the space below, similar to that in Exercise 2, for a solution to the same difference equation, but starting from $P(0) = 3.8$.

t	P	ΔP
0	3.8	
1		
2		
3		
4		
5		
6		
7		
8		

Exercise 4. (a) Rewrite equation (26),

$$\Delta P \; = \; kP\Delta t. \tag{26R}$$

with $\Delta t = 1$, in the form

$$P_{n+1} - P_n = kP_n$$

at time $t = n$. Show that

$$P_{n+1} = (1+k)P_n \quad \text{for} \quad n = 0, 1, 2, \cdots. \tag{27}$$

[Formula (27) is a recursive formula; it gives each value of P if you know the previous one. However, if you want to know P_{30}, you have to do 30 multiplications.]

(b) Write out the first several cases of formula (27), starting at $n = 0$. Find a pattern from which you can deduce an explicit *formula for P_n, one that would enable you to find P_{30} from P_0 and k with only a few keystrokes on your calculator.*

(c) Use your formula, with $k = 0.02$, to check your answer to Exercise 1 for the world population at the start of the 21st century.

We can solve equation (18) algebraically by following the same pattern of steps you carried out numerically in Exercises 2 and 3 and algebraically in Exercise 4. In order to proceed, we have to know a value of P (an "initial value") at some particular time t (the "initial time"). Since the time scale is under our control, we choose to make the initial time $t = 0$, and we denote the initial value by P_0. Thus, we are really solving the **"difference-equation-with-initial-value problem"**[31] consisting of equation (18) together with the specification of an initial value:

$$\frac{\Delta P}{\Delta t} \;=\; kP \quad and \quad P = P_0 \quad \text{when} \quad t = 0. \tag{28}$$

First, we recall that equation (26), $\Delta P = kP\Delta t$,   is equivalent to equation (18), $\frac{\Delta P}{\Delta t} = kP$; we rewrite equation (26), as in Exercise 4 (but without specifying a value for Δt):

$$P_{n+1} \;-\; P_n \;=\; k\,P_n\,\Delta t. \tag{26R}$$

Then we add P_n to both sides of the equation in order to express P_{n+1} (the "next P" at time n) in terms of P_n (the "current P" at time n):

$$P_{n+1} \;=\; P_n \;+\; k\,P_n\,\Delta t \;=\; (1 + k\Delta t)\,P_n.$$

This recursive form of the difference equation tells us how to get each P_{n+1} from its predecessor, for $n = 0, 1, 2,$ and so on. Thus, we can make a table of algebraic results, similar to the tables of numbers in Exercises 2 and 3:

---

[31]*Every* function that satisfies the difference equation (18) must have *some* initial value, i.e., a value at $t = 0$. Since we have not actually specified the initial value, our procedure will find *all* solutions of the difference equation. It is conceptually simpler to find solution functions "one-at-a-time," that is, to assume we know an initial value — whether we do or not.

n	t	P
0	0	P_0
1	Δt	$(1 + k\Delta t)\, P_0$
2	$2\Delta t$	$(1 + k\Delta t)\, P_1 \;=\; (1 + k\Delta t)^2\, P_0$
3	$3\Delta t$	$(1 + k\Delta t)\, P_2 \;=\; (1 + k\Delta t)^3\, P_0$
4	$4\Delta t$	$(1 + k\Delta t)\, P_3 \;=$
5	$5\Delta t$	

Exercise 5. *(a) Explain the algebraic step following each equals sign. Fill in the blanks in lines 4 and 5 of the table. What pattern do you see emerging?*
(b) Explain why $P_n = a^n P_0$ for some constant a. What is a?

Your explanation shows that, given a starting population, each later population value can be expressed explicitly as a function of the step number n. What we really want as a solution to the difference-equation-with-initial-value problem (28) is P expressed as a function of t. But t and n are clearly related by the first two columns of the table: $t = n\,\Delta t$, so $n = \dfrac{t}{\Delta t}$. Furthermore, $a^n = a^{t/\Delta t} = b^t$, where $b = a^{1/\Delta t}$. Thus, we have our desired form for the solution to the difference-equation-with-initial-value problem:

$$P(t) \;=\; P_0\, b^t, \quad \text{where } b \;=\; (1 + k\Delta t)^{1/\Delta t}. \tag{29}$$

Exercise 6. *Continue your explanation in Exercise 5 to explain the steps leading to the solution in equation (29), filling in any steps we may have left out.*

Exercise 7. *This spring, after cubs were born, the bear population of Black Bear Natural Wilderness Area was 317, half of them female. Each spring, 70% of the females each produce two cubs, and 50% of all the cubs are female.*
(a) Assuming no deaths, estimate the bear population in 5 years. Do your counting with whole bears, not with formula (29), which may produce "fractional bears." [Hint: You may find it simpler to count the females year by year, and then double to get the total population.]
(b) Now use formula (29) to estimate the population in 5 years. How close is this estimate to the one in part (a)?

SUMMARY

We began our discussion of population growth with an assumption about "natural" biological growth: In the absence of any significant factors other than reproductive "urges," a biological population will tend to grow at a rate proportional to its size. For discrete time steps Δt, the symbolic formulation of that assumption is the *difference equation*

$$\frac{\Delta P}{\Delta t} \;=\; \text{constant} \cdot P, \tag{30}$$

where the "constant" depends on the species, on the location,[32] and on Δt, but *not* on t or P.

Many populations grow more or less continuously, *not* in discrete time steps, so early in the chapter we introduced the idea of *derivative* — the *instantaneous* rate of change — by letting the time step Δt shrink to zero. For continuously growing populations, the model analogous to (30) is the *differential equation*

$$\frac{dP}{dt} \;=\; \text{constant} \cdot P. \tag{31}$$

Our primary objective in this chapter has been to determine what functions $P = P(t)$ could be solutions of (31). But, in order to make sense of the question, we had to study the object called "derivative," the instantaneous rate of change that is represented by the left-hand side of (31).

For that purpose, we changed the subject to a physical problem, the motion of a falling object, which gives rise to a familiar *power* function and for which "derivative" has a familiar interpretation as *velocity*. Through numerical, graphical, and algebraic studies of linear, quadratic, and other polynomial functions, we saw that we could reduce the study of rates of change of such functions to a small set of rather simple formulas.[33]

On our way back to population growth problems, we did some numerical calculations with another familiar situation — compound interest

---

[32]A species may have different natural growth rates in different locations because of different food sources or climatic conditions, for example.

[33]A summary of definitions and formulas appears at the end of this summary section.

— that formally has the same structure as the natural growth equation $\frac{\Delta P}{\Delta t} = kP$, i.e., growth rate proportional to the amount present. Those calculations suggested that exponential functions might be appropriate models for savings account balances, hence also for natural population growth.[34] Thus, we turned our attention to the problem of calculating derivatives of exponential functions.

We first determined graphically that exponential functions of the form $f(t) = b^t$ did indeed appear to satisfy equation (31), that is, to have derivatives proportional to themselves, at least approximately: Using a difference quotient approximation $g(t)$ to the derivative $f'(t)$, we saw that $g(t)/f(t)$ appeared to be constant for every choice of b. We called the proportionality constant $L(b)$ (because it varied with the choice of b), and we did some approximate numerical calculations of several values of $L(b)$: $L(2) = 0.6931...$, $L(3) = 1.0986...$, $L(4) = 1.3862...$. Those calculations suggested the existence of a special base, universally known as e, for which $L(e) = 1$; in a classroom activity, you found that $e = 2.71828...$. The significance of the special ("natural") base e is that the derivative of e^t is the *same* function e^t:

$$\frac{de^t}{dt} = e^t. \tag{11R}$$

Next, we saw that "changing the scale" of the independent variable from t to kt had the effect of scaling the rate of change by the same factor:

$$\frac{d}{dt}\, e^{kt} = ke^{kt}. \tag{32}$$

(This scale-change statement is true for all functions, not just exponential functions.) Scaling the output or dependent variable by a constant factor A has a similar effect (for all functions, not just exponential functions) of scaling the derivative by the same factor:

$$\frac{d}{dt}\, Ae^{kt} = kAe^{kt}. \tag{13R}$$

---

[34]In a lab associated with this chapter, or in problems at the end of the chapter, you may have pursued interest rate calculations further, seeing that the same conclusion is appropriate when interest is compounded more frequently — including daily compounding, which is "nearly continuous."

Formula (13) provided us with a sufficiently large family of exponential functions to solve any natural growth problem of the form $\frac{dP}{dt} = kP$: If $P(t) = Ae^{kt}$, then $P'(t) = kP(t)$ for every choice of the constant A. If we choose A to be the population P_0 at time $t = 0$, then we can match any initial condition for a differential-equation-with-initial-value problem.

While exponential functions with base e turned out to be enough to solve *all* continuous natural growth problems, the equation

$$\frac{d}{dt}\, e^{kt} \;=\; ke^{kt} \tag{32R}$$

also told us indirectly how to complete the calculation of the derivative of an exponential function with any base b, that is, how to find $L(b)$. If we set $b = e^k$ (that is, choose k to be $\ln b$), then equation (32) becomes

$$\frac{d}{dt}\, b^t \;=\; (\ln b)\, b^t. \tag{15R}$$

After developing all the necessary tools and carrying out the solution of our instantaneous growth rate problem, we turned to the question of *whether given data actually fits one of our theoretical models.* We found that a semilog plot could reveal whether the function sampled by the data grows exponentially: If the plot is straight, it does; otherwise, it does not. Similarly, we found that a log-log plot could reveal whether the function grows like a power function: If the plot is straight, it does; otherwise, it does not.

If we are starting with data of unknown origin and have no idea what type of model it might fit, we can follow a multi-step strategy:

1. Make an ordinary scatter plot (see Chapter 1). If the data points fall roughly along a straight line, the relationship between variables is approximately linear.

2a. If the relationship is clearly not linear, make a semilog plot. If the points *now* fall roughly along a straight line, the relationship is approximately exponential.

2b. If 2a does not produce a straight line, reverse the variables and try the semilog plot again. If the relationship is now exponential, then the original relationship was logarithmic. (Why?)

3. If nothing has worked yet, try a log-log plot. If that produces straightness, the relationship is that of a power function.[35]

Of course, we are just at the beginning of our study of possible models for real-world data; there is no reason to think that "linear," "exponential," "logarithmic," and "power" are the only possibilities.

In Chapter 1 we began classifying our methods, models, and "realities" in several different ways, for example, "discrete" vs. "continuous." Our methods, both conceptual and computational, can also be classified as "approximate" or "exact," which is not the same as "discrete" vs. "continuous." A discrete method may be an approximation to an exact, continuous method, or it may be the other way around: A continuous method may be an approximation to an exact, discrete method. Even when the method or model is "exact" (whether discrete or continuous), we may still find ourselves doing "approximate" calculations simply because the exact calculations would be incredibly difficult or even impossible. The developments in this chapter provide examples of all of these possibilities.

Conceptually, "real" population growth (say, of humans in a given country) has a continuous independent variable (time) and a discrete dependent variable (population, changing in unit steps with each birth, death, immigration, or emigration). On the other hand, the fine detail of unit changes in very large populations is unimportant for studying large-scale trends, so we simplify things by imagining that $P = P(t)$ changes continuously as a function of continuous time. Only then does it make sense to speak of its instantaneous rate of change, which we imagine to be another continuously varying function. But, in order to formulate the concept of instantaneous rate of change, we began by approximating by difference quotients (average rates of change over small time steps), an inherently discrete idea. This led to *two* (related, but different) models for

---

[35]If this doesn't work out, why is there no point in reversing the variables as we did in Step 2?

continuous population growth, the discrete model embodied in a difference equation, and the continuous one represented by a differential equation. Each is, *at best*, an approximation to "reality."

As it happened, both the discrete model and the continuous model could be solved *exactly*, that is to say, by formal manipulation of the symbols.[36] Both procedures led to the same conclusion: A population whose growth rate (average or instantaneous) is proportional to the population itself must grow exponentially.

In both our calculus solution of the continuous model (Section 2.5) and our algebraic solution of the discrete model (Section 2.8), we found it convenient to talk about "initial values," that is, starting points for our unknown population functions. The idea, illustrated by our direction field pictures of the growth rate equation, is this: If you know where you are, and you always know what direction you are going, then you always know where you will be at any future time. Differential (and difference) equations always have *infinitely many* "solutions," functions that satisfy the growth rate condition or curves that fit the direction field. But, once you pick a starting point, that usually determines *exactly one* solution. Thus, once a starting time is identified, the possible starting values of the dependent variable determine the possible solutions — *all* the possible solutions.

Our "initial value" approach to these problems comes with a price: the extremely unwieldy description "differential-equation-with-initial-value problem" (or "difference-equation-with-initial-value problem"). It should come as no surprise that mathematicians, scientists, and engineers usually refer to these problems by much shorter names. In fact, both names are conventionally shortened to "initial value problem." We will use this phrase in Chapter 3 and thereafter, but you should not allow its apparent simplicity to fool you. "Initial value problem" *does not* mean, as English usage would suggest, "the problem of finding an initial value." Indeed, the initial value of P usually is known in advance or is easy to find. Rather,

---

[36]For the discrete case, this process was not completed until the optional Section 2.8.

you should always interpret the phrase **initial value problem** to mean **the problem of finding a function that simultaneously satisfies both the differential equation and the initial condition.** When we need to distinguish between *difference* equations and *differential* equations, we will speak of, respectively, "discrete initial value problems" and "continuous initial value problems." If neither adjective is present, and the context does not indicate otherwise, "initial value problem" will mean a *differential* equation with a specified starting point.

Definitions and Formulas

Definition of Derivative

If y is a function of x, then

$$\frac{dy}{dx} \text{ is the limiting value of } \frac{\Delta y}{\Delta x} \text{ as } \Delta x \text{ approaches zero.}$$

Derivative Formulas

Derivatives of power functions (the Power Rule):

$$\frac{d}{dt}\, t^n = n\, t^{n-1}. \tag{5R}$$

The Constant Multiple Rule:

$$\frac{d}{dt}\, cf(t) = c\,\frac{df}{dt}. \tag{6R}$$

The Sum Rule:

$$\frac{d}{dt}\, [f(t) + g(t)] = \frac{df}{dt} + \frac{dg}{dt}. \tag{7R}$$

Derivatives of polynomial functions:

Combine the Power Rule, the Constant Multiple Rule, and the Sum Rule.

Derivatives of exponential functions:

$$\frac{d}{dt}\, e^t = e^t, \text{ where } e \text{ is the natural base, } 2.71828\ldots. \tag{11R}$$

$$\frac{d}{dt}\, e^{kt} = k\, e^{kt}, \text{ for any constant } k. \tag{32R}$$

$$\frac{d}{dt}\, b^t = (\ln b)\, b^t, \text{ for any constant exponential base } b. \tag{15R}$$

Scale change formula for the independent variable (special case of the Chain Rule):

$$\text{If } u = kt, \text{ where } k \text{ is constant, then } \frac{d}{dt} f(kt) = k \frac{d}{du} f(u). \tag{12R}$$

Scale change formula for the dependent variable (the Constant Multiple Rule):

$$\frac{d}{dt} cf(t) = c \frac{df}{dt}. \tag{6R}$$

Limiting Values

Formula (33) follows from computations of the "effective yield" from interest compounded daily or continuously (in lab or in solving Problem 3 at the end of the chapter) or from the observation that solutions of the discrete natural growth problem approximate solutions of the continuous natural growth problem (see "Problems and Solutions" below). The other two formulas follow from our definition of $L(b)$ and our subsequent determination that $L(b) = \ln b$.

$$\text{The limiting value of } (1 + k\Delta t)^{1/\Delta t} \text{ as } \Delta t \to 0 \text{ is } e^k. \tag{33}$$

$$\text{The limiting value of } \frac{b^{\Delta t} - 1}{\Delta t} \text{ as } \Delta t \to 0 \text{ is } \ln b. \tag{34}$$

$$\text{The limiting value of } \frac{e^{\Delta t} - 1}{\Delta t} \text{ as } \Delta t \to 0 \text{ is } 1. \tag{35}$$

Problems and Solutions

$$\text{IF} \qquad \frac{dP}{dt} = kP \quad and \quad P = P_0 \text{ when } t = 0, \tag{36}$$

$$\text{THEN} \quad P = P_0\, e^{kt}. \tag{37}$$

$$\text{IF} \qquad \frac{\Delta P}{\Delta t} = kP \quad and \quad P = P_0 \text{ when } t = 0, \tag{28R}$$

$$\text{THEN} \quad P(t) = P_0\, b^t, \text{ where } b = (1 + k\Delta t)^{1/\Delta t}. \tag{29R}$$

Practice with Calculations

*Exercises in this category include (a) calculations that can be done by machines and (b) practice on important topics from courses that precede calculus. You need to develop **facility** with both categories — not because such calculations are a central feature of the course, but because they should not frustrate you or keep you from concentrating on the more important parts of the course by occupying a lot of your time or by leading to lots of mistakes. Even though routine calculations can be done quickly and accurately by computer or calculator, you need to develop judgment about when to use a machine and when not to, and you need to know how to tell when you might have pressed the wrong button. These skills are acquired and sharpened by **practice**.*

You should expect to see exercises like these as some portion of your homework assignments, quizzes, and tests.

1. On the first set of coordinate axes below, draw a line with slope $-\frac{2}{3}$. On the second, draw a line with slope $\frac{3}{5}$. On the third, draw a line with slope $-\frac{3}{4}$; on the fourth, a line with slope -2. (Each block is one unit square.)

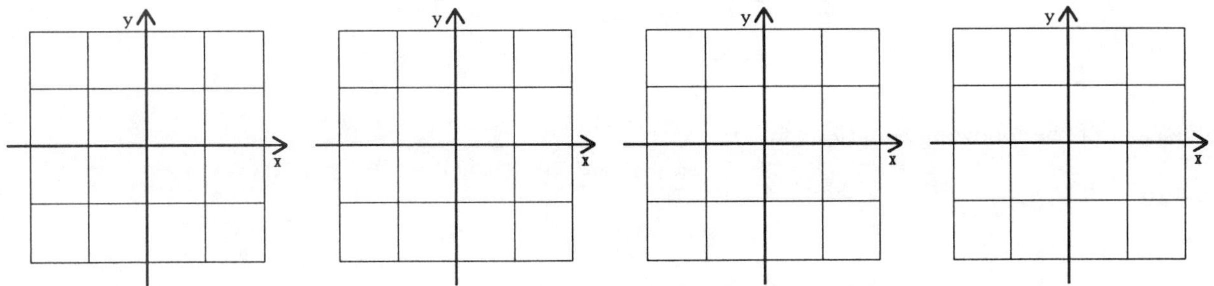

2. A line with slope $\frac{1}{2}$ passes through the point $(5, -2)$.

   (a) Sketch the line.

   (b) Find an equation of the line.

3. (a) Sketch the line that passes through the points $(3, -7)$ and $(1, -3)$.

   (b) Find an equation of the line.

4. A line with slope 1.7 passes through the point $(1.5, -2.3)$.

   (a) Sketch the line.

   (b) Find an equation of the line.

5. (a) Sketch the line that passes through the points $(1.3, -0.7)$ and $(1.5, -3.2)$.

   (b) Find an equation of the line.

6.   Put your calculator in *radian* mode. Find the average rate of change of the cosine function (use the *cos* button) over each of the intervals $[0, \pi/4]$, $[0, \pi/2]$, $[0, \pi]$.

7.   Put your calculator in *degree* mode. Find the average rate of change of the cosine function (use the *cos* button) over each of the intervals $[0°, 45°]$, $[0°, 90°]$, $[0°, 180°]$.

8.   Solve the equation $100 = 27e^{.07t}$ for t.

9.   Solve the equation $P = P_0 e^{kt}$ for t.

Rewrite each of the following equations in exponential form:

10. $\log_9 3 = \frac{1}{2}$ 11. $\log_{10} 1000 = 3$ 12. $\log_{10} 0.1 = -1$

Given that $\log_b 2 = 0.6309$ and $\log_b 5 = 1.465$, find

13. $\log_b 10$ 14. $\log_b \frac{5}{2}$ 15. $\log_b \frac{1}{5}$

16. $\log_b \sqrt{b^3}$ 17. $\log_b 5b$ 18. $\log_b 8$

Solve each of the following equations for t:

19. $t^8 = 20$ 20. $e^t = 20$ 21. $10^t = 20$

Solve each of the following equations for x:

22. $2^{-x} = 3^x$ 23. $2^{-x} = 2^x + 2$ 24. $2^x = 3^x + 2$

25. $2^{-x} = 3^x + 2$ 26. $3^x = 2^x + 2$ 27. $3^{-x} = 2^x + 2$

Calculate each of the following derivatives:

28. $\dfrac{d}{dt} e^{2t}$ 29. $\dfrac{d}{dt} e^{-t}$ 30. $\dfrac{d}{dt} 3^t$

31. $\dfrac{d}{dt} (3t - 5)$ 32. $\dfrac{d}{dt} 6\,t^2$ 33. $\dfrac{d}{dt} 7$

34.   Express $17\,e^{0.07t}$ in the form cb^t.

35.   Express $23 \cdot 2^t$ in the form ce^{kt}.

Find the derivative of each of the following functions:

36. $t^5 - 7t^4 + 2t^2 - 6t + 1$

37. $t^5 + 2t^4 - 3t^2 - 8$

38. $4t^4 - t^3 + 2t^2 - t + 7$

39. $13 - 27t + 5t^2 + t^3$

40. $t^5 - 7e^{4t} + 2t^2 - 6e^{-t} + 1$

41. $t^5 + 2t^4 - 3t^2 - 8 + e^{-t/2}$

42. $4e^{4t} - e^{3t} + 2e^{2t} - e^t + 7$

43. $13e - 27t + 5t^2 + e^{3t}$

Solve each of the following differential-equation-with-initial-value problems:

44. $\dfrac{dP}{dt} = 0.12P$ with $P = 500$ at $t = 0.$

45. $\dfrac{dP}{dt} = -0.12P$ with $P = 500$ at $t = 0.$

46. $\dfrac{dP}{dt} = 0.12P$ with $P = 500$ at $t = 2.$

47. $\dfrac{dP}{dt} = -0.12P$ with $P = 500$ at $t = 2.$

On each of the following direction fields, sketch three solution functions.

48.

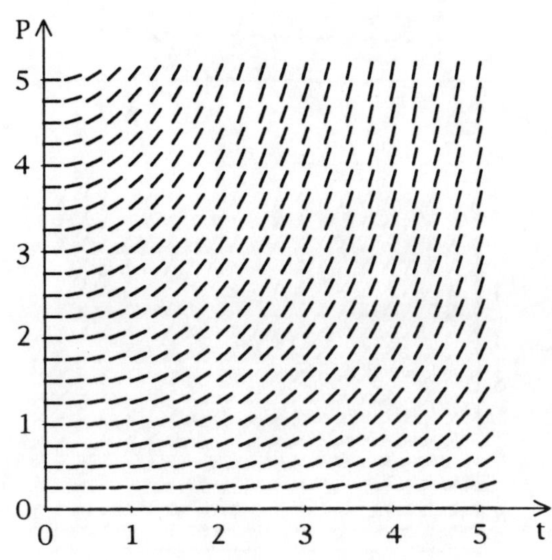

49.

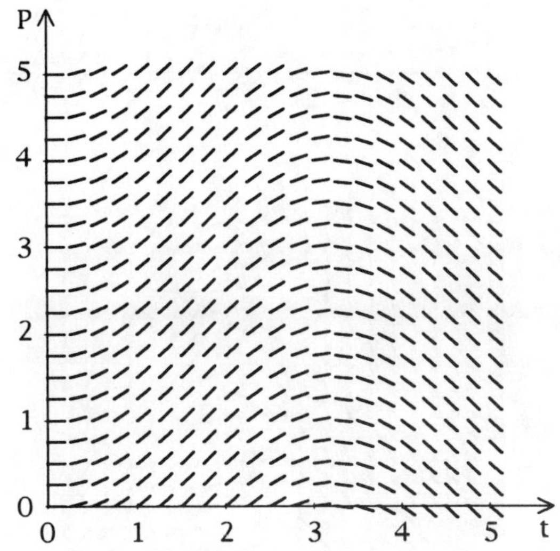

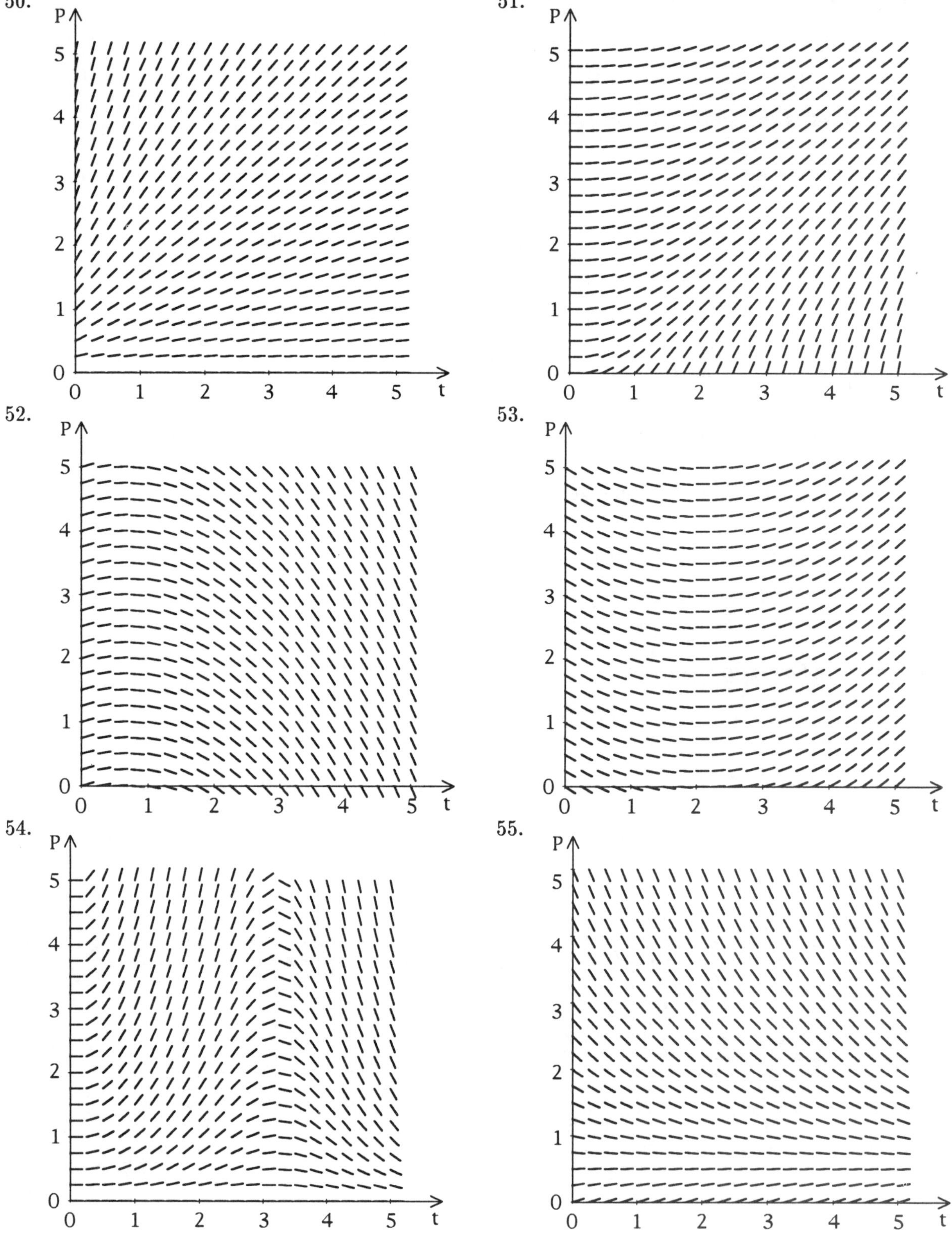

Conceptual Exercises

Exercises in this category use and/or further develop the concepts introduced in this and earlier chapters. The level of difficulty is similar to that of the exercises embedded in the text and at the ends of sections. While many of these exercises are presented in "realistic" contexts, they are not representative of real problems. Rather, these exercises are to help you get ready for tackling real problems.

You should expect to see exercises like these as a major portion of your homework assignments, quizzes, and tests.

1. It follows from your calculation of derivatives of polynomials that
$$\frac{d}{dx}\,(mx+b)\ =\ m.$$
That is, the *instantaneous* rate of change of a linear function (at every x) is the same thing as its *average* rate of change, i.e., its slope. Explain why that is true.

2. (Cf. Conceptual Exercise 9 at the end of Chapter 1)

   (a) What is the rate of change of temperature in degrees Fahrenheit with respect to temperature in degrees Celsius?

   (b) What is the rate of change of temperature in degrees Celsius with respect to temperature in degrees Fahrenheit?

   (c) What relationship do you observe between these two rates?

3. (Cf. Problem 9 at the end of Chapter 1) Suppose a weight of 10 ounces stretches a spring one inch.

   (a) For arbitrary weights hung from the same spring (within the elastic limit), what is the rate of change of weight with respect to displacement?

   (b) What is the rate of change of displacement with respect to weight?

   (c) What relationship do you observe between these two rates?

4. A town with a population of 10,240 at the start of a year has an average growth rate of 5% per year.

   (a) What is the population at the end of that year?

   (b) What is the population half-way through the next year?

   (c) What is the population half-way through the third year?

   (d) Explain your strategy for answering these questions.

5.   (a) A biologist has a colony of bacteria that contains 1.5 biomass units when first weighed and 1.95 biomass units an hour later. What is the biomass of the colony t hours later?

(b) Her first measurement was made at 8:30 A.M. and her last just before she went home at 4:30 P.M. Estimate the last measurement.

6.   Suppose $f(t)$ is any function that has a derivative, and C is any constant. Show that the function $g(t) = f(t) + C$ has the same derivative as $f(t)$.

7.   Suppose $f(t)$ and $g(t)$ are any functions that have derivatives. Show that

$$\frac{d}{dt}\,[f(t) + g(t)] \;=\; \frac{df}{dt} + \frac{dg}{dt}\,.$$

8.   Fill in the tables below. Use your results to estimate ln 8 and ln 10. Check your estimates by using the $\boxed{\ln}$ key on your calculator.

Δt	$8^{\Delta t}$	$\dfrac{8^{\Delta t} - 1}{\Delta t}$	Δt	$10^{\Delta t}$	$\dfrac{10^{\Delta t} - 1}{\Delta t}$
.01			.01		
.001			.001		
.0001			.0001		
.00001			.00001		

9.   Suppose gasoline sells for $1.19 a gallon in June and for $1.25 a gallon in July. What can you say about the average price of a gallon of gasoline during those two months?

10.[37] A colony of bacteria living on a Petri dish under optimal conditions doubles in size every ten minutes. At noon on a certain day, the Petri dish is completely covered with bacteria. At what times (to the nearest hour, minute, and second) was the covered percentage of the dish

(a) 50%,   (b) 25%,   (c) 5%,   (d) 1%?

---

[37]From *Calculus Problems for a New Century*, edited by Robert Fraga, MAA Notes Number 28, 1993.

11. "In recent years there has been a rise in the concentration of atmospheric methane of more than 1 percent per year. The increase is both rapid and significant because ... methane is 20 times as effective as carbon dioxide in trapping heat."[38] How long will it take to increase the concentration of atmospheric methane by 50%? [The answer is not "50 years."]

12. Table 6 gives the population of Mexico for each of the years 1980 through 1986. Decide whether the population was growing exponentially during those years. If so, find an exponential function that approximately fits this data. By what percentage did the population grow each year?

Year	1980	1981	1982	1983	1984	1985	1986
Population (millions)	67.38	69.13	70.93	72.77	74.66	76.60	78.59

Table 6. Population of Mexico.

---

[38]From "Global Climatic Change" by R. A. Houghton and G. M. Woodwell, *Scientific American*, April 1989, pages 36-44.

Problems and Projects

The problems presented here are not intended for individual homework assignments or for tests. These problems should be attempted by groups of two to four students sharing ideas, whether in the classroom or elsewhere. Some of the more extensive problems are suitable for projects with time frames ranging from a class period to a week.

1. Figures 41, 41, and 42 show successive zoom-ins on the graph of $f(t) = |t|$ in the vicinity of the point $(0,0)$.

    (a) Is this function locally linear at that point? Why or why not?

    (b) What is the average slope of the graph from $-\Delta t$ to Δt for a small value of Δt?

    (c) Can you use such an average slope to estimate the rate of change at 0? Why or why not?

    (d) What is the average slope of the graph from 0 to Δt for a small value of Δt? From $-\Delta t$ to 0?

    (e) What do you conclude about instantaneous rate of change at $t = 0$?

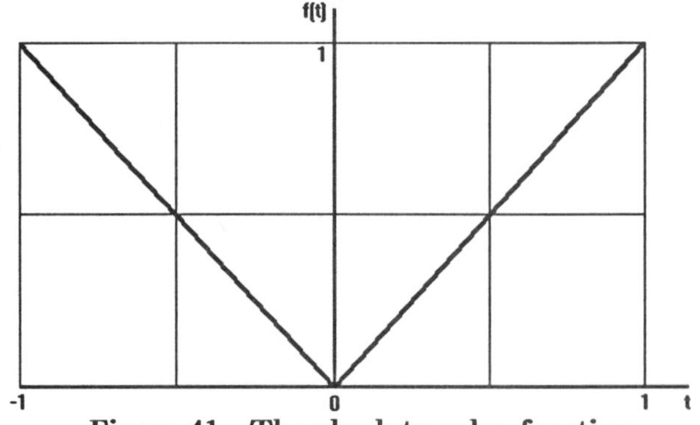

Figure 41. The absolute value function.

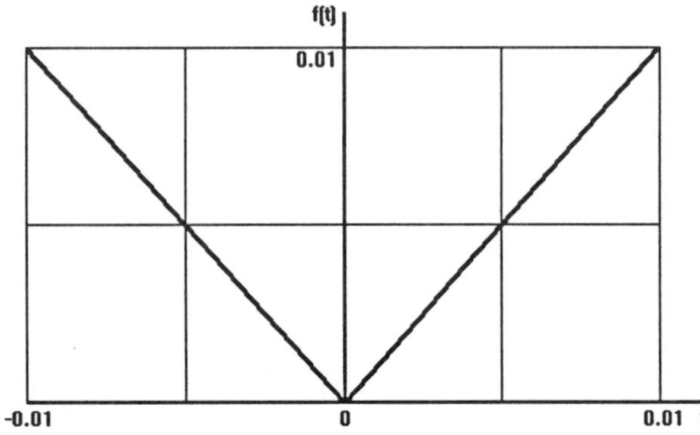

Figure 42. The absolute value function.

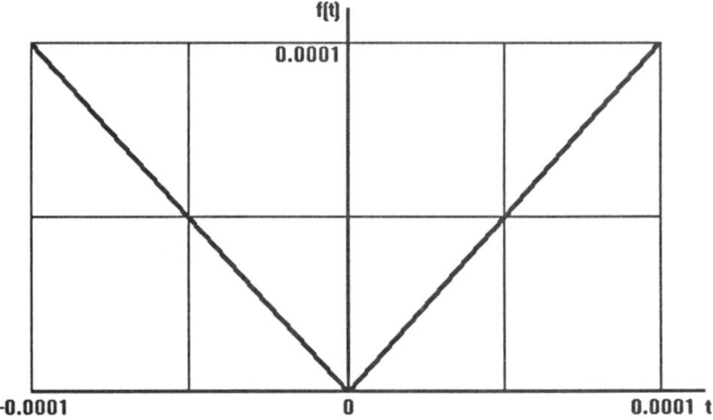

Figure 43. The absolute value function.

2. Use properties of exponentials and the definition of logarithms,

$$y = \log_b x \quad \text{if and only if} \quad x = b^y,$$

[see Section 1.6] to establish the following properties of logarithms:

$$\log_b(AB) = \log_b A + \log_b B$$

$$\log_b A^B = B \log_b A$$

$$\log_b \frac{A}{B} = \log_b A - \log_b B$$

3.[39] Suppose you deposit $1000 in a savings account that pays 8% interest compounded annually, and you leave it (and the interest) there indefinitely.

(a) What is your balance one year after you made the deposit?

(b) By what factor has your money increased?

(c) What is your balance two years after the initial deposit?

(d) Twenty years after the initial deposit?

(e) What is your balance t years after the initial deposit?

4.[39] Answer the same questions as in the previous problem

(a) if the interest is compounded quarterly,

(b) if the interest is compounded monthly,

(c) if the interest is compounded daily.

(d) The bank offering 8% compounded daily might advertise an "effective yield" of 8.33%. What does that mean?

5.[39] Some banks advertise continuous compounding of interest. This means (or should mean) that the instantaneous rate of change of your balance is the advertised interest rate times the balance.

(a) Answer the same questions as in Problem 3 under the assumption of continuous compounding.

(b) How much difference is there between daily compounding and continuous compounding? In particular, what is the effective yield in this case?

---

[39]Adapted from D. A. Smith, *Interface: Calculus and the Computer*, 2nd Ed., Saunders College Publishing, 1984.

6.[39] Relate your computations in the three preceding problems to the statement:

$$\text{The limiting value of } (1 + k\Delta t)^{1/\Delta t} \text{ as } \Delta t \to 0 \text{ is } e^{k}. \qquad (33R)$$

What specific values of k and Δt were you using?

7.[39] The First National Bank offers $8\frac{1}{4}\%$ interest on deposits compounded daily, the Merchants and Farmers Bank offers $8\frac{3}{8}\%$ compounded quarterly, and the Central Savings and Loan offers $8\frac{1}{2}\%$ compounded annually.

(a) Which bank would you choose if you planned to deposit a substantial amount of money in a long-term savings account?

(b) Can you think of a reason to split your deposit between two (or more) of the banks?

8.[40] An important concept in business and banking is the time required for an investment to double. A common rule of thumb, called the "rule of 72," is that the doubling time in years is approximately 72 divided by the interest rate r (as a percentage). For example, at 8% interest, it takes about 9 years to double an investment.

(a) If the interest is compounded continuously at $r\%$, find the exact doubling time.

(b) Make a table showing the doubling time in years, for $r = 2\%$, 4%, 8%, 12%, and 18%, calculated three ways: from the exact formula in part (a), from the "rule of 72," and from the similar "rule of 69." Which rule of thumb comes closest to the exact answers?

(c) Can you think of a reason why people in business would prefer the "rule of 72" to the "rule of 69"?

9.[40] Table 7 contains data on the number of lawyers in a certain town and the total annual income of those lawyers. Assume that, during the five year period represented by the table, the rate of inflation was 5% per year.

(a) What statistic best describes the financial state of a typical lawyer in a given year? What formula computes it? Construct a table corresponding to this statistic.

---

[40]Adapted from *Calculus Problems for a New Century*, edited by Robert Fraga, MAA Notes Number 28, 1993.

(b)   When were lawyers most well off?   Which year saw the greatest improvement? Which year saw the worst deterioration?

Year	Number of Lawyers	Total Annual Income
1982	40	$2,500,000
1983	45	$2,700,000
1984	50	$3,400,000
1985	40	$3,000,000
1986	50	$3,600,000

Table 7. Lawyers and their income.

10.   (a) The data in the following table comes from either an exponential function or a power function. Decide which. Find a formula that fits the data. Your formula should have one of the forms $y = cb^t$, $y = ce^{kt}$, or $y = ct^k$, with explicit values for the constants.

t	y
1	24.7
5	17.2
10	11.0
20	4.5
35	1.2
50	0.3

(b) The data in the following table comes from either an exponential function or a power function. Decide which. Find a formula that fits the data. Your formula should have one of the forms $y = cb^t$, $y = ce^{kt}$, or $y = ct^k$, with explicit values for the constants.

t	y
1	25.0
5	7.5
10	4.4
20	2.6
35	1.7
50	1.3

11.   It may have occurred to you while working Exercise 5 in Section 2.6 that you don't really need logarithms to tell whether the data in Table 6 fits $s = ct^2$; taking square roots would do as well.

(a) Verify this by adding a column for $\sqrt{s}$ to Table 6 and testing for linearity as you did in that Exercise.

(b) Discuss the relative merits of these two procedures. Suppose you wanted to test whether $s = ct^p$ for some unknown power p. Could you do this by taking roots? Could you do it by taking logarithms? In the process, could you find out what p must be? Explain.

12. Suppose you are testing your car's speedometer against the odometer. (You assume the odometer is accurate.) You drive along the interstate and try to hold your speed constant at 50 mph, while your friend takes down distance readings at regular one-minute intervals. (You do this early Sunday morning when there is little traffic, so that you do not get run off the road by drivers wanting to go 65!) Here is the data recorded by your friend:

Time (minutes)	Distance traveled (miles)
1	0.8
2	1.6
3	2.4
4	3.2
5	4.0
6	4.8
7	5.7
8	6.4
9	7.3
10	8.1

(a) Were you able to hold the speed relatively constant? Explain briefly.

(b) What is your best estimate of the speed at which the car was traveling?

(c) How far had the car traveled in the first 7.3 minutes?

(d) If the car continued moving in this manner, how far would it have gone in 12 minutes?

(e) Find an algebraic expression for a function $s(t)$ that gives the distance the car has traveled as a function of time t for values of t under consideration.

13. As in the previous problem, we have data on the distance a car has traveled; in this case the measurements were made every half-minute.

Time (minutes)	Distance traveled (miles)
0.1	0.06
1.0	0.25
1.5	0.56
2.0	1.00
2.5	1.56
3.0	2.25
3.5	3.06
4.0	4.00
4.5	5.06
5.0	6.25

(a) What can you say about the average speeds?

(b) Find an algebraic expression for a function that approximates the average speeds.

(c) Using this function for speed, find a function that approximates the distance traveled at each time t.

(d) Compare your approximation to the actual data given. How accurate is your formula? Where is it most accurate? Where is it least accurate?

(e) Use your approximation to estimate the distance the car had traveled at 3.3 minutes.

(f) Assuming that the car continued traveling in the same manner, estimate how far it would have gone in 8 minutes.

(g) How realistic is this data for cars you know about?

14.[41] At the start of a [not very realistic] trip, an automobile is moving at 10 miles per hour. It accelerates at a constant rate of 10 miles per hour per hour for five hours. [Acceleration is the instantaneous rate of change of velocity.]

(a) Find an equation for the velocity of the car as a function of the number of hours after the start of the trip, and graph the velocity as a function of time.

(b) What is the slowest velocity of the car for $1 \le t \le 3$? Use that slowest velocity to estimate the distance the car travels between $t = 1$ and $t = 3$.

---

[41]Adapted from *Calculus Problems for a New Century*, edited by Robert Fraga, MAA Notes Number 28, 1993.

(c) What is the fastest velocity of the car for $1 \le t \le 3$? Use that fastest velocity to estimate the distance the car travels between $t = 1$ and $t = 3$. What have you learned from your two estimates?

(d) What is the slowest the car travels for $1 \le t \le 2$? What is the slowest the car travels for $2 \le t \le 3$? What is the minimum distance the car can travel in the second hour of the trip? What is the minimum distance the car can travel in the third hour of the trip? Use these results to estimate the distance the car travels between $t = 1$ and $t = 3$.

(e) What is the fastest the car travels for $1 \le t \le 2$? What is the fastest the car travels for $2 \le t \le 3$? What is the maximum distance the car can travel in the second hour of the trip? What is the maximum distance the car can travel in the third hour of the trip? Use these results to estimate the distance the car travels between $t = 1$ and $t = 3$.

(f) What have you learned from the two estimates in parts (d) and (e)? How does this compare with what you learned from parts (b) and (c)?

(g) Suggest a technique based on this procedure that might lead to an exact answer (or a very precise estimate) for the total distance the car travels between $t = 1$ and $t = 3$.

15. We consider the growth of fruit flies in a favorable laboratory environment: unlimited food, unlimited space, and no predators. Our objective is to find an *approximating* function that we can describe by a formula and that we can use to estimate the population at any (reasonable) time, without actually counting the flies. The "real" population function has only integer values — we do not count pieces of flies! — but we allow our approximating function to assume fractional values. When we interpret our estimates from our approximating function, we have to remember not to be too impressed by a prediction of, say, 788.025 flies on day 15.

Table 7 shows the data obtained by counting the flies at the same time each day for ten days:

Day Number	Number of Flies
0	111
1	122
2	134
3	147
4	161
5	177
6	195
7	214
8	235
9	258
10	283

Table 7. Fruit fly data.

(a) Calculate the growth in the population for each of the ten days. (Add a column to the table to show the daily growth.) Determine the *rate* of growth for each of the ten days. What can we say about the rate of growth of the population? In particular, is it constant? What else can you say about the rate of growth?

(b) Biologists argue that, for populations of this type, the rate of growth should be proportional to the population.   How can we test whether the data in hand supports this theory?

(c) Carry out your test. If the data supports the theory, how can you estimate the proportionality constant? Assuming there is one, what is your best estimate of the proportionality constant?

(d) We need to know what sort of a function has a rate of change proportional to the function itself. Decide which of the following functions have rates of change proportional to their own values at times $t = 0, 1, 2, \ldots$.   For those that do, find the constant of proportionality. (Show your calculations and conclusions for each of the six cases.)

    (i) $f(t) = t^2$ (ii) $f(t) = 2^t$ (iii) $f(t) = 7 \cdot 2^t$

    (iv) $f(t) = 10^t$ (v) $f(t) = t^3$ (vi) $f(t) = 7 \cdot t^3$

(e) Can you find a formula for a function $f(t)$ whose *rate of change* (for one-unit time steps) is k times $f(t)$, where k is the constant of proportionality you obtained in Part (c)? Can you find one that *also* has the value 111 at $t = 0$? (Explain how to find such a function, and state your conclusion about such a function.)

(f) Check the values of your formula-defined function from Part (e) against the fruit fly data in Table 7. (Add another column to the table.) Does the formula approximate the data reasonably well? If not, can you adjust the formula to fit better — without violating the conditions of Part (e)?

(g) What does your formula predict for the fruit fly population on day 20? How confident are you of this prediction?

16.   Figure 45 shows data on the numbers of AIDS cases in the early years of the AIDS epidemic in the U.S.

(a) What is the significance of the approximately straight lines shown in the graph? [Look carefully at what is being plotted on the vertical axis.]

(b) For reasons we will see later in the course (Chapter 5), many epidemics start out with roughly exponential growth in the number of cases; is the AIDS data exponential? Why or why not?

(c) What sort of model (formula) would you expect to fit the "total" data in Figure 45?

(d) If you had the numbers of cases in each year (instead of the graph), how would you test the data graphically to see what sort of model might fit? What would cause you to accept or reject each of the possible models you can think of?

(e) Knowing what you know from Figure 45, what would you expect to see in each of your graphical tests?

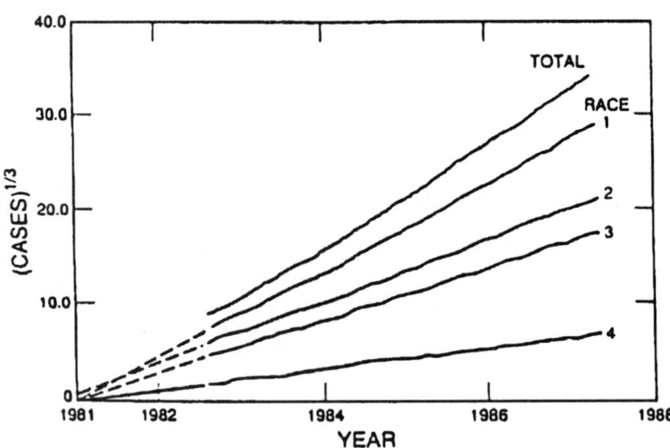

Figure 45. Data on AIDS cases reported by
  the Centers for Disease Control.[42]
  Breakdown by race: 1. White,
  2. Black, 3. Hispanic, 4. Unknown.

---

[42]From "Modeling the AIDS Epidemic" by Allyn Jackson, *Notices of the American Mathematical Society*, vol. 36 (1989), pages 981-983.

17. (Cf. Exercises 3 and 4 in Section 2.6 and the discussion of the Math Pipeline) What would have had to be different between 1972 and 1985 to double the number of Ph.D. recipients in the mathematical sciences in 1985?

18. Figure 46 shows electromagnetic field (EMF) exposure from three different sources as "functions" of distance from source.

(a) Explain why none of the relationships shown is really a function.

(b) We could, nevertheless, replace each of the three non-function graphs by an "average" function. Draw on Figure 46 what you think each of those average functions would look like.

(c) The graphs for high-voltage transmission lines and for distribution lines each have a "horizontal" part and a "sloping" part. What does it mean for a part of the graph to be horizontal?

(d) On the type of graph shown, what does it mean for a part of the graph to be sloping and straight?

(e) For each of the sloping parts, estimate the slope, and then determine an appropriate model to fit that part of the data, using the graph type, the straightness, and the slope.

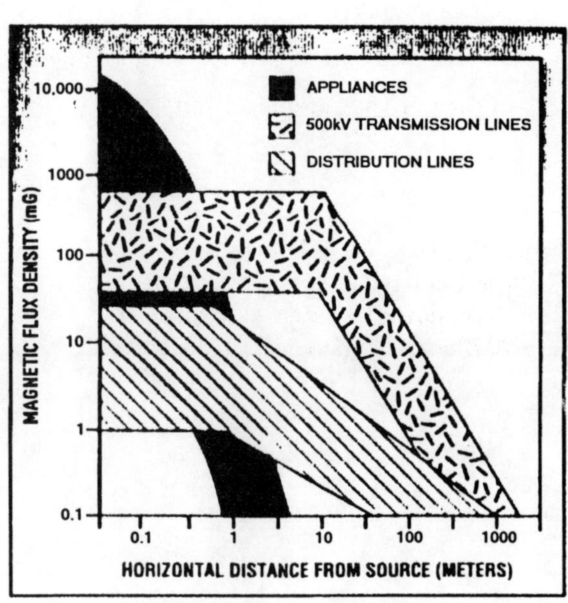

Figure 46. Electromagnetic field exposure from three sources.[43]

---

[43]From a "News and Comment" article on EMF exposure and cancer risk in *Science*, Sept. 7, 1990, page 1097. This figure is typical of data displays in the scientific literature.

CHAPTER 3

Initial Value Problems

In Chapter 2, we encountered the concept of *initial value problem* — a differential equation together with an initial condition — as the key to our study of "natural" population growth. Indeed, this concept is key to the study of many problems of growth and decay, of motion, of all sorts of change. In this chapter, we will study more problems that can be modeled as initial value problems: the change in temperature of a body cooling in the air, the growth of a population with immigration, and the speed of an object falling under the force of gravity. Along the way, we will encounter additional key concepts and methods: the utility of changing the independent variable in a differential equation (additive scaling), a technique for solving algebraic equations, and some general problem-solving advice.

3.1 Differential Equations and Initial Values

General Forms and Examples

The general family of differential equations of immediate interest to us consists of those that specify, at every point in some region of the plane, the *slope* of an unknown function. That is, if t is our independent variable (think "time"), and y is our dependent variable (think any function of time), then we might want to study equations of the form

$$\frac{dy}{dt} = g(t, y), \tag{1}$$

where g can be a function of *both* t and y. In this course, however, we will restrict our attention to differential equations that specify the slope *either* as a function of t alone (in general, the independent variable alone) *or* as a function of y alone (in general, the dependent variable alone). Examples of both types are readily available from Chapter 2.

Our dependent variable in Chapter 2 was often population, so we called it P instead of y. The only "slope function" we considered for the population growth rate was $kP,$[1] which did *not* depend on $t,$ so the exponential (or "natural") growth equation

$$\frac{dy}{dt} = ky$$

is an example of a differential equation of the form

$$\frac{dy}{dt} = g(y), \tag{2}$$

with $g(y) = ky.$

When we used the falling body model in Chapter 2, we took for granted the formula $s = ct^2,$ from which we calculated the velocity as $\frac{ds}{dt} = 2ct.$ On the other hand, if we had started from the velocity formula as "known,"[2] then the formula $\frac{ds}{dt} = 2ct$ would have specified the position function $s = s(t)$ as the unknown dependent variable in a differential equation of the form

$$\frac{dy}{dt} = g(t), \tag{3}$$

where $g(t) = 2ct.$ We will return to our study of the falling body problem later in the chapter; for purposes of discussing direction fields and solutions, we will concentrate mostly on equations of the form (2).

Solutions and Direction Fields

A "**solution**" of the differential equation, whether of the form (2) or (3), is a *function*, $y = f(t),$ that "satisfies" the equation, that is, a function that can be substituted for y in the differential equation to produce an *identity* for all time values t in some interval.

For equation (3), satisfying the equation means nothing more than $f'(t) = g(t).$ So, for example, $f(t) = t^2$ is a solution of the differential

---

[1] In an early lab on constrained population growth, we considered a more complicated right-hand side, $cP(M - P),$ which happened also not to depend on $t.$

[2] Physically, it is much more likely that our *a priori* knowledge would be of velocity rather than of position.

equation $\dfrac{dy}{dt} = 2t$, because $f'(t) = 2t$. The function $h(t) = t^2 + 5$ is another solution. However, $\phi(t) = t^3$ is *not* a solution, because $\phi'(t) = 3t^2 \neq 2t$.

For an equation of type (2), a function f satisfies the equation if

$$f'(t) \; = g(f(t)) \tag{4}$$

for all numbers t in some interval. Notice that both sides of (4) are functions of t alone; the requirement for f to be a solution of (2) is that its derivative, $f'(t)$, be precisely the function $g(f(t))$. So, for example, $f(t) = e^{2t}$ is a solution of the differential equation $\dfrac{dy}{dt} = 2y$, because $f'(t) = 2e^{2t} = 2f(t)$. The function $h(t) = 5e^{2t}$ is another solution. However, $\phi(t) = e^t$ is *not* a solution, because $\phi'(t) = e^t = \phi(t) \neq 2\phi(t)$.

In general, a differential equation has an infinite number of solutions. For example, we have already noted that both e^{2t} and $5e^{2t}$ are solutions of $\dfrac{dy}{dt} = 2y$. Other solutions of this differential equation are $3e^{2t}$, $-7e^{2t}$, and $1.432e^{2t}$. In fact, if C is *any* constant, then the function Ce^{2t} is a solution.

Exercise 1. (a) In the case of the natural growth equation,

$$\frac{dy}{dt} = ky, \tag{5}$$

what is $g(y)$?

(b) Describe one solution function $f(t)$ explicitly.

(c) Express the infinite family of solutions in terms of $f(t)$.

[Hint: We are only asking you to write down things you have done already, but we are asking that you think about them in a slightly different way.]

Our Assumption About Solutions

As we did in Section 2.5, we consider the family of solutions of the differential equation $\dfrac{dP}{dt} = 0.25P$. Figures 1 through 4 (repeated from Chapter 2) illustrate the process of relating solution functions to the direction field that is a visual representation of the differential equation.

In Figure 1, we show the graphs of several functions that satisfy the differential equation. Each of these functions has a different value at $t = 0$, i.e., they satisfy different initial conditions. Figure 2 shows the same curves, but drawn with short, almost-straight-line segments. Figure 3 picks up on the idea of short line segments by plotting, at each point of a regular grid, a line segment whose slope is determined by the differential equation itself. And Figure 4 show how solutions can be recreated by following the directions of the direction field, once a starting point has been chosen.

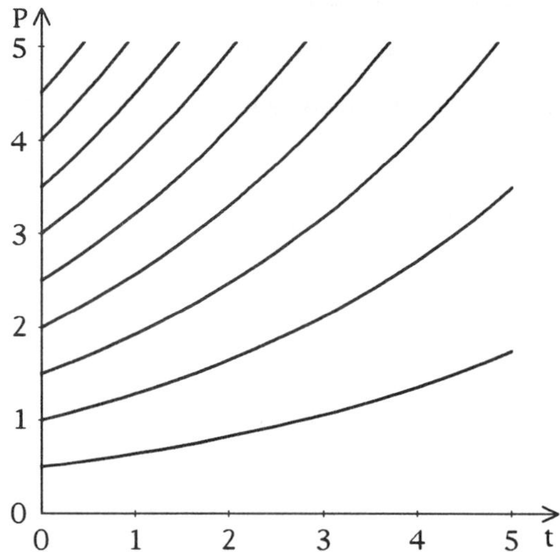

Figure 1. Solution curves for $\frac{dP}{dt} = 0.25P$.

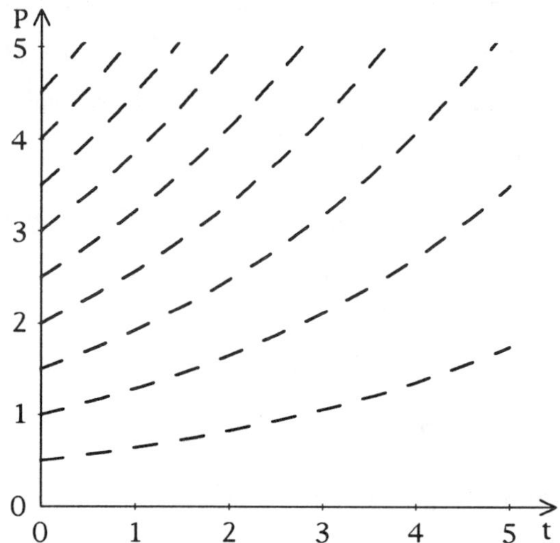

Figure 2. Segments of solution curves.

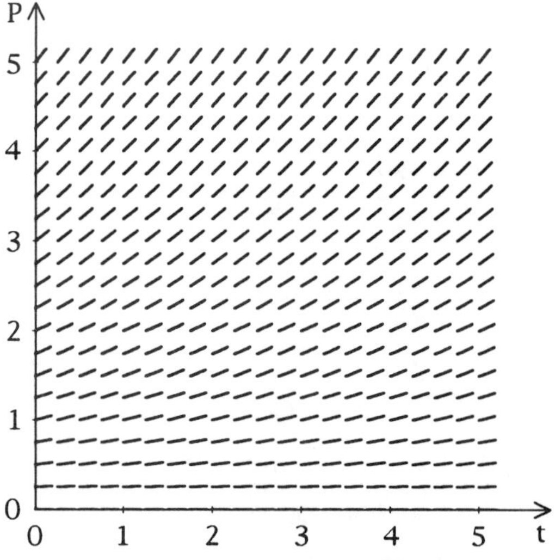

Figure 3. Direction field for $\frac{dP}{dt} = 0.25\,P$.

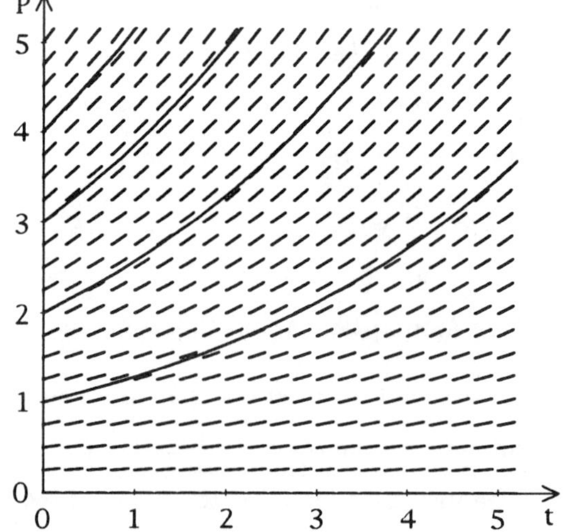

Figure 4. Direction field with solution curves.

Figures 3 and 2 remind us that we do not need to know solutions in advance in order to construct the direction field; we can do that directly from the differential equation, because it contains the information we need for determining the slope at each point. And, given the direction field, we can construct visual representations (graphs) of solutions, whether we know formulas for them or not.

It is easy to imagine a solution of the differential equation through *any* point in the direction field. Thus, the infinite family of solutions must actually fill up the field, in the sense that *every* point lies on *some* solution, namely the one for which it is an initial point. Less obvious is the observation that every point lies on *just one* solution, that is, solution curves cannot coalesce or cross. That certainly seems to be the case in Figure 2. In fact, this is a correct observation about direction fields under very general circumstances — but *not always*. The "very general circumstances" include practically all physically interesting differential equation models, and we will encounter no difficulties if we simply *assume* that each point in a direction field lies on a *single* solution curve. Thus, given a differential equation and an initial point (t_0, y_0), we will speak of *the* function determined by the differential equation whose graph passes through the given point.

The assumption just stated is a form of physical "determinism": No matter where we are, if we always know (a) where we are and (b) the direction we are headed, then we always know exactly where we will go.

That brings us to the "initial value problem," which, as we have seen, consists of a differential equation *plus* an "initial condition" — a statement about one point that must lie on the solution curve, such as $y = y_0$ when $t = t_0$. Under very general and physically reasonable (but as yet unstated) conditions, every initial value problem,

$$\frac{dy}{dt} = g(t, y) \text{ with } y = y_0 \text{ when } t = t_0,$$

has a *unique* solution, that is, a function $f(t)$ such that $f'(t) = g(t, f(t))$ *and* $f(t_0) = y_0$.

Fundamental Assumption: In this *Reader*, we assume that every initial value problem has a unique solution.

Growth is not always "natural," and relatively few physical processes behave like "continuously compounded interest."   In order to cope with other differential equations, we need to know more about derivatives and about the interactions of the differentiation process with the variables in our functional relationships.   We will explore those interactions in two basic types of initial value problem, corresponding to our two types of differential equation,

$$\frac{dy}{dt} = g(y), \tag{2R}$$

and

$$\frac{dy}{dt} = g(t), \tag{3R}$$

It will be helpful to have names by which to refer to these types. The first we call (informally) "Type 1," or (more formally) **autonomous**;[3] these names refer to any initial value problem of the form

$$\frac{dy}{dt} = g(y), \quad \text{with} \quad y = y_0 \text{ at } t = t_0. \tag{6}$$

The second we call (informally) "Type 2," or (slightly more formally) **explicit-derivative type**;[4] these names refer to any initial value problem of the form

$$\frac{dy}{dt} = g(t), \quad \text{with} \quad y = y_0 \text{ at } t = t_0. \tag{7}$$

---

[3]"Autonomous" is a technical term that means the description of the *derivative* does not depend on the independent variable, usually time.

[4]There is no standard technical term for this type.   Our made-up adjective, "explicit-derivative," signifies that the derivative is described explicitly as a function of the independent variable.

Section Summary

Write summaries of each of the following concepts:

Autonomous Differential Equation

Explicit-Derivative Differential Equation

Initial Value Problem

Direction Field

Exercises

2. For each of the following differential equations, decide whether the equation is of autonomous or explicit-derivative type:

(a) $\dfrac{dy}{dt} = 3t^2 + 2$ (b) $\dfrac{dy}{dt} = 2 - y$

(c) $\dfrac{dy}{dt} = 2y$ (d) $\dfrac{dy}{dt} = e^{2t}$

3. Which of the following functions are solutions of the differential equation $\dfrac{dy}{dt} = 4y$?

(a) $f(t) = e^{4t}$ (b) $f(t) = e^{2t}$

(c) $f(t) = 5e^{4t}$ (d) $f(t) = e^{4t} + 5$

4. Which of the following functions are solutions of the differential equation $\dfrac{dy}{dt} = 4t$?

(a) $f(t) = 2t^2$ (b) $f(t) = 2t^2 + 3$

(c) $f(t) = e^{4t}$ (d) $f(t) = 3t^2$

5. Which of the following functions is **the** solution of the initial value problem

$$\frac{dy}{dt} = 4t \;\; \text{with} \;\; y = 4 \;\; \text{at} \;\; t = 1 \; ?$$

(a) $f(t) = 2t^2$ (b) $f(t) = 2t^2 + 1$

(c) $f(t) = 2t^2 + 2$ (d) $f(t) = 2t^2 + 4$

6. Which of the following functions is **the** solution of the initial value problem

$$\frac{dy}{dt} = 4y \;\; \text{with} \;\; y = 3 \;\; \text{at} \;\; t = 0 \; ?$$

(a) $f(t) = e^{4t}$ (b) $f(t) = 2e^{4t}$

(c) $f(t) = 3e^{4t}$ (d) $f(t) = 4e^{4t}$

7. Find three solutions of the differential equation $\dfrac{dy}{dt} = 8y$.

8. Find three solutions of the differential equation $\dfrac{dy}{dt} = 8t$.

9. Find *the* solution of the initial value problem

$$\frac{dy}{dt} = 8y \quad \text{with} \quad y = 2 \quad \text{at} \quad t = 0.$$

10. Find *the* solution of the initial value problem

$$\frac{dy}{dt} = 8t \quad \text{with} \quad y = 2 \quad \text{at} \quad t = 0.$$

3.2 The Homicide Detective — A Chilling Tale[5]

Negative Exponentials

In this section we investigate an initial value problem that describes heat loss from a cooling body. A feature of this cooling body problem will be the appearance of exponential functions such as $e^{-1.7t}$ or $e^{-2.5t}$ or $e^{-3.7t}$, i.e., functions of the form e^{-kt} where the constant k is positive. In Figure 5 we graph the sample functions just mentioned.

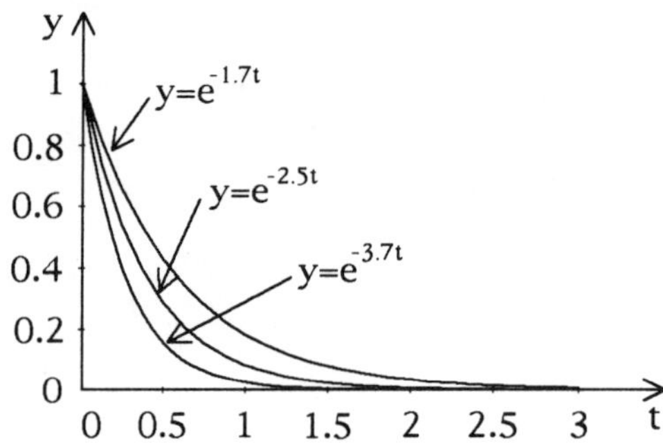

Figure 5. Graphs of functions of the form $y = e^{-kt}$.

Notice that all these functions have the value 1 at $t = 0$ and all decrease to 0 as t increases to infinity. Also notice that the larger the value of k, the faster the function decreases.

We already know how to differentiate functions of this type: In general,

$$\frac{d}{dt}\, e^{rt} = re^{rt} \text{ for any constant } r.$$

In particular, this is true for $r = -1.7$ or $r = -2.5$ or $r = -3.7$. For example,

$$\frac{d}{dt}\, e^{-1.7t} = -1.7e^{-1.7t} \ .$$

More generally,

$$\frac{d}{dt}\, e^{-kt} = -ke^{-kt} \quad \text{for all constants } k.$$

---

[5]Based on "An Application of Newton's Law of Cooling," by James F. Hurley, *Mathematics Teacher* 67 (1974), pp. 141-142. See also "The Homicide Problem Revisited," by David A. Smith, *The Two-Year College Mathematics Journal* 9 (1978), pp. 141-145.

The Crime and the Suspects

Imagine yourself a script writer for your favorite TV detective. Imagine further that this detective has studied calculus and remembers it. Here's the scene: On Monday morning, a wealthy industrialist is found murdered in his office in the climate-controlled building he owns (sorry, owned). The discovery is made by his loyal and devoted secretary when she reports for work as usual at 8:30 A.M.; she promptly calls the police, and our hero is on the scene in minutes. The secretary has not seen her boss since Friday afternoon, and (we will learn later in the show) no one else is willing to admit having seen him over the weekend. Everyone but the secretary hated this man, so there are lots of suspects. Solving the case may depend critically on accurate determination of the *time* of the crime, as that may show that some (all but one?) of the suspects could not have been present.

The office building is located in Crystal City, a development in Arlington, Virginia, just outside of Washington, DC. Here are some of the people who had or may have had access to the building between Friday afternoon and Monday morning:

- The janitorial crew that works an 11 P.M. to 7 A.M. shift each night, Sunday through Thursday.

- The industrialist's ex-wife, who, unknown to everyone (except the detective, who has ways of finding out), still has a key to the building. Her whereabouts over the weekend cannot be accounted for, but she is known to have reported for work at 9 A.M. Monday at her regular job — in Cleveland.

- The industrialist's trusted but incompetent nephew, who had been given a managerial job in the firm, and who would inherit a large share of the fortune. His tennis partner confirms his presence on the tennis court most of Saturday; he (the nephew) says he went sailing alone all day Sunday. His wife claims he was in bed with her from midnight Sunday to about 9 A.M. Monday. He normally showed up for work between 10 A.M. and noon.

- The vice-president who is the real brains and the principal operating officer of the corporation. She is known to have rejected advances from the industrialist (when he was still married) and still to have advanced to the top on sheer determination and talent. She is also known to have a fixed work schedule of 6:30 A.M. to 6:30 P.M. six days a week — every day but Sunday, which she devotes to church and social activities. She was at work two hours before the secretary came in and discovered the body, but her usual practice was not to go near the owner's office, because he would always waste her time with off-the-wall ideas that would have been detrimental to the firm.

The Investigation: Body Temperature

What would you have our thoroughly modern detective do? There are some easy-to-measure physiological characteristics that distinguish the living from the dead. For example, no matter how obviously dead the victim is, our detective will probably check for a pulse. However, once the activity of the heart ceases, the absence of a pulse gives no clue as to the time of death. Another measurement that separates living from dead, and that *does* tell us something about time of death, is body temperature.

As we all learned in high school biology, metabolic activity maintains the human body in a state of "homeostasis," that is, at a constant body temperature (except for occasional fever and chills) of 37°C.[6] When death occurs, metabolic activity ceases, and the body begins to lose heat to its surroundings — if the surrounding (or "ambient") temperature is lower. Of course, the temperature of a climate-controlled office would be nowhere near normal body temperature. Our detective can easily determine the ambient temperature by looking at the office thermostat; let's suppose it is set at 21°C.[7]

---

[6] If you are not accustomed to thinking of body temperature in degrees Celsius, this is a good time to try out your formulas from Conceptual Exercise 17 of Chapter 1. Convert 37°C to Fahrenheit, and see if it agrees with your recollection of normal body temperature.

[7] Convert that one to Fahrenheit also. Is it a reasonable room temperature?

The Initial Value Problem

Now, at what rate does a body lose heat? Elementary physics provides us a theoretical model (verifiable by simple experiments) called Newton's Law of Cooling: The change in temperature of a cooling body in a small time interval Δt is jointly proportional to Δt and to the *difference* between the temperature of the body and the ambient temperature.[8] That looks complicated; it gets simpler when we convert it to mathematical symbols.

Let's write $T = T(t)$ for temperature of the cooling body at time t. Then Newton's Law of Cooling, applied in the setting of constant ambient temperature of 21°C, says that $\Delta T = -k(T - 21)\Delta t$. We have chosen to call the joint proportionality constant "$-k$" because it must be negative (why?). Now we can divide by Δt to find an average rate of change over a small interval, then let the interval shrink to zero to get a description of the instantaneous rate of change:

$$\frac{dT}{dt} \;=\; -k(T - 21). \tag{8}$$

That's our differential equation (except we don't know k yet); now we need to turn it into an initial value problem. We have deliberately postponed a decision on the time scale (in particular, what to call "time $= 0$") in order to keep our computations as simple as possible. Your first thought might be: "Of course! Set $t = 0$ at the time of the murder, and then $T_0 = 37$ at $t = 0$." However, our detective may not think that's such a good idea. We know the body temperature at time of death, all right, but we don't know the clock time at that instant — indeed, that's the very problem we're trying to solve. Thus, we have no direct way to relate that time to anything our detective can measure. On the other hand, he can measure body temperature any time he wants to (after his arrival on the scene) and match that temperature with the time on his watch. The temperature function is an open book *after* the body has been discovered!

Let's suppose the first measurement of temperature is taken at 8:50 A.M. and found to be 30°C. If we declare $t = 0$ to mean 8:50 A.M., then

---

[8]The "law" is usually stated in terms of a "cooling body" even if no homicide has occurred.

our initial value problem is

$$\frac{dT}{dt} = -k(T - 21), \quad \text{with} \quad T = 30 \quad \text{at} \quad t = 0. \tag{9}$$

Now we have to (a) solve the initial value problem, (b) determine k (to make the solution an explicit function of time), and (c) use the solution to figure out the time of death.

Solution of the Initial Value Problem

Our differential equation is not exactly like the problems considered in Chapter 2 because the rate of change is proportional to $T - 21$, not to T. That suggests a change of dependent variable to, say $y = T - 21$. Since y differs from T by a constant, we know that $\frac{dy}{dt} = \frac{dT}{dt}$, so the transformed initial value problem is

$$\frac{dy}{dt} = -ky, \quad \text{with} \quad y = 9 \quad \text{at} \quad t = 0. \tag{10}$$

Now we have something familiar, namely, the natural growth equation — except our "growth" factor is negative. The problem has exactly the form of equation (36) in Chapter 2,

$$\frac{dy}{dt} = ry \quad and \quad y = y_0 \quad \text{when} \quad t = 0,$$

with $r = -k$, so the solution is given by equation (37) in Chapter 2:

$$y = y_0 \, e^{rt}.$$

That is,

$$y = 9 \, e^{-kt}. \tag{11}$$

Of course, y is not the quantity we wanted to know about — it was merely a computational convenience. But we can reverse the change of the dependent variable by substituting $y = T - 21$ and solving for T:

$$T = 21 + 9 \, e^{-kt}. \tag{12}$$

Determination of the Proportionality Constant

We're ready for step (b). The constant k has something to do with how fast the body is losing heat, and our detective has no knowledge of that yet, because he has taken only one measurement. A second measurement will give us another (t,T) pair that we can substitute in (12) to find an equation in which k is the only variable. When should we take that measurement?

That brings us to the question of an appropriate *unit* for t, a decision we have postponed until we had some good reason to make it. If we want to solve for k, the most convenient value of t to substitute into the solution formula,

$$T = 21 + 9\, e^{-kt}, \tag{12R}$$

is $t = 1$ (why?). Thus, we should choose our unit of time to be a good time interval between temperature measurements. There's no big hurry to move the body — we could certainly wait an hour.[9] If we used one second or one minute as the time unit, our thermometer would probably record "no change" in temperature — thermometers just aren't that sensitive. If we waited a whole day for the second measurement — with the body still in the office (to keep the ambient temperature constant) — we might create some unnecessary problems. So one hour looks like a good time unit and a good time to take the second measurement.

At precisely 9:50 A.M., our detective takes his second measurement and finds that the body temperature has dropped to 28°C, that is, $T = 28$ at $t = 1$. We substitute this information into (12), or better yet, into (11), and we find

$$7 = 9\, e^{-k}.$$

Now we take natural logs, and we get

$$\ln 7 = \ln 9 - k,$$

or

---

[9]The hour would have to pass much faster, of course, in a half-hour TV show.

$$k = \ln 9 - \ln 7 = 0.2513144.$$

The numerical value comes from a calculator, which any modern detective always has at hand.

Determination of the Time of Death

Now we're ready for step (c) — we want to know the time on the clock when the decaying temperature function given by

$$T = 21 + 9\, e^{-kt} \tag{12R}$$

had the value 37. The procedure is similar to step (b), except now we know k and want to solve for t. We substitute $T = 37$ in (12) [or (11)], and we find

$$16 = 9\, e^{-kt}.$$

Taking logs again, and substituting our known value for k, we find

$$\ln 16 = \ln 9 - kt = \ln 9 - t\,(\ln 9 - \ln 7).$$

We solve for t and find the time of death to be

$$t = \frac{\ln 9 - \ln 16}{\ln 9 - \ln 7} = -2.29.$$

This still needs interpretation. The minus sign means the time of death was *before* the time of the first measurement at 8:50 A.M. — could it be otherwise? In fact, the time of death was 2 hours, 17 minutes before 8:50, or 6:33 A.M.

Conclusion (of the calculation and of the TV show): The ex-wife didn't have time to commit the murder and drive to Cleveland, nor could she have gotten a plane that early. The nephew was probably in bed. So that leaves the vice-president, who was probably in the building at the right time. Or did the janitor do it?

Section Summary

List the steps involved in solving the initial value problem

$$\frac{dT}{dt} = -k(T - 21), \quad \text{with} \quad T = 30 \quad \text{at} \quad t = 0, \qquad \text{(9R)}$$

and determining the time of death. Add more step numbers as necessary.

1. Change the dependent variable to $y = T - 21$.

2. Rewrite the initial value problem in terms of y.

3. Solve the initial value problem for y.

4.

5.

Exercises

1. Suppose the detective's two measurements had been 26°C and 25°C.

    (a) What would the time of death have been?

    (b) Whom would you have the detective arrest in this case?

2. Follow that first instinct to set $t = 0$ at time of death.

    (a) How much more complicated are the computations?   Where do the complications arise?

    (b) You should be able to overcome these complications; can you?

    (c) Assume the same measurements of $30°$ and $28°$; do you get the same answer for the time of death? Why or why not?

3. In the space below, label the axes and sketch the graph of the victim's body temperature function from midnight Sunday to noon Monday. (Your calculator may help.)

Normal body temperature ⸺

Room temperature ⸺⸺⸺⸺

4. Find three solutions of the differential equation $\dfrac{dy}{dt} = -8y.$

5. Find three solutions of the initial value problem

$$\frac{dy}{dt} = -8y \quad \text{with} \quad y = 2 \quad \text{at} \quad t = 0.$$

6. If $z = 10 + 8e^{-kt}$ for all $t,$ and $z = 14$ at $t = 1,$ find $k.$

7. If $z = 10 + 8e^{-0.693t}$ for all $t,$ find t when $z = 12.$

3.3 Immigration

The United States is "a nation of immigrants" — over its 200 year history, some 50,000,000 people have come here to live from someplace else. How is it, then, that we could model our population growth in Chapter 2 and completely ignore immigration, not to mention wars, the Great Depression, and the post-war baby boom? We couldn't, of course; we found (in Exercise 1 of Section 2.6) that the natural growth model is adequate from 1790 to 1860, but the effect of the Civil War was evident in the census data immediately after that.

In this section we formulate and solve an initial value problem that attempts to account for immigration. In the process of analyzing this model, we will have to solve a rather complicated nonlinear equation. We pause here to examine the problems involved in solving such an equation.

Nonlinear Equations

Suppose we want to find a real number x such that

$$7x - 6 = 3x + 4.$$

Since this equation is linear, we can find an equivalent equation in which the variable x appears only once:

$$4x = 10$$

or

$$x = 2.5.$$

Now suppose we want to find a real number x such that

$$x = e^{-x}. \tag{13}$$

Again we might hope that we can combine both occurrences of the variable x into one. However, we cannot find any algebraic operations that will do this. We will have to use some other approach.

For that matter, how can we be sure there actually is an x that satisfies (13)? One way is to graph the two functions, $y = x$ and $y = e^{-x}$ together and see if their graphs cross. If they do, the x-coordinate of the point of intersection is a solution. We show the two graphs in Figure 6.

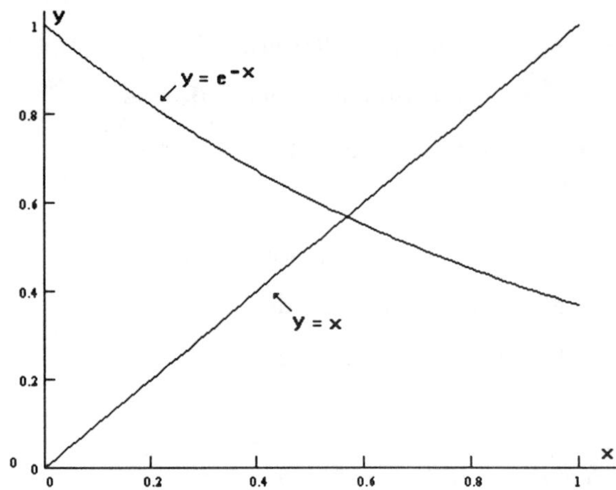

Figure 6. Graphs of the functions x **and** e^{-x}**.**

Exercise 1. (a) *Estimate the number* x *such that* $x = e^{-x}$ *from Figure 6.*

(b) *Use your calculator to see how close your estimate is to a solution.*

Exercise 2. (a) *What happens to equation (13),* $x = e^{-x}$, *if you take the natural logarithm of both sides? Is this any easier to solve than (13)?*

(b) *Use a computer or graphing calculator to graph both sides of your new equation. Estimate the solution* x *from this graph. Is your estimate the same as in Exercise 1?*

In Section 3.4, we will examine a systematic way to solve nonlinear equations by graphing. In Chapter 4 we will return to this problem and use calculus to find another way to solve nonlinear equations. Now let's look at a model for immigration; keep an eye out for the nonlinear equation.

A Proposed Model

A model that takes immigration into account might have the form

$$\frac{dP}{dt} = kP + (\text{immigration rate}).$$

That is, the growth rate would be made up of two terms or "components," one for natural biological growth, the other for immigration. To keep things simple (and because we have the tools at hand for a solution), let's assume that the immigration rate is constant, say, that m people per year are added to the population through immigration.[10] Then our differential equation is

$$\frac{dP}{dt} = kP + m. \tag{14}$$

As with our previous population models, we can turn this into an initial value problem by declaring $t = 0$ to mean a time at which we know the population P_0.

Equation (14) is similar to our cooling body equation,

$$\frac{dT}{dt} = -k(T - 21), \tag{8R}$$

in that the right-hand side has a term proportional to P and a constant term. It is different in that k is not a factor of both terms, as it was in (8). But we can fix that [i.e., make equation (14) look like a problem we have already solved] by factoring out k:

$$\frac{dP}{dt} = k\left[P + \frac{m}{k}\right].$$

If we compare this with the cooling body problem (8), we know exactly what to do: Shift the dependent variable by substituting $y = P + \frac{m}{k}$ and setting $y_0 = P_0 + \frac{m}{k}$ at $t = 0$. Now the modified initial value problem is

$$\frac{dy}{dt} = ky \quad \text{with} \quad y_0 = P_0 + \frac{m}{k} \quad \text{at} \quad t = 0. \tag{15}$$

Exercise 3. *(a) Explain why the solution of (15) is*

$$y = \left[P_0 + \frac{m}{k}\right] e^{kt}.$$

(b) Show that the solution of (14) is

$$P = \left[P_0 + \frac{m}{k}\right] e^{kt} - \frac{m}{k}. \tag{16}$$

Analysis of the Model

Is the model reasonable? How can we tell whether it fits any real data? Well, a step along the way to the solution was to show that y behaves like a natural growth function: $y = y_0\, e^{kt}$. In principle, we could test that just as we did in Chapter 2 with a semilog plot of y as a function of t.

---

[10]Actually, m has to be measured in the same units as P. Thus, if population is measured in millions, then m represents m million people per year.

However, in the natural growth case, we did not need to know k in order to carry out the plot — indeed, we could find k from the plot. Here, we need to know both k and m just to find out what values of y correspond to the known values of P.

The situation is not hopeless, if we formulate our problem in a reasonable way. The growth constant k usually cannot be inferred with any accuracy directly from the data, which is never as smooth as the exponential solution. But the immigration parameter m is observable: Every immigrant is counted. Unfortunately, what these counts show (in the U. S. and everywhere else) is that nobody's immigration rate is constant.[11]

Suppose we take m to be the *average* immigration rate over a decade in which immigration was relatively constant, namely the 1950's. For that decade, about 250,000 immigrants arrived each year.[12] Then we can ask whether there is a constant k such that the immigration model reasonably fits the census data over any substantial period of time. If so, we can conclude that natural growth and immigration together more or less explain our population growth. If not, then we know we have to look for other factors to provide an adequate explanation. Either way, we learn something.

If k really is constant (as our model requires for our "solution" to make sense), then we can estimate it by using a second measurement of population, just as we did in the previous section with the second temperature measurement. Specifically, we can substitute $P = P_1$ at time $t = t_1$ to obtain an equation involving k alone. For example, we could take $t = 0$ in 1790, and let t_1 be 50 (which would represent 1840). Referring back to Table 4 in Chapter 1 (or Table 5 in Chapter 2), we see that P_0 is 3.929 (in millions), and P_1 is 17.069. Since we are measuring population in millions, our assumed value for m is 0.25. When

---

[11]In principle, it could be — if a rigid quota system were rigidly enforced.

[12]That also happens to be the long-term average immigration rate. Source: Table 1 in the *U. S. Immigration and Naturalization Service Annual Report*, 1961, reproduced in M. T. Bennett, *American Immigration Policies: A History*, Public Affairs Press, Washington, DC, 1963.

we substitute all these values into equation (16), we get

$$17.069 = \left[3.929 + \frac{0.25}{k}\right] e^{50k} - \frac{0.25}{k}. \tag{17}$$

The promised nonlinear equation has appeared! The unknown is k, not x, and (17) is considerably more complicated than

$$x = e^{-x}, \tag{13R}$$

but our approach to solving both problems will be the same.

We will solve both (17) and (13) in the next section. When we solve (17), we will find that a reasonable approximation to k is 0.001227.

Exercise 4. *Use your calculator to check that $k = 0.001227$ is an approximate solution to (17).*

We have our proposed solution to the immigration problem:

$$P = \left[P_0 + \frac{m}{k}\right] e^{kt} - \frac{m}{k}, \quad \text{where}\ \ m = 0.25\ \ \text{and}\ \ k = 0.001227. \tag{18}$$

But how can we tell whether the solution makes any sense? There are at least two aspects of the solution that need to be checked. First, in general, it's a good idea to check whether the result of your calculations is an object that satisfies the mathematical conditions of the problem. No matter how good we are at calculation, we make mistakes some of the time. And "answers" corrupted by even "small" mistakes are of no particular value to anyone. Second, if the solution turns out to mathematically correct, we still need to know whether it actually fits the data for which it was proposed as a model. These two aspects of post-solution analysis are quite different, and we will take them up one at a time.

Checking the Solution of the Initial Value Problem

In the Section "A Proposed Model," we set out to solve the differential equation

$$\frac{dP}{dt} = kP + m, \tag{14R}$$

where m is the immigration rate, with the initial condition $P(0) = P_0$. In Exercise 3, you found the solution to be

$$P = \left[P_0 + \tfrac{m}{k}\right] e^{kt} - \tfrac{m}{k}.$$
(16R)

If you did the calculation correctly (and if we stated the conclusion correctly), we should find that the function $P(t)$ described in (16) actually satisfies *both* the differential equation and the initial condition.

Checking the initial condition is easy:

$$\begin{aligned}
P(0) &= \left[P_0 + \tfrac{m}{k}\right] e^{k \cdot 0} - \tfrac{m}{k} \\
&= \left[P_0 + \tfrac{m}{k}\right] - \tfrac{m}{k} \\
&= P_0.
\end{aligned}$$

To check whether $P(t)$ satisfies the differential equation, we have to calculate the derivative, $\dfrac{dP}{dt}$, of our proposed solution function (16) and compare the result with the expression on the right in (14). The derivative calculation is not difficult, but we need to recall some of the differentiation formulas that appeared in Chapter 2. We will need:

the **Sum Rule**,

$$\frac{d}{dt}\left[f(t) + g(t)\right] = \frac{df}{dt} + \frac{dg}{dt},$$
(19)

the **Difference Rule**,

$$\frac{d}{dt}\left[f(t) - g(t)\right] = \frac{df}{dt} - \frac{dg}{dt},$$
(20)

and the **Constant Multiple Rule**,

$$\frac{d}{dt} \, Af(t) = A \, \frac{df}{dt}, \text{ for any constant } A.$$
(21)

Now we are ready to differentiate our solution function. Next to each step, fill in a reason for the step.

<u>Step</u>   <u>Reason</u>

$$\begin{aligned}
\frac{dP}{dt} &= \frac{d}{dt}\left(\left[P_0 + \tfrac{m}{k}\right] e^{kt} - \tfrac{m}{k}\right) \\[2mm]
&= \frac{d}{dt}\left[P_0 + \tfrac{m}{k}\right] e^{kt} - \frac{d}{dt}\,\frac{m}{k} \qquad \text{We applied the Difference Rule.} \\[2mm]
&= \left[P_0 + \tfrac{m}{k}\right] \frac{d}{dt}\, e^{kt} \\[2mm]
&= \left[P_0 + \tfrac{m}{k}\right] k\, e^{kt}
\end{aligned}$$

$$= (kP_0 + m)\, e^{kt}$$

On the other hand, if we substitute the definition of $P(t)$ from (16) into the right-hand side of the differential equation (14), we get

$$kP(t) + m = k\left(\left[P_0 + \frac{m}{k}\right]e^{kt} - \frac{m}{k}\right) + m$$

$$= k\left[P_0 + \frac{m}{k}\right]e^{kt} - m + m$$

$$= (k\,P_0 + m)\, e^{kt} .$$

Since substitution of $P(t)$ into each side of (14) gives the same result, $P(t)$ does indeed satisfy the differential equation.

Exercise 5. In section 3.2, we solved the initial value problem

$$\frac{dT}{dt} = -k(T - 21), \quad \text{with } T = 30 \text{ at } t = 0. \tag{9R}$$

Our solution was

$$T = 21 + 9\, e^{-kt}, \tag{12R}$$

where $k = 0.2513144$. Check this solution by showing that T and its derivative actually satisfy the differential equation in (9) and that $T(0)$ is indeed 30.

Comparing the Solution to the Data

Next we ask whether the solution (18) of the proposed immigration model,

$$P = \left[P_0 + \frac{m}{k}\right]e^{kt} - \frac{m}{k}, \quad \text{where } m = 0.25 \text{ and } k = 0.001227, \tag{18R}$$

actually fits the census data for the U. S. population. (We repeat the data in Table 1.) The solution in (18), rewritten in the form

$$P + \frac{m}{k} = \left[P_0 + \frac{m}{k}\right]e^{kt},$$

suggests that we should look at a semilog plot of the shifted data values

$$P_0 + \frac{m}{k}, \ P_1 + \frac{m}{k}, \ \dots, \ P_{19} + \frac{m}{k},$$

versus time t.

Exercise 6. Explain why the semilog plot of shifted data should be a straight line if the function in (18) fits the data.

t (years A.D.)	P (millions)	$P + m/k$	t (years A.D.)	P (millions)	$P + m/k$
1790	3.939		1890	62.948	
1800	5.308		1900	75.996	
1810	7.240		1910	91.972	
1820	9.638		1920	105.711	
1830	12.866		1930	122.775	
1840	17.069		1940	131.669	
1850	23.192		1950	150.697	
1860	31.443		1960	179.323	
1870	38.558		1970	203.185	
1880	50.156		1980	226.546	

Table 1. U. S. Census Data, 1790–1980.

Exercise 7. Fill in the $P + \frac{m}{k}$ columns in Table 1.

Figure 7 shows the semilog plot of $P + \frac{m}{k}$ versus t.

Exercise 8. Write a short discussion (not more than a page) of the extent to which the semilog plot in Figure 7 confirms or refutes the reasonableness of the immigration model as a description of U. S. population growth.

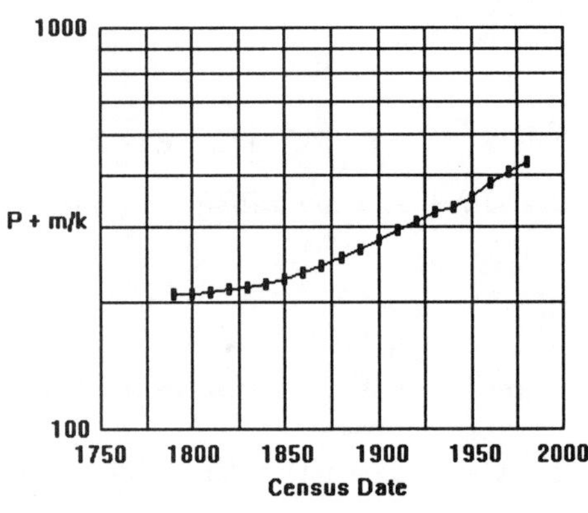

Figure 7. Semilog plot of $P_i + \frac{m}{k}$ versus time.

Section Summary

List the steps in solving the initial value problem

$$\frac{dP}{dt} \;=\; kP + m \quad \text{with} \quad P(0) = P_0$$

and checking the solution.

Exercises

9. Consider the initial value problem

$$\frac{dz}{dt} = 2z + 1 \quad \text{with} \quad z = 3 \quad \text{at} \quad t = 0. \tag{22}$$

(a) Make a change of variable to rewrite the differential equation in (22) in the form

$$\frac{dy}{dt} = 2y.$$

(b) Rewrite the initial condition in terms of y.

(c) Find y as a function of t so that $y(t)$ satisfies the initial value problem described in (a) and (b).

(d) Find the solution $z = z(t)$ of the original initial value problem (22).

10. Solve the initial value problem

$$\frac{dz}{dt} = 3z + 2 \quad \text{with} \quad z = 4 \quad \text{at} \quad t = 0.$$

3.4 Solving Nonlinear Equations by Zooming

The Simpler Problem

Before we work on the determination of the constant k for the immigration problem, let's look at the simpler problem posed at the beginning of Section 3.4. We want to find the value of x such that

$$x = e^{-x}. \tag{23}$$

As a first step, we rewrite equation (23): We want to find an x that satisfies

$$e^{-x} - x = 0. \tag{24}$$

This changes the problem from finding the point of intersection of two graphs to finding a zero[13] of one function, $f(x) = e^{-x} - x$. We can see from the graph of $f(x)$ in Figure 8 that the function $e^{-x} - x$ does have a zero and that it lies between 0.5 and 0.6.

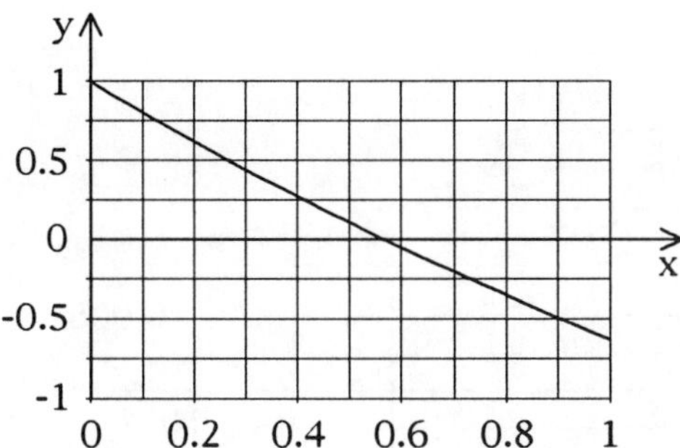

Figure 8. Graph of $y = x - e^{-x}$ from $x = 0$ to $x = 1$.

Now let's zoom in on the portion of the graph between 0.5 and 0.6 (Figure 9):

---

[13]A note on the proper use of words: A **zero of a function** is a number in the domain (a value of x in this case) at which the function value is zero. A **root of an equation** [such as equation (24)] is a number that satisfies the equation. Thus, our problem can be correctly stated as either "Find a zero of the function $f(x) = e^{-x} - x$" or "Find a root of the equation $e^{-x} - x = 0$." The terms "zero" and "root" are often confused; we will try to help you keep them straight.

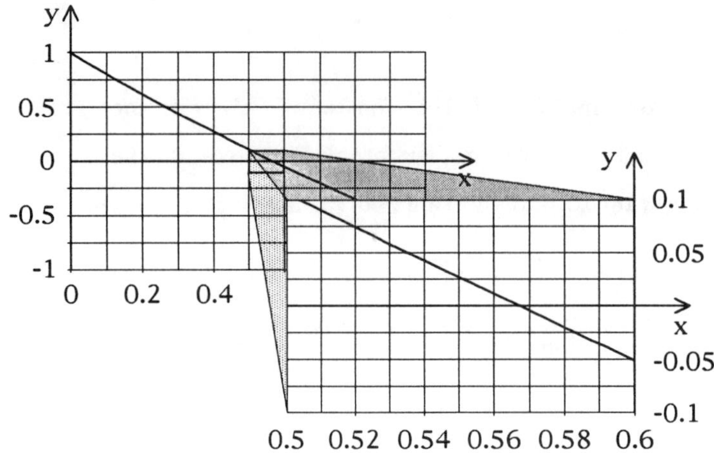

Figure 9. Graph of $y = x - e^{-x}$ from $x = 0.5$ to $x = 0.6$.

From Figure 9 we can see that the zero is between 0.56 and 0.57. We can zoom in again to obtain an even better estimate:

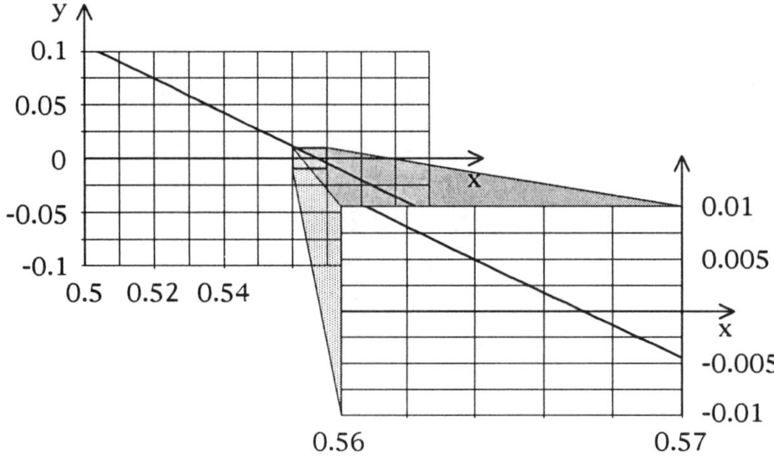

Figure 10. Graph of $y = x - e^{-x}$ from 0.56 to 0.57.

In Figure 10 we can see that the zero is between 0.567 and 0.568, but closer to 0.567. We can continue zooming to obtain any desired degree of accuracy.

Does this successive zooming process recall something we did in Chapter 2? Do you see something familiar in the emerging "straightness" as we look at smaller and smaller segments of the curve? Hold that thought — we will put it to work in the next chapter when we take a deeper look at the problem of solving a nonlinear equation by reducing it to the much simpler problem of solving a linear equation.

Exercise 1. If we take logs of both sides of equation (23), we obtain the equivalent problem of finding a number x that satisfies

$$-x = \ln x. \tag{25}$$

(a) Rewrite this problem as one of finding the zero of a single function.

(b) Use a graphing calculator or computer to zoom in to a solution accurate to three significant digits.

Exercise 2. Use the zooming method to find all values of x such that

$$x^2 = -10 \cos x.$$

Find each zero (there are four) to three significant digit accuracy.

The More Complicated Problem

We return to the problem encountered in our investigation of population growth with immigration,

$$17.069 = \left[3.929 + \frac{0.25}{k}\right] e^{50k} - \frac{0.25}{k}. \tag{17R}$$

To put the problem in slightly more tractable form, we clear the fractions (i.e., multiply through by k) and bring all terms to one side of the equation (i.e., subtract $17.069\,k$ from both sides). The equivalent equation is

$$(3.929\,k + 0.25)\, e^{50k} - 0.25 - 17.069\,k = 0. \tag{26}$$

This is not a simple equation, but it is typical of equations that turn up in real problem solving. As we will see, our methods for solving such equations are almost independent of how complex they are, except for the number of keys we have to press on our calculator or computer.

Equation (26) has some special problems we attend to before discussing how to solve it. First, it is simple to check that $k = 0$ is a solution. (Why is that a solution? Do the substitution.) However, $k = 0$ is *not* a solution of equation (17)! In this case, 0 is a "spurious" solution, one that was introduced by our algebraic manipulation (which manipulation?). Having a spurious solution is not usually a serious problem, as long as we are aware of it and are not misled by it. However, the form of equation (26) suggests that any positive number k that is also a solution must be *very small*. Specifically, the term involving e^{50k} is very large even for fairly small values of k, so there is no way that this term can be offset by the negative terms in the equation unless k is really tiny. (Check it out. See how big that term is for $k = 0.2$, say.) This means that the solution we want must be very close to the solution we already know — and don't want.[14] And that's a problem.

There is a standard tool for spreading out values of the independent variable: multiplicative scaling. For example, we can spread out k values by a factor of 100 if we change to a new independent variable, $x = 100\,k$. Each place k occurs, we substitute $\frac{x}{100}$, and we get

$$(0.03929\,x \;+\; 0.25)\,e^{x/2} \;-\; 0.25 \;-\; 0.17069\,x \;=\; 0. \qquad (27)$$

This is the equation we will actually solve for x; then we will divide the answer by 100 to find k.

Zooming In

The left-hand side of (27) is a function of x, which we name $f(x)$:

$$f(x) \;=\; (0.03929\,x \;+\; 0.25)\,e^{x/2} \;-\; 0.25 \;-\; 0.17069\,x. \qquad (28)$$

We'll apply our-zoom in approach to this problem exactly as we did for the simpler problem. In Figure 11 we see that the solution we want is between $x = 0.1$ and $x = 0.2$. In Figure 12 we zoom in on the interval $[0.1, 0.2]$.

---

[14]If we keep the original problem firmly in mind, we will find no surprise in the fact that k must be close to zero. We are looking for the proportional growth coefficient for a real (human) population on a time scale with one year as the unit. That alone makes k small — typically not more than 0.02. But we are also attributing a significant part of the growth to immigration, which should make k even smaller.

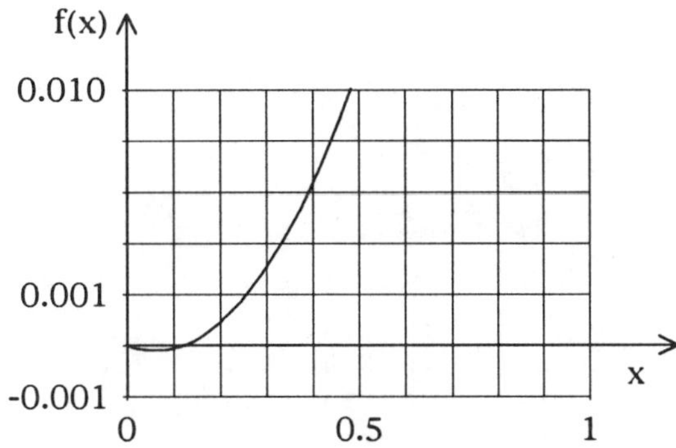

Figure 11. $f(x) = (0.03929\,x + 0.25)\,e^{x/2} - 0.25 - 0.17069\,x.$

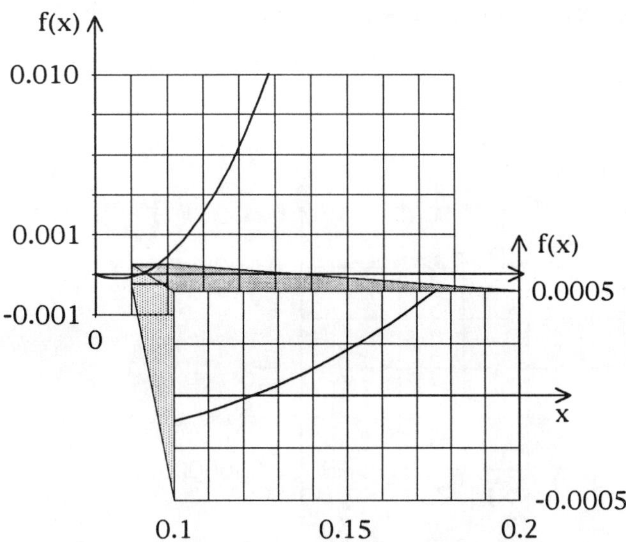

Figure 12. First zoom-in.

Now we repeat the divide-and-conquer strategy that we used in the simpler problem. The grid in Figure 12 shows at a glance that the solution must lie in the interval [0.12,0.13]. Thus, we now know the first two decimal places of the solution. (What are they?)

In Figure 13, we zoom in on the interval [0.12,0.13]. What does this figure tell us about the third decimal place of the solution? In Figure 14, we zoom in once more. Now you should know the solution to four decimal places. Write the answer here: ______________________

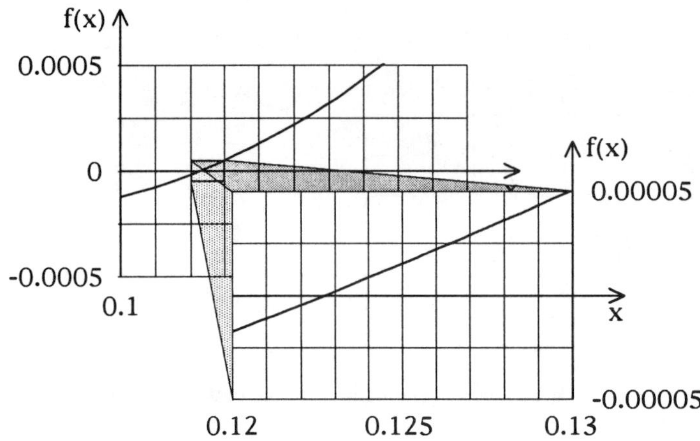

Figure 13. Second zoom-in.

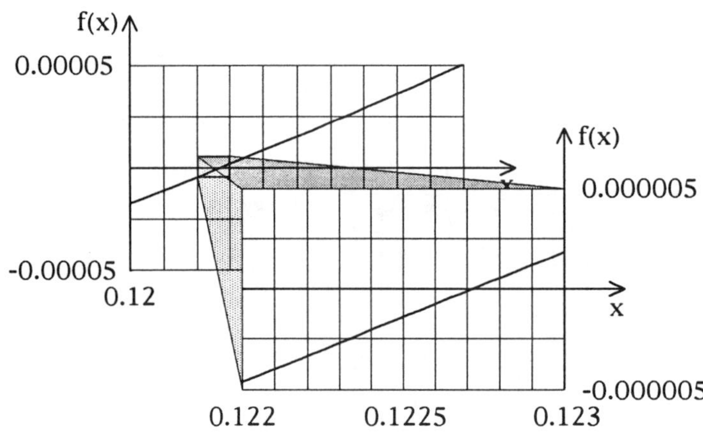

Figure 14. Third zoom-in.

You now know x to four decimal places, but x was merely a "variable of convenience," introduced to make the solution process easier. The quantity we really wanted to know was k — remember k? Recall how x and k were related; then write the value of k here: $k =$ __________ Compare this answer with the value we used in Section 3.3.

Zooming Without Graphing

The solution we just carried out doesn't absolutely require a computer; a graphing calculator would do as well. But what if we have nothing to work with but a non-graphing calculator (one without a $\boxed{\text{Solve}}$ key that would automate the solution process)? In principle, we could carry out the same process, subdividing an interval containing the solution, and computing the value of f at each point from left to right until we see a sign change. However, *every* evaluation of f is a significant problem in its own right, so this could be a very tedious process. (Try it with your calculator — find the value of f at 0.12.)

On the other hand, if f is relatively simple to evaluate, you can easily find a solution of $f(x) = 0$ by a numerical version of zooming in that can be carried out on your calculator. The next exercise will guide you through this procedure. Don't be intimidated by the apparent length of the exercise — this is not a difficult procedure, and you will find it very useful to know how to solve equations with only a calculator.

Exercise 3. *We will attempt to solve the equation* $x + 2 = e^x$.

(a) First make a rough sketch of $y = x + 2$ and $y = e^x$ on the same axes, and explain why our equation can have only one positive solution. (It also has a negative solution, which we will ignore for now.) Estimate from your rough sketch about how big the positive solution is.

(b) Rewrite the equation in the form $f(x) = 0$. (You get to choose f; the choice is not uniquely determined.) Now, using your calculator, find two positive values of x, reasonably close together, so that f is positive at one value and negative at the other. You now know an interval in which the solution must lie (why?).

(c) From here on, there are many ways you can proceed. For example, you could proceed in small steps from your left end point toward your right end point, checking the value of f at each point until you see a sign change. Or, you could use the sizes of your f values at the end points to estimate about how far across the interval the sign change must occur. However you proceed, your objective is to find a smaller interval at each stage that still contains the solution. You may not be able to reduce the size of the interval by a factor of 10 at each step, but that is not required. With a few minutes calculation (and careful record keeping), you should know the solution to three or four decimal places. Write your answer here: The only positive number x that satisfies the equation $x + 2 = e^x$ is ___________________

Section Summary

Describe the process of solving nonlinear equations by zooming-in.

Exercises

4. Use zooming to find **all** solutions of the equation $3x^2 = e^x$.

5. Find the solution of $e^{-x} = x$ without graphics, in the manner described in Exercise 3.

3.5 Falling Bodies

An "Experiment"

Here's a "thought experiment." (Please don't try to carry it out —
someone might get hurt, and you would certainly get arrested.) Suppose
you dropped a marble out of the window at the top of the Washington
Monument. How fast do you think it would be going just before it hits the
ground, about 535 feet below?[15] Write your guess in the margin, for
reference after we have solved the problem.

Guess: ___________

In Chapter 2, you examined data (repeated below in Table 2) giving the
position s of a falling object as a function of time t. In particular, by
constructing a log-log plot of the data (Exercise 5 of Section 2.6 and Figure
39), you showed the plausibility that s might be a power function of t,
specifically, a constant times the square of t.

Time (seconds)	Distance (meters)	Time (seconds)	Distance (meters)
0.5	1.237	5.5	147.661
1.0	4.962	6.0	175.769
1.5	11.052	6.5	207.346
2.0	19.398	7.0	239.973
2.5	30.579	7.5	276.944
3.0	44.113	8.0	312.798
3.5	60.258	8.5	354.793
4.0	78.024	9.0	396.130
4.5	99.216	9.5	443.067
5.0	122.757	10.0	489.007

Table 2. Position of a falling object.

We can *estimate* the velocity v (which means the time derivative or
rate of change of s) of the falling object by calculating difference quotients
of the data in Table 1, that is, by calculating $v_i = \dfrac{s_i - s_{i-1}}{t_i - t_{i-1}}$ for each i
from 1 to 20. In Figure 15, we show the results of this calculation; the
approximate velocity function is approximately linear. (The obvious lack of
linearity near the end of the fall may be the result of measurement error

---

[15]One of the tidbits of information we recall from high school (it may or may not
be correct) is that the Washington Monument is 555 feet, $5\frac{1}{5}$ inches tall. The window
is not at the very top, of course.

when the object is falling that fast, or it may mean that half a second is not a "small" time step when the object is falling that fast.) Therefore, the acceleration (which means the time derivative or rate of change of v) is approximately constant (why?). Is that physically reasonable? Should it apply to all falling objects, some falling objects, or just the one measured in this particular experiment?

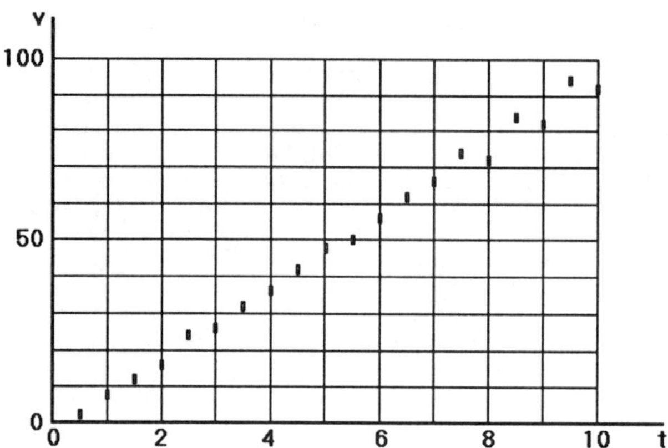

Figure 15. Approximate velocity (in meters per second) of a falling object.

Gravity and Newton's Second Law of Motion

In section 1.2, we discussed briefly the theoretical framework for considering questions such as those just raised; this framework dates back more than 300 years to Newton's formulation of "laws of motion." These "laws" are very concise statements that encapsulate many generations of observations about moving objects, such as Simon Stevin's experiment a century earlier, which showed that objects of very different weights would fall the same distance in almost identical times. In particular, Newton's Second Law of Motion (as applied to an object of unchanging mass, such as our marble) states that the net force acting on an object is the product of its mass and its acceleration.[16] In symbols, this says

$$F \;=\; ma \;=\; m\,\frac{dv}{dt},$$

---

[16]In other words, the net force is proportional to acceleration; "mass" is then defined to be the proportionality constant. The law has to be stated differently if the object's mass is changing, as would be the case, say, for a rocket, whose mass decreases as it consumes its fuel. We will take up this case in Chapter 7.

where F is the net force, m is the (constant) mass of the object, and a is the acceleration.

Our falling marble is subject to a gravitational force, called its *weight*, which is the result of its interaction with another (very large) object, the Earth. Newton also formulated a Universal Law of Gravitation, which says that the gravitational attraction between two particles ("point masses") is jointly proportional to their masses and inversely proportional to the square of the distance between them. When the objects are moving toward each other under gravitational attraction, this law says, in particular, the gravitational force is *not* constant. The Earth hardly qualifies as a "point mass," but it is a very large collection of point masses, and one can deduce from the Law (as Newton did) that, *very roughly*, you get the right answers about Earth's gravity if you assume the entire mass of the Earth is concentrated at its center point — that is, treat the Earth as a point mass. The Law still says the gravitational force is *not* constant.

However, the marble falls about one-tenth of a mile, and the distance to the center of the Earth is some 40,000 times farther, so the distance between the two centers is *essentially* constant. Therefore the gravitational attraction is also. In fact, most of our human existence takes place so close to the surface of the earth[17] that we can safely assume the gravitational force acting on any particular object is constant. We do precisely that when we say that an object has a "weight" that we expect not to depend on where the object happens to be. What is really constant for a given object is its *mass*, which, as we noted, is the proportionality constant between weight (force) and the gravitational acceleration. Careful experiments have shown that the gravitational acceleration at or near the surface of the earth is approximately 32.17 feet per second per second; this constant is usually abbreviated "g."

Exercise 1. *The distance unit for the problem at hand happens to be feet, but we will often use metric units as well — as we did in Table 2 and Figure 15. Convert $g = 32.17$ ft/sec^2 to meters per second per second.*

---

[17]Some exceptions: air and space travel, deep diving and mining, and serious mountain climbing.

What we have seen so far is that "constant acceleration" can account for gravity near sea level. However, it certainly cannot account for other possible forces acting on the marble, such as air resistance (which depends on its velocity) and wind currents (which are mostly unpredictable). Thus, in order for our "constant acceleration" model to apply, we have to *assume* these other forces are not present or are so small as not to matter. For now, we make that assumption; later, we will consider the effects of wind and air resistance.

An Initial Value Problem for Velocity

We come at last to an initial value problem for v as a function of t:

$$\frac{dv}{dt} = g, \quad \text{with} \quad v = 0 \quad \text{at} \quad t = 0. \tag{29}$$

(Where does the initial condition come from?) This is the *inverse* of a problem we have studied already: If a function is *linear*, then we have seen that its average rate of change is constant (that's the slope of the linear graph), so its instantaneous rate of change is also constant. Having studied the inverse problem, we know a possible solution to (29): Any linear function with slope g, say, $v = gt + C$, will satisfy the differential equation. And by substituting $v = 0$ and $t = 0$ simultaneously, we see that the initial condition can be satisfied if $C = 0$. Thus,

$$v = gt \tag{30}$$

is *a* solution of (29). Since we have assumed that each initial value problem has only one solution, it follows that (30) must be *the* solution of (29).

We now know velocity as a function of time; that would answer the "how fast" question about the marble if we knew *how long* it takes to fall 535 feet. But that's a question about the functional relationship between *distance* and time, which we don't know yet. However, we can read (30) as a differential equation relating distance and time (because $v = \frac{ds}{dt}$), and we can set up another initial value problem:

$$\frac{ds}{dt} = gt, \quad \text{with} \quad s = 0 \quad \text{at} \quad t = 0. \tag{31}$$

(Where does the initial condition come from this time?)

Again, we happen to have studied the appropriate inverse problem already. Where did we see a function whose derivative was a constant times t? Oh, yes — that was one of our earliest examples of a derivative calculation in Chapter 2. For any function of the form $s = ct^2$, we found that its derivative was $2ct$. The constant factor c was arbitrary, so we can set $2c = g$, and we will match the derivative in (31). As with the previous initial value problem, we now know a whole family of solutions of the differential equation: $s = \frac{1}{2}gt^2 + C$, where C could be any constant (see Conceptual Exercise 6 in Chapter 2). But when we substitute the initial condition, $s = 0$ when $t = 0$, we see again that C has to be zero. Thus,

$$s = \frac{1}{2}gt^2 \tag{32}$$

is a solution of (31), and therefore (by uniqueness) *the* solution.

Now we can determine the time of fall from (32) and then the velocity from (30).

Section Summary

We list formulas that appeared in the process of obtaining formula (32) from Newton's Second Law of Motion together with the assumptions that $v = 0$ and $s = 0$ at time $t = 0$. Next to each formula, write a sentence explaining this step of the development.

$$F = ma$$

$$mg = m\frac{dv}{dt}$$

$$\frac{dv}{dt} = g$$

$$\frac{dv}{dt} = g \ \text{ with } \ v = 0 \ \text{ at } \ t = 0$$

$$v = gt + C$$

$$v = gt$$

$$\frac{ds}{dt} = gt \ \text{ with } \ s = 0 \ \text{ at } \ t = 0$$

$$s = \frac{g}{2}\,t^2 + C$$

$$s = \frac{g}{2}\,t^2$$

Exercises

2. *Finish the marble problem:*

    (a) How long does the marble fall, and how fast is it going just before it hits the ground? Express your answer first in feet per second; then convert it to meters per second and miles per hour.

    (b) Compare your answer for speed at impact with the guess you wrote in the margin at the start of this section. Were you close? Can you see why you would be arrested if you tried the experiment?

3. *Suppose the marble were dropped from a height of 100 feet (approximately the height of a ten-story building).*

    (a) What would be its velocity (in feet per second) just before impact?

    (b) How long would it take to hit the ground?

4. *Suppose the marble were dropped from a height of 4 feet (i.e., from arm's height above the ground).*

    (a) What would be its velocity (in feet per second) just before impact?

    (b) How long would it take to hit the ground?

5. *Recall where in the calculation we used the assumption that the marble was dropped, as opposed to being thrown.*

    (a) Suppose the marble was tossed straight up (from a height of 535 feet) with an initial speed of 40 feet per second. How does this affect the time of fall and velocity at impact?

    (b) Suppose it was tossed straight down at 40 feet per second. How does that affect the time of fall and velocity at impact?

3.6 Problem-Solving Strategies

Our discussions of the homicide, immigration, and falling marble problems suggest some observations about problem solving. Other thoughts about problem solving will appear from time to time throughout the course. However, when you find yourself stuck with no ideas about what to do next, you may want to return to this section. As you consciously put these strategies into practice, you will become a better problem solver.

First, be aware that even moderately realistic problems often have many "steps" and require pulling together many ideas. Furthermore, the steps are not stated as part of the problem — you have to work out the divide-and-conquer strategy. In this sense, *problems* are quite different from "exercises." We saw this, for example, in the homicide problem (Section 3.2), which could be stated quite simply: Take temperature measurements on the body, and determine the time of death. That simple demand led us into Newton's Law of Cooling, setting up an initial value problem, rescaling the problem in terms of a convenient variable, relating decay to negative growth, using what we learned in Chapter 2 about exponential solutions, doing a reverse substitution to get back to the temperature variable, using measurements to determine the constants in the model, substituting a known temperature to find an unknown time, and using logarithms and a calculator to complete the calculation.

We (the authors) were the problem solvers in the previous sections of this chapter, and we provided an exposition of our thought processes as a model of what you should do when you tackle the problems posed as in-class or take-home projects. We are *not* providing "recipes" that you can simply follow step-by-step. No two problems are alike, and we are not asking you to mimic our specific steps.

Second, be aware of the choices that are yours to make, especially the options for scaling and re-scaling variables and the closely related options for choosing positive directions, signs of quantities, units, and location of the origin. Your first instincts about these matters may not be the best choices, but there may be no way to know until you get into the computations.

We actually shielded you from our best example of the need to make choices when we suggested (again in Section 3.2) that you set $t = 0$ at the time of first measurement, not at the time of death. And then we asked you (in Exercise 2 of that section) to see what would happen if you had made the other choice. The homicide problem had another good example: Recall that we held off on deciding what unit to use for time until we had a reason to make that choice. And then we saw that very small units (seconds or minutes) and very large units (days or years) would not be sensible for the problem at hand.

If you find a computation getting messy, ask yourself if there might be a better set of choices, and don't shy away from making those choices.

Third, keep your goal clearly in focus. Don't pursue calculations just because they are there to do. Ask yourself whether they will actually get you closer to your goal. If not, you should probably be thinking in some other direction.

Our most important examples of this principle are yet to come, but here is one that might have happened in Section 1.6 on Inverse Functions. We considered there the following table of data, in which x represented time and y represented drug sales in billions of dollars:

x (Year)	y (Sales)
1985	22
1986	25
1987	28
1988	31
1989	35
1990	38

Let us suppose you are asked to determine the year in which sales reached $30 billion. The question is trivial, but if you respond to such questions by jumping immediately into "calculate" mode — without thinking — you might do something like this: (1) observe that the sales are growing roughly linearly (about $3 billion/year); (2) write a linear formula for sales, such as

$$y = 3\,(x - 1985) + 22;$$

(3) invert the function by solving for x as a function of y; (4) substitute $y = 30$ into your new formula, and solve for x. If you actually did all this, your answer would be $x = 1987\frac{2}{3}$ — and you still wouldn't have a sensible answer to the question that was asked. The problem is not that the calculation is *wrong*, just that it is unnecessary.

Fourth, try to relate what you are doing to problems you have already solved. For example, when we found (Section 3.2 again) that Newton's Law of Cooling could be transformed into a natural growth problem — with temperature decay thought of as "negative growth" — we knew the solution of the initial value problem without repeating any of the labor from Chapter 2. Then we found (Section 3.3) that the immigration problem could be transformed into something very like Newton's Law of Cooling.

Fifth, be aware that abstract symbols are sometimes a labor-saving device. We saw a modest example in the cooling body problem: When we found k, we wrote it down with seven decimal places — and could have written more from a 12-place calculator. But we had no idea how many places we needed for subsequent calculation. Rather than risk miscopying long strings of digits or introducing rounding errors too early, we kept the symbol k in subsequent calculations, knowing that we could substitute its value whenever we needed to.

Sixth, feel free to manipulate your algebraic quantities in any ways that look like they might be profitable. We did this, for example, in changing

$$\frac{dP}{dt} = kP + m \tag{14R}$$

into

$$\frac{dP}{dt} = k\left[P + \frac{m}{k}\right],$$

and in changing

$$17.069 = \left[3.929 + \frac{0.25}{k}\right]e^{50k} - \frac{0.25}{k} \tag{17R}$$

to

$$(3.929 \, k \, + \, 0.25) \, e^{50k} \, - \, 0.25 \, - \, 17.069 \, k \, = \, 0 \qquad \text{(26R)}$$

and then to

$$(0.03929 \, x \, + \, 0.25) \, e^{x/2} \, - \, 0.25 \, - \, 0.17069 \, x \, = \, 0. \qquad \text{(27R)}$$

You can always multiply and divide by the same (nonzero) quantity to produce a new factor; you can always add and subtract the same quantity to produce a new term; you can always rescale your variables if there is a good reason to do so. Rearrange, combine, expand, factor, substitute until your expression has the form needed for whatever you have decided will be the next step.

In your previous experience with algebra, your were probably told often to "simplify" — which may have meant contradictory things, depending on the context. You may have been given the impression that there is only one "right" form for each expression. In truth, simplification often helps, but sometimes "unsimplification" does, too. We saw a good example of that in Exercise 4 of Section 1.7, in which you showed that $\sqrt{x^2+1}-x$ is algebraically the same as $1/\!\left(\sqrt{x^2+1}+x\right)$. The first of these was virtually useless for numerical evaluation at a large value of x, and the second had no numerical problems at all. The right form for an expression is the one that works for whatever comes next. The one thing your manipulations must *not* do is change the value of the expression you are manipulating.

Seventh, be aware that the solution process for one problem may generate an equally substantial problem of an entirely different nature, one that must be solved before any more progress can be made on the first problem. That's what happened when we found (in Section 3.3) that the immigration problem was not *identical* to the cooling body problem, that we would need to know something about solving nonlinear equations before we could finish the calculation.

SUMMARY

We remind you again — because it's easy to forget — that "initial value problem" is not very descriptive; it's just jargon to shorten the unwieldy phrase "differential-equation-with-initial-value problem."   However, it is *standard* jargon, so you must get used to using it, and you must associate the right kind of problem with it.

In this chapter, we have considered (and will continue to consider throughout the course) just two kinds of differential equations. The first kind ("autonomous") expresses the derivative (rate of change) of an unknown function in terms of the dependent variable only. The second ("explicit-derivative") expresses the derivative as an explicit function of the independent variable. Either kind of differential equation can be expected to have infinitely many solutions; adding the specification of a single point on the solution curve (an initial condition) usually singles out a unique solution function — so usually that we simply *assume* that **every initial value problem has a unique solution.**[18]   When you have mastered the meanings of five of the last seven words in the previous sentence — "initial," "value," "problem," "unique," "solution" — you will be well on your way to understanding what this course is about and, in particular, how calculus is used to model the real world.

We recalled at the beginning of the chapter the useful visual device called "direction field." To be candid, one would not want to draw very many direction fields by hand; that job is tedious unless done by a computer. The most important role of the direction field is as a *conceptual* tool. Whether you actually draw the field or not, you should imagine each differential equation being represented by such a field, by a large number of "direction markers" whose slopes are determined by the differential equation. Solving the differential equation then means finding curves (graphs of functions) that pass through the direction field always following

---

[18]We hasten to add that this statement is FALSE. There are initial value problems that have more than one solution, and there are initial value problems that have no solutions. However, we won't meet any of those in this course, so the statement is "true enough" to be our working hypothesis.

the direction markers. And solving an initial value problem means picking out the solution curve that goes through the right initial point.

So far, our principal technique for solving differential equations has been to first solve the easier *inverse* problem — calculating derivatives of known functions — and then to guess the *form* of a solution and make the unknown constants match up with the statement of the problem.

There are two kinds of unknown constants: (i) those that appear as part of the problem (or as part of the model), and (ii) one (per problem) that arises from the fact that every derivative is the derivative of a whole family of functions.

The value of the second kind of constant is invariably determined by making sure that the initial condition is satisfied. Indeed, that was our purpose in focusing on initial value problems: Satisfying the initial condition has the effect of singling out exactly one solution from an infinite family of possibilities. Furthermore, most "real" problems either come with an initial condition already specified or are amenable to having one specified as part of the problem solving process.

The other kind of unknown constants (sometimes called "parameters") may be more troublesome, in that there are no rules on how many there are or on how to determine their values. Each new situation may be different from the ones you have seen before. You have to match up known information with the unknown constants and use appropriate algebraic and numeric calculations to find the appropriate values.

You will quickly run out of differential equations for which you can guess the form of a solution — unless you "leverage" this tool by adding other supplementary tools to your toolbox. The leveraging tools we considered in this chapter are collectively called *scaling*.

There are four possibilties: additively scaling either independent or dependent variables, and multiplicatively scaling either independent or dependent variables. Our primary use for these tools will be to turn apparently new problems involving derivatives into ones we have seen before and know how to solve.

Three of the four scaling possibilities actually came up in Chapter 2. The first to appear (Section 2.3) was the Constant Multiple Rule,

$$\frac{d}{dt}\, cf(t) \;=\; c\,\frac{df}{dt},$$

which told us that when the *dependent* variable, $y = f(t)$, is scaled by a factor of c, the rate of change is scaled by the same factor.

Next to appear (Section 2.4) was a multiplicative scale change of the *independent* variable, which led to a similar conclusion about derivatives. Specifically, we saw that knowing $\frac{d}{dt}e^t = e^t$ also told us how to differentiate e^{kt}: Multiplying the independent variable by k multiplies the rate of change by k. That is, $\frac{d}{dt}e^{kt} = ke^{kt}$.

Then, in Section 2.5, we saw an example of *additive scaling* (or shifting) *of the independent variable.* Specifically, the solution of the initial value problem

$$\frac{dP}{dt} \;=\; kP \quad and \quad P = P_0 \ \ when \ \ t = t_0 \tag{33}$$

is

$$P \;=\; P_0\, e^{k(t-t_0)}; \tag{34}$$

in particular, if t_0 happens to be 0, then the solution of

$$\frac{dP}{dt} \;=\; kP \quad and \quad P = P_0 \ \ when \ \ t = 0 \tag{35}$$

is

$$P \;=\; P_0\, e^{kt}. \tag{36}$$

The distinction between the problem-solution pair (33)-(34) and the pair (35)-(36) is whether the initial condition is specified at 0 or somplace else. The second pair is simpler to understand — and the result is easier to remember. It's often the case that initial value problems are simpler if we can *choose* the initial point to be at $t = 0$. Fortunately, we usually *can* make that choice. If the problem is stated with an initial point at some t_0 that is not 0, we can change the independent variable to, say, $x = t - t_0$, solve the problem as in (36) with x replacing t, and then make the reverse substitution to get the solution (34).

The fourth type of scaling, *additive scaling of the dependent variable* (or vertical shifting), appeared for the first time in Section 3.2 and was just what we needed for the cooling body and immigration problems. As you will see, it works in much the same way for a number of other problems at the end of the chapter. Here the idea was that the shift $y = T - 21$ could turn the initial value problem

$$\frac{dT}{dt} = -k(T - 21), \quad \text{with } T = 30 \text{ at } t = 0. \tag{9R}$$

into the more familiar (and already solved) initial value problem

$$\frac{dy}{dt} = -ky, \quad \text{with } y = 9 \text{ at } t = 0. \tag{10R}$$

We worked through two apparently similar problems, cooling and immigration, both using additive scaling in much the same way, to point out a difficulty that can and did arise (with the second problem, but not the first) in determining a value for an unknown constant. We will often find ourselves with nonlinear algebraic equations we need to solve.[19] This is an algebraic problem that algebra alone usually fails to solve!

We saw that, for a single equation in a single unknown, this problem is conceptually simple — we just need to find where the graph of a function crosses the horizontal axis. And with a powerful enough tool — a graphing computer or calculator — even the practical part of the problem is not difficult, although we found it necessary to use a scaling tool to visually separate two solutions that were very close together.

Even without a graphing device, the *concept* (crossing the axis) leads to a numeric procedure that can be carried out on any calculator, although the procedure may be tedious. In the next chapter, we will put calculus to work on this equation-solving problem and find a method that is much less tedious to carry out. We hinted at the method when we pointed out the

---

[19]In Chapter 1 we stressed the point that "solve" in this course would usually mean "find a function" that satisfies the conditions of the problem, whereas in your previous courses in mathematics it often meant "find a number" (the value or values of a variable determined by an equation) or "extract a variable" from an equation. However, we still need to find numbers as well, and often in situations different from any you may have seen before.

straight-line appearance of our graphs as we zoomed in on the crossing points — we're going to find another use for "local linearity."

In order to tie up a loose end left dangling in Chapter 2, we used the falling body problem as our principal example of a "Type II" (explicit derivative) differential equation. Rather than start from a position formula from a physics book (a "solution," not a "problem"), we (re)started the discussion from the physical principles embodied in Newton's Second Law of Motion and Law of Universal Gravitation. That approach led us to the observation that the textbook "solution" is just the end result of solving two simple initial value problems, one to get the velocity formula from the assumption of constant acceleration, the other to get the position formula from the velocity formula.

Derivative Formulas[20]

General formulas

Scaling the independent variable:

If $u = kt$, where k is constant, then $\dfrac{d}{dt} f(kt) = k \dfrac{d}{du} f(u)$.

The Constant Multiple Rule:

$\dfrac{d}{dt} Af(t) = A \dfrac{df}{dt}$, for any constant A.

The Sum Rule:

$\dfrac{d}{dt} [f(t) + g(t)] = \dfrac{df}{dt} + \dfrac{dg}{dt}$.

The Difference Rule:

$\dfrac{d}{dt} [f(t) - g(t)] = \dfrac{df}{dt} - \dfrac{dg}{dt}$.

Derivatives of exponential functions

$\dfrac{d}{dt} e^{kt} = k \, e^{kt}$, for any constant k.

$\dfrac{d}{dt} b^{t} = (\ln b) \, b^{t}$, for any constant exponential base b.

Derivatives of power functions (the Power Rule)

$\dfrac{d}{dt} t^{n} = n \, t^{n-1}$.

Derivatives of polynomial functions

Combine the Power Rule, the Constant Multiple Rule, and the Sum Rule.

---

[20]This summary repeats formulas from Chapter 2, but with some reorganization and renaming.

Practice with Calculations

*Exercises in this category include (a) calculations that can be done by machines and (b) practice on important topics from courses that precede calculus. You need to develop **facility** with both categories — not because such calculations are a central feature of the course, but because they should not frustrate you or keep you from concentrating on the more important parts of the course by occupying a lot of your time or by leading to lots of mistakes. Even though routine calculations can be done quickly and accurately by computer or calculator, you need to develop judgment about when to use a machine and when not to, and you need to know how to tell when you might have pressed the wrong button. These skills are acquired and sharpened by **practice**.*

You should expect to see exercises like these as some portion of your homework assignments, quizzes, and tests.

Calculate the derivative of each of the following functions:

1. $f(t) = e^{3t}$ 2. $g(t) = e^{-2t}$ 3. $P(t) = 1.2^t$

4. $h(t) = 3t - 5t^2$ 5. $\phi(t) = 6 - e^{3t}$ 6. $q(t) = 2t + e^{-t}$

7. $Q(t) = 7t^4 + 5t^3 - 4t^2 + 3t - 10$ 8. $w(t) = t^3 + \sqrt{2}\,e^{2t}$

Each of the next four exercises shows a direction field. In each case, sketch the graphs of at least three different solutions of the corresponding differential equation.

9.

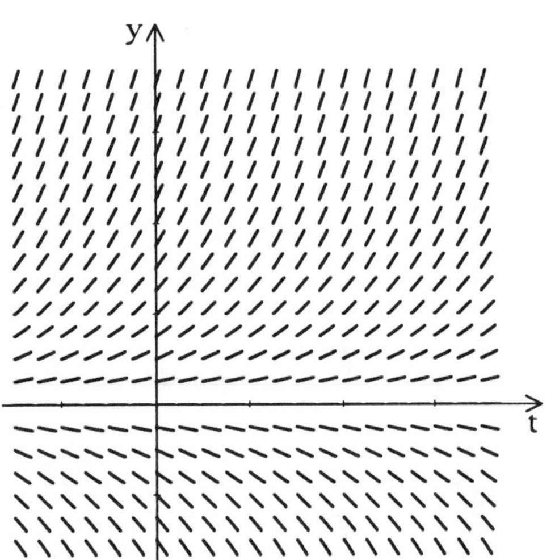

Figure 16. Direction field for $\dfrac{dy}{dt} = \dfrac{1}{2}\,y$.

10.

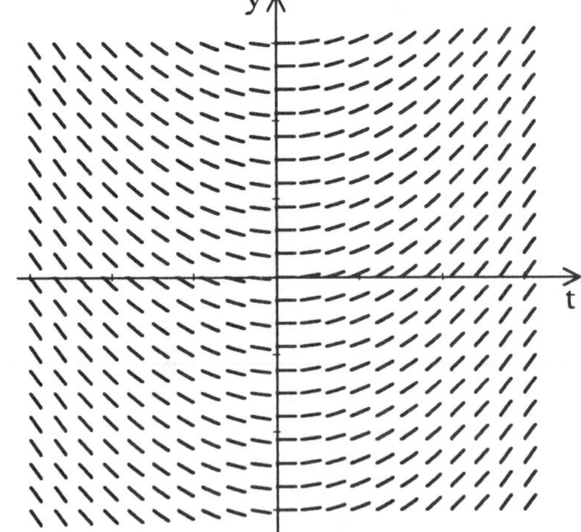

Figure 17. Direction field for $\dfrac{dy}{dt} = \dfrac{1}{2}\,t$.

11.

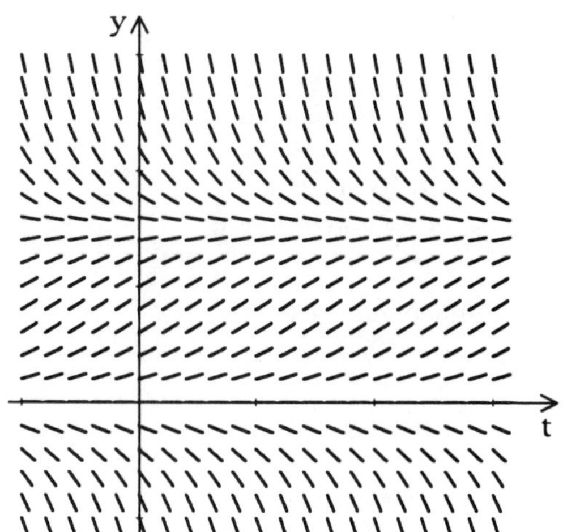

Figure 18. Direction field for $\dfrac{dy}{dt} = \dfrac{3}{10}\, y\,(3-y).$

12.

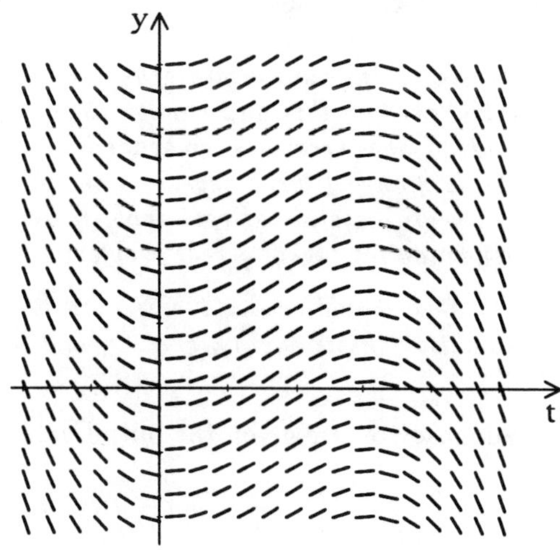

Figure 19. Direction field for $\dfrac{3}{10}\, t\,(3-t).$

13. Use zooming to solve the equation $x^2 = e^{-x}$. (How many solutions does this equation have?)

14. Solve the equation $z^2 + 3z - 4 = e^{-2z}$.

15. Find the only positive number x that satisfies the equation $x + 2 = e^x$.

16. (a) Let $y = 2 - 2\,e^{-3t}$. Find $\dfrac{dy}{dt}$.

    (b) Show that $y = 2 - 2\,e^{-3t}$ is a solution to the differential equation $\dfrac{dy}{dt} = 6 - 3y$.

17. Solve $\dfrac{dy}{dt} = 4 - 2y$, where $y = 1$ when $t = 0$.

18. Solve $\dfrac{dy}{dt} = 3t^2$, where $y = 5$ at $t = 2$.

Conceptual Exercises

Exercises in this category use and/or further develop the concepts introduced in this and earlier chapters. The level of difficulty is similar to that of the exercises embedded in the text and at the ends of sections. While many of these exercises are presented in "realistic" contexts, they are not representative of real problems. Rather, these exercises are to help you get ready for tackling real problems.

You should expect to see exercises like these as a major portion of your homework assignments, quizzes, and tests.

1.   Suppose you are given the initial value problem

$$\frac{dy}{dt} = y(1 - y), \quad \text{with} \quad y = 4 \text{ at } t = -1.$$

Change the independent variable and formulate an equivalent initial value problem for which the initial value is specified at time 0.

2.   Suppose you are given the initial value problem

$$\frac{dy}{dt} = t(1 - t), \quad \text{with} \quad y = 4 \text{ at } t = -1.$$

Change the independent variable and formulate an equivalent initial value problem for which the initial value is specified at time 0.

3.   Suppose you are given the initial value problem

$$\frac{dy}{dt} = y(1 - y), \quad \text{with} \quad y = 4 \text{ at } t = -1.$$

Change the scale of the independent variable multiplicatively and formulate an equivalent initial value problem for which the initial value is specified at time -2.

4.   Suppose you are given the initial value problem

$$\frac{dy}{dt} = y(1 - y), \quad \text{with} \quad y = 4 \text{ at } t = -1.$$

Make the substitution $Y = 2y$, and formulate an equivalent initial value problem with Y as the dependent variable.

5.   (a) Sketch a direction field for $v = \dfrac{ds}{dt} = ct$, where c is a positive constant.

(b) Sketch a solution.

(c) What feature of the field corresponds to the fact that s does not appear in the derivative?

6. Here again is the solution of our immigration model:

$$P = \left[P_0 + \frac{m}{k}\right] e^{kt} - \frac{m}{k}. \tag{16R}$$

(a) How much difference would it have made in the immigration model if we assumed $m = 0.24$ instead of 0.25?

(b) What would k be in that case?

7. Suppose your car is capable of accelerating smoothly from a standing start at the rate of 1.4 feet per second per second.

(a) Find the acceleration, the velocity, and the distance traveled, all as functions of time, over the first minute of travel.

(b) What is the speedometer reading at the end of one minute?

(c) What fraction of a mile does the car travel in that time?

8. If someone dropped a baseball out of the Washington Monument, do you think a professional catcher could catch it? Why or why not?

9. In Exercise 5 at the end of Section 3.5, you modified the falling marble problem by giving the marble an initial velocity of 40 feet per second (up or down).

(a) Solve the problem for an arbitrary initial velocity v_0 (positive or negative) and an arbitrary starting height s_0.

(b) What are the time of fall and the velocity at impact?

10. Although many people know of the famous golf ball experiment, conducted during the Apollo 11 mission to the moon, most people are unfamiliar with the experiments performed by the Frenchman, Forget Menot. Having calculated the acceleration due to gravitational force on the moon to be 5.4 feet per second per second, Forget threw a ball straight up with an initial velocity of 5 feet per second. Unfortunately, Forget neglected to bring his watch and was unable to estimate the total time it took for the ball to rise to its peak and return to the ground. Help Capitaine Menot by calculating this time and also by calculating the height of the ball at its peak.

11. A company is considering two ways to depreciate a piece of capital equipment that originally cost $14,000 and is worth $10,000 after one year:

> Method I assumes the equipment depreciates at a rate proportional to the difference between its value and and its scrap value of $400.

> Method II assumes the equipment depreciates linearly, i.e. at a constant rate.

(a) Using Method I, find the value of the equipment at the end of 2 years and at the end of 3 years.

(b) Using Method II, find the value of the equipment at the end of 2 years and at the end of 3 years.

(c) Which method produces "faster" depreciation? Explain.

Problems and Projects

The problems presented here are not intended for individual homework assignments or for tests. These problems should be attempted by groups of two to four students sharing ideas, whether in the classroom or elsewhere. Some of the more extensive problems are suitable for projects with time frames ranging from a class period to a week.

1. Radioactive substances tend to decay at a rate proportional to the amount present at any given time. Let y_0 be the amount of a radioactive substance present at time $t = 0$, and let $y = y(t)$ be the amount present at any time t.

   (a) Write an initial value problem whose solution is $y(t)$.

   (b) Solve your initial value problem to find a formula for $y(t)$.

   (c) The "half-life" of a radioactive substance is the time it takes for a given amount to decay to half that amount. Find an expression for half-life that depends only on the proportionality constant in your differential equation.

2.[21] Superman has a violent reaction to red kryptonite, which fortunately has a half-life of only 15 hours when it decays into green kryptonite. It is no longer dangerous to Superman when 90% of the red kryptonite has decayed. If Superman is exposed to pure red kryptonite, for how long is he in danger?

3.[21] (a) What is the relation between the half-life H of a radioactive substance and the "third-life" T of the same substance?

   (b) What is the relation between the doubling time D of a colony of rabbits and the tripling time T of the same colony?

4. A physician decides to give a patient an infusion of glucose at a rate of c grams per hour. The body of the patient simultaneously converts the glucose and removes it from the bloodstream at a rate proportional to the amount present in the bloodstream, say at r grams per hour per gram of glucose present.

   (a) Explain why the amount $G = G(t)$ of glucose present at time t can be modeled by a differential equation of the form $\dfrac{dG}{dt} = c - rG$.

---

[21] Adapted from *Calculus Problems for a New Century*, edited by Robert Fraga, MAA Notes Number 28, 1993.

(b) Given an initial amount G_0 of glucose in the bloodstream at time 0, find an explicit formula for the amount present at any time t. [Hint: Compare with Newton's Law of Cooling.]

(c) What rate of infusion would keep the glucose level in the bloodstream constant?

(d) A physician orders an infusion of 10 grams of glucose per hour for a patient. Laboratory technicians determine that the patient has 2 grams of glucose in his bloodstream and that his body will remove glucose from the bloodstream at a rate of 3 grams per hour per gram of glucose. How much glucose will be in the bloodstream t hours after the infusion is started? In particular, how much after two hours? How long will it take for the glucose level in the bloodstream to reach 3 grams?

(e) Suppose a patient's bloodstream has 2 grams of glucose, and her physician wants to raise this amount to 3.5 grams in three hours. It is determined that her system removes glucose from the bloodstream at a rate of 4 grams per hour per gram of glucose. How fast should the physician order the glucose to be infused into the patient's body?

5. At a certain instant, just before lifting off, a plane is traveling down the runway at 285 kilometers per hour. The pilot suddenly realizes something is wrong and aborts the takeoff by cutting power and applying the brakes. Assume that the effect of this action is a deceleration proportional to time t. After 28 seconds the plane comes to a stop. How far does the plane travel after the takeoff is aborted?

6. Figure 20 is a diagram of an electrical circuit with a resistance of R ohms, an inductance of L henries, and a battery (i.e., a constant voltage source) of V volts. If the current (in amperes) at time t (in seconds after closing the switch) is $i = i(t)$, then the voltage drop at the resistor is $R\,i$, and the voltage drop at the inductor is $L\,\dfrac{di}{dt}$. According to Ohm's Law, the voltage input V must balance the voltage drops, i.e.,

$$L\,\frac{di}{dt} + R\,i \;=\; V.$$

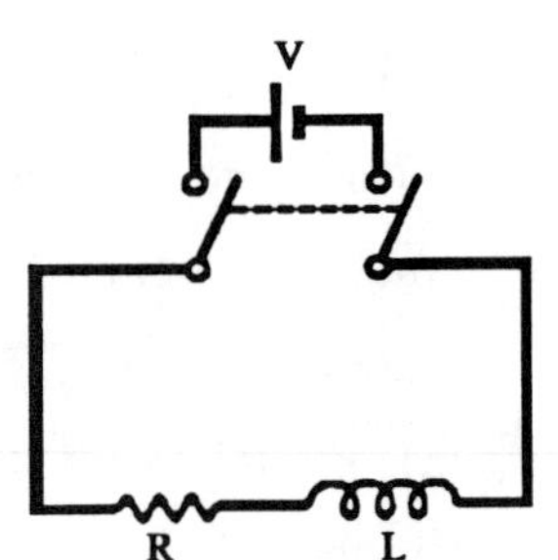

Figure 20. An RL circuit.

(a) Find the current as a function of time. [Hint: Compare with Newton's Law of Cooling.]

(b) If the switch is left closed for a long time, what is the limiting value of the current?

7. Suppose we consider air resistance in our model for the falling marble. Then the total force F acting on the marble has two components, the gravitational force and the retarding force. If we write F_g for the gravitational force and F_r for the retarding force (air resistance), then $F = F_g + F_r$. For purposes of this problem, we assume that F_r is proportional to the velocity, say $F_r = -kv$.

(a) Why is the coefficient of v negative?

(b) Use Newton's Second Law of Motion to argue that v must satisfy a differential equation of the form $\frac{dv}{dt} = g - cv$ for some constant c. How is c related to k and the mass of the object?

(c) Under the assumptions stated above, set up an initial value problem for the velocity of the marble dropped from the Washington Monument.

(d) Assume further that the numerical values of k and of the mass are such that c turns out to be 0.08. Solve the initial value problem to find an explicit formula for velocity as a function of time. [Hint: Compare with Newton's Law of Cooling. In particular, note that the right-hand side of the differential equation is now of Type I (autonomous) rather than Type II (explicit-derivative).]

(e) Your explicit formula for velocity in part (d) also provides a differential equation — of the explicit-derivative type — for the position function $s = s(t)$. Set up and solve an initial value problem for the position function. [Hint: See the summary of derivative formulas at the end of the chapter.]

(f) Under the assumptions stated above, how long would it take a marble dropped from the Washington Monument to reach the ground? How fast would it be going when it hit the ground?

8. A tank contains 25 pounds of salt dissolved in 200 gallons of water. There is a pipe at the top of the tank to bring in additional solution, and a pipe at the bottom of the tank to drain off solution. The tank itself

contains a large stirrer that keeps the solution in the tank thoroughly mixed. Starting at time $t = 0$, water containing $\frac{1}{2}$ pound of salt per gallon enters the tank at the rate of 4 gallons per minute, and the well-stirred solution leaves the tank at the same rate.

(a) If this process continues a long time, how much salt do you think will eventually be in the tank? Does your answer about the eventual salt content depend on how much salt there was in the tank at the start of the process? Why or why not?

(b) Find a differential equation that describes the rate of change of the amount of salt in the tank. Supply an initial condition for your differential equation, and solve for an explicit formula that gives the amount of salt as a function of time. How much salt is in the tank after 10 minutes? after 2 hours? What does your formula tell you about how much salt there will be in the tank after a long time? Does your answer agree with what you determined in part (a)? Why or why not? [Hints: Let $S = S(t)$ be the *amount* of salt in the tank at time t, and let $C = C(t)$ be the *concentration* of the salt at time t. How are S and C related? The flow of salt out of the tank depends on the concentration; use the relationship between S and C to express that flow in terms of S.]

9. The cooling system in an old truck holds about 10 liters of coolant. Last summer, the system was flushed by running tap water into a tap-in on the heater hose while the engine was running, and simultaneously draining the thoroughly mixed fluid from the bottom of the radiator. Water flowed in at the same rate that the mixture flowed out — about 2 liters per minute. The system was initially 50% antifreeze. Let W be the amount of water in the system after t minutes; then it follows that

$$\frac{dW}{dt} = 2 - 2\frac{W}{10}.$$

(a) Explain why this differential equation is the correct one.

(b) Find W as a function of time.

(c) How long should water have run into the system to ensure that 95% of the mixture was water?

Radiocarbon Dating[23]

In 1960, Willard Libby won a Nobel Prize in chemistry for his work in the 40's and 50's that included discovery of a radioactive isotope of carbon and development of the technique of "radiocarbon dating" to determine the ages of once-living objects.

"Ordinary" carbon, found in abundance in all living things, has an atomic weight of 12; its unstable isotope has an atomic weight of 14 and is therefore called "carbon 14" (C^{14}) or "radiocarbon." This isotope is created in the upper atmosphere by interaction of cosmic ray neutrons with nitrogen. It is then oxidized to a radioactive form of carbon dioxide, which is mixed by winds with the stable carbon dioxide already present in the atmosphere. Because this process — formation of carbon 14, oxidation, mixing, and decay of the radioactive carbon dioxide back to nitrogen — has been going on for eons, the ratio of C^{14} to ordinary carbon in the atmosphere has long since reached a steady state. That is, the proportion of radiocarbon in the air is constant.

All air-breathing plants take in carbon dioxide with this constant proportion of carbon 14 (relative to carbon 12), and thus the carbon in their tissues has the same proportion of radiocarbon. Animals that eat these plants (for example, humans) also incorporate carbon in their tissues with this constant proportion of radiocarbon. However, when plant or animal dies, it ceases to take in any more carbon (or anything else!). The radioactive carbon then in its tissues continues to decay, but the ordinary carbon does not, so the proportion of radiocarbon in its tissues decreases.

Radioactivity is observed and measured by devices, such as the Geiger counter, that count "events" or "disintegrations" of the isotope that emit a subatomic particle. From a count of disintegrations per minute (dpm), one can infer the proportion of C^{14} in a given quantity of carbon.

Facts about the decay rate, coupled with observation of the actual proportion of carbon 14 present, suffice to determine the approximate time of death of organisms that lived many thousands of years ago:

Fact 1. Any radioactive isotope tends to decay at a rate proportional to the amount of that isotope present.

Fact 2. C^{14} has a half-life of approximately 5730 years. That is, in 5730 years a given quantity of C^{14} will have decreased to half as much.

Fact 3. The radiocarbon in living tissue decays at a rate of about 15.30 (measured) dpm per gram of contained carbon.

---

[23]The remaining problems and this background text are adapted from *Calculus with Analytic Geometry* by George F. Simmons, McGraw-Hill, 1985, pp. 226-229, and from *Some Applications of Exponential and Logarithmic Functions* by W. Thurmon Whitley, UMAP Module 444, COMAP, 1989.

10. (a) Suppose $y = y(t)$ is a function whose value at time t is the proportion of carbon 14 in a once-living tissue, where $y_0 = y(0)$ is the proportion at time of death. Explain how Fact 1 leads to a formula for $y(t)$ that involves y_0 and the proportionality constant.

    (b) Explain how knowledge of the half-life determines the proportionality constant, whether or not you know y_0. What is the proportionality constant for carbon 14? Does your answer depend on the unit in which time is measured? Explain why it does or does not.

    (c) Invert the function you found in part (a) to find time as a function of amount of C^{14} present.

11. (a) A fossilized bone of a man found in 1980 in Western Pennsylvania contained approximately 17% of its original C^{14}. Estimate the year the man died.

    (b) A bone uncovered in Kenya was found to contain only 10% of its original C^{14}. Approximately how long ago did death occur?

12. The objects listed in Table 3 were tested in 1950 for radioactivity, with the indicated results expressed in disintegrations per minute per gram of carbon.

    (a) Explain why knowing the *rate* of C^{14} disintegration is enough to determine the *proportion* of C^{14} present at time t. In particular, for the chair leg in Table 3, why is the ratio $10.14{:}15.30$ the same as the ratio $y(t){:}y_0$?

    (b) Estimate the age of each object in Table 3.

Object	dpm per gram
A chair leg from the tomb of Tutankhamen	10.14
A house beam from Babylon in the reign of Hammurabi	9.52
Giant sloth dung from Gypsum Cave in Nevada	4.17
A hardwood atlatl[24] from Leonard Rock Shelter in Nevada	6.42

Table 3. Objects measured for C^{14} decay rate.

---

[24]A device for throwing a spear.

13.　(a)　The objects in Table 4 have been dated by radiocarbon dating, with the indicated results.　Estimate the measured disintegrations per minute per gram of carbon in each object, and explain your estimate.

Object	Age in years
Linen wrappings from the Dead Sea scrolls	$1,917 \pm 200$
Charcoal from the Lascaux caves in France	$15,516 \pm 900$
Charcoal from Stonehenge in England	$3,798 \pm 275$
Charcoal from the Crater Lake volcano eruption, Oregon	$6,453 \pm 250$

Table 4. Objects dated by C^{14} disintegrations.

(b)　Think about the "uncertainties" in the estimated ages; what do they say about possible uncertainties in the procedure for counting disintegrations?

CHAPTER 4

Differential Calculus and Its Uses

This chapter, which is approximately at the center of our first course in calculus, has a title that might well be the title of the course itself. In this chapter we will consolidate and build on what we have accomplished already and then marshal our forces for the rest of the course.

By first addressing problems that appeared naturally as differential-equation-with-initial-value problems, we have found a need for also addressing the (usually easier) inverse problem of calculating derivatives of known functions. The problems we will tackle in this chapter lead *directly* to the need for calculating derivatives — in most cases, derivatives of functions we have not yet differentiated. As we find out how to differentiate these new functions, we will expand our repertoire of derivative formulas and thereby enhance our ability to solve differential equations as well.

We begin this chapter by considering what derivatives tell us about the graphs of functions and what graphs tell us about derivatives. In this context we introduce the second derivative of a function. We find that both first and second derivatives have an intimate connection with problems of *optimization*, that is, of finding the best (or worst) way to do something. That connection brings us back to a problem only partially (and crudely) solved in Chapter 3 — that of solving nonlinear algebraic equations — and we put the derivative to work in another way to find a more elegant and practical solution, one we associate, again, with the name "Newton."

We next turn to two fundamental physical properties of light — reflection and refraction — and we find that our mathematical descriptions of the optimizing properties of light force us to expand our repertoire of derivative formulas once more. By the time we have conquered reflection and refraction, our list of important derivative formulas will include the Product Rule, the General Power Rule, and the most important of these formulas, the Chain Rule for differentiating compositions of functions. By the end of the chapter we will have all the formulas we need except for two very simple ones that will appear in Chapter 6, when we study oscillating and rotating phenomena.

4.1 Derivatives and Graphs

Analyzing Graphs

We have seen on several occasions the value of graphing a function in order to visualize the relationship it represents.   Now we examine what the derivative of a function $f(t)$ tells us about the graph of that function — and what the graph tells us about the derivative.

If $f'(t)$ is *positive* for values of t between a and b, then every tangent line to the graph of f in that interval slopes upward (from left to right); we say the function is **"increasing"** in the interval $a < t < b$. Similarly, if the derivative is negative over an interval, the function is **"decreasing."** (See Figures 1 and 2.)

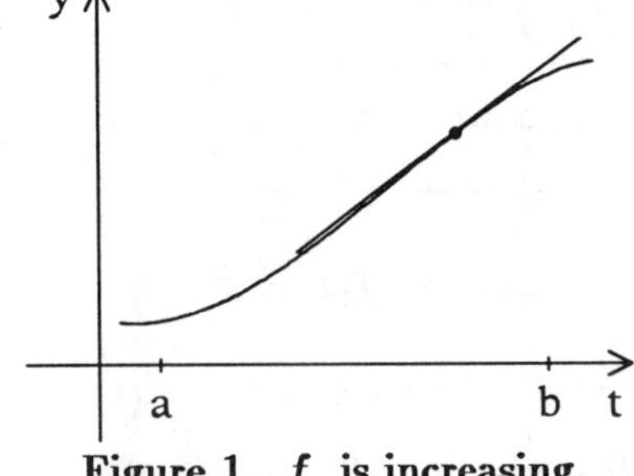

Figure 1. f is increasing.

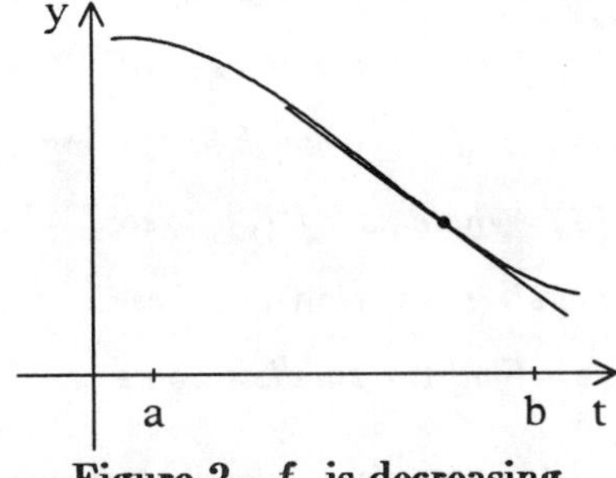

Figure 2. f is decreasing.

Exercise 1. (a) What happens on a graph if the derivative changes from positive to negative?   Specifically, what happens at $t = a$ if the derivative is positive to the left of a and negative to the right of a? Sketch a picture on the left set of axes below.

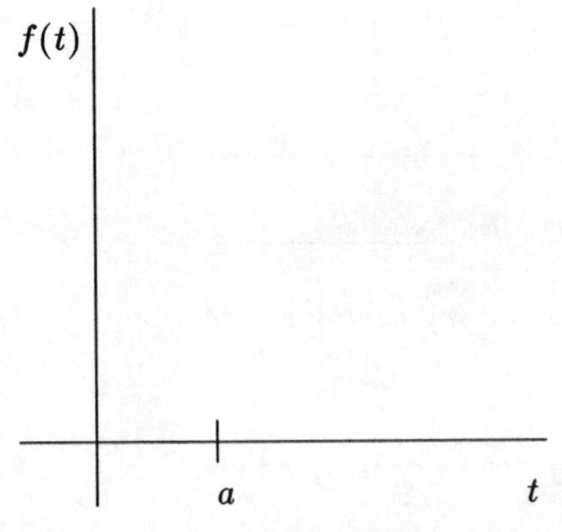

$f'(t)$ changes from positive to negative at $t = a$.

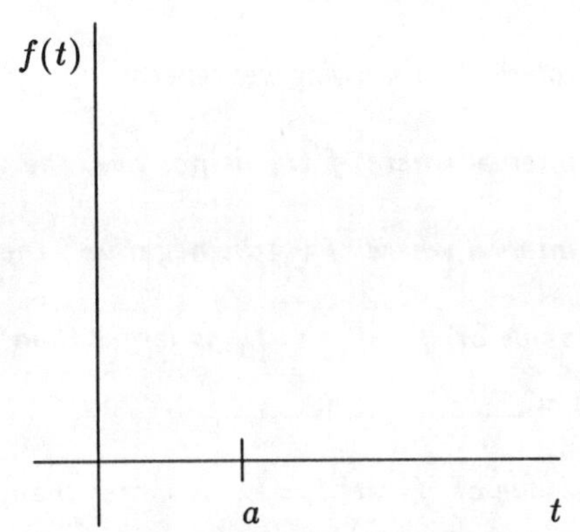

$f'(t)$ changes from negative to positive at $t = a$.

(b) What happens if the derivative changes from negative to positive (going from left to right)? Sketch a picture on the right set of axes.

Exercise 2. *In Figures 3 and 4, we graph the function $f(t) = t^3 - t$ and its derivative.*

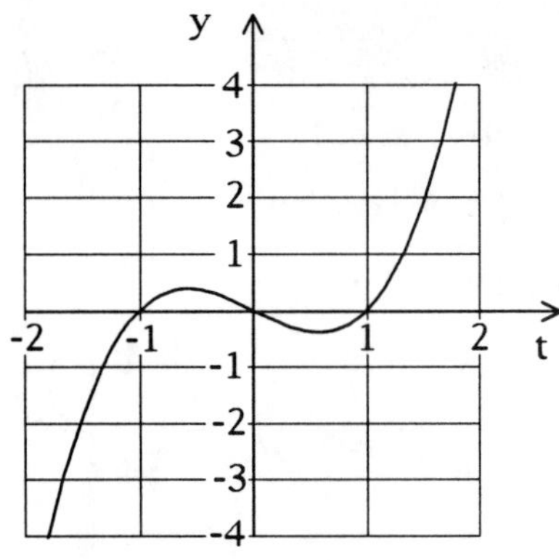

Figure 3. $f(t) = t^3 - t.$ Figure 4. Graph of $f'(t).$

(a) Find a formula for $f'(t)$.

(b) Identify the range of t values for which $f'(t)$ is positive. For these values of t, what is happening on the graph of f?

(c) Identify the range of t values for which $f'(t)$ is negative. For these values of t, what is happening on the graph of f?

(d) Where is $f'(t)$ zero? What is the largest value $f(t)$ assumes for negative t? How are these two questions related?

(e) Find the smallest value $f(t)$ assumes for positive t.

In the next exercise we ask you to summarize the relationship between the graph of a function and the values of its derivative.

Exercise 3. *Complete the following sentences:*

(a) In an interval where $f'(t)$ is positive, the graph of $f(t)$ is ________________________ .

(b) In an interval where $f'(t)$ is negative , the graph of $f(t)$ is ________________________ .

(c) If the value of f at $t = t_0$ is larger than the value of f at any nearby points, then $f'(t) =$ ________________________ .

(d) If the value of f at $t = t_0$ is larger than the value of f at any nearby points, then $f'(t) =$ ________________________ .

In the next exercise, you will use the graph of a function to draw conclusions about the values of the derivative.

Exercise 4. In Figure 5, we graph the function $g(t) = -t^3 + 12t$.

(a) From the graph, decide for what values of t the function $g'(t)$ is positive.

(b) From the graph, decide for what values of t the function $g'(t)$ is negative.

(c) From the graph, decide for what values of t the function $g'(t)$ is zero.

(d) Calculate $g'(t)$, and use algebra to check your answers to (a) – (c).

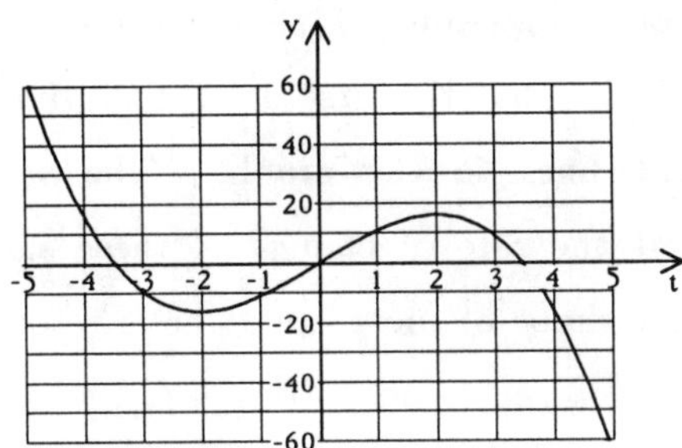

Figure 5. Graph of $g(t) = -t^3 + 12t$.

The Second Derivative

We get more information about the graph of a function if we differentiate the function twice. The derivative of a function is a new function $\frac{d}{dt} f(t)$. For example, if $f(t) = t^3 - t$, then $\frac{d}{dt} f(t) = 3t^2 - 1$. This new function also has a derivative, which we may denote $\frac{d}{dt}\left[\frac{d}{dt} f(t)\right]$; for $f(t) = t^3 - t$, this new function is $6t$. It is common practice to discard the brackets in the notation $\frac{d}{dt}\left[\frac{d}{dt} f(t)\right]$ and to replace the symbols "dd" with "d^2" and "$dtdt$" with "dt^2". This yields the notation $\frac{d^2}{dt^2} f(t)$; for our example, $\frac{d^2}{dt^2} f(t) = 6t$. This function, obtained by differentiating f twice, is called the "**second derivative**" of f. Parallel to the f' notation for the first derivative, the second derivative is also denoted $f''(t)$. So, for our example, we could also write $f'(t) = 3t^2 - 1$ and $f''(t) = 6t$.

We don't have to stop with first and second derivatives; we could talk of third derivatives, fourth derivatives, and so on. For now, we have no need for these so-called "higher derivatives," $\frac{d^3}{dt^3} f(t)$, $\frac{d^4}{dt^4} f(t)$, ... , and we will confine our attention to first and second derivatives. However, higher derivatives will play an important role in Chapter 11.

Second derivatives are not entirely new in our development; we just didn't point them out before. When we were discussing falling bodies, we let $s(t)$ represent the distance the object had fallen at time t. The velocity $v(t)$ was the derivative of $s(t)$, and the acceleration $a(t)$ was the derivative of $v(t)$; that is, $a(t) = s''(t) = \dfrac{d^2 s}{dt^2}$.

Graphs and the Second Derivative

Now we consider how the graph of a function $f(t)$ and its **second** derivative $f''(t)$ are related. We know that the first derivative gives us information about slopes of tangent lines to the graph.   The second derivative gives us information about the rate of change of these slopes. How is that information related to the shape of the graph? Let's return to original example.

Exercise 5. *We repeat the graph of* $f(t) = t^3 - t$, *and we add the graph of its second derivative in Figure 6.*

(a) Where is $f''(t)$ *positive? What does the graph of* f *look like in this range?*

(b) Where is $f''(t)$ *negative? What does the graph of* f *look like in this range?*

(c) Where is $f''(t)$ *zero?*

(d) Trace the graph of f *with your finger. How does the curve change when you pass over the point where* $f''(t)$ *is zero?*

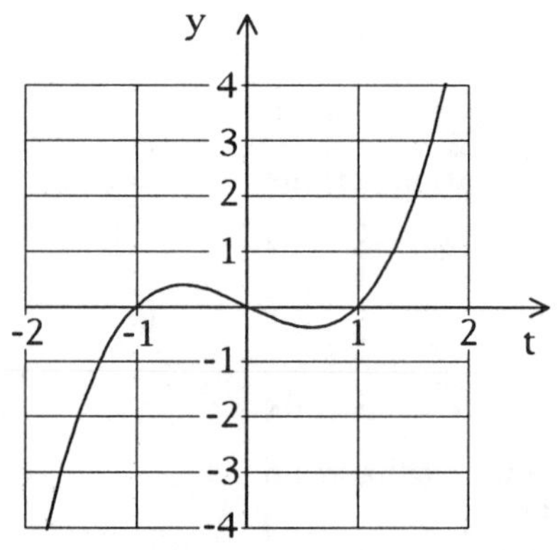

Figure 3R. $f(t) = t^3 - t.$

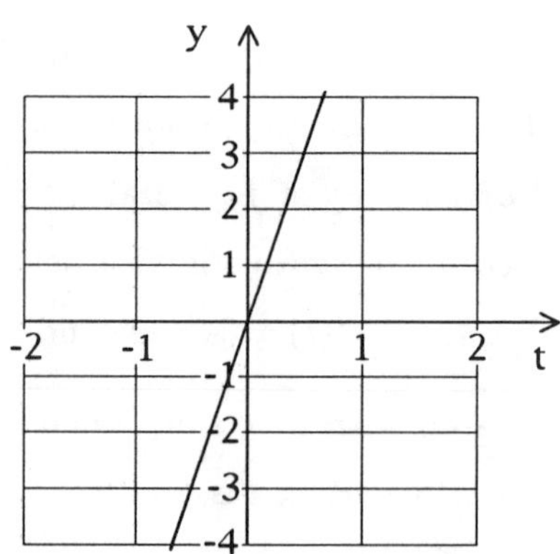

Figure 6. Graph of $f''(t) = 6t.$

The second derivative f'' measures the rate of change of f' (i.e., of *slope*). In Exercise 5 you found that the slope of f is decreasing when t is negative and increasing when t is positive. If you imagine walking along the graph of f from left to right, "decreasing slope" means you are turning right, and "increasing slope" means you are turning left. We describe the graph in the first case as **"concave down"** and in the second case as **"concave up."** Thus, our function $f(t) = t^3 - t$ has a graph that is concave down for negative values of t and concave up for positive values of t.

Exercise 6. *We repeat the graph of the function* $g(t) = -t^3 + 12t$:

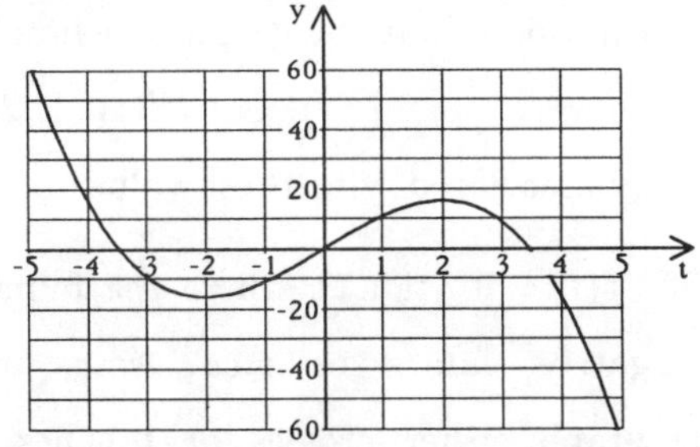

Figure 5R. Graph of $y = -t^3 + 12t$.

(a) For what range of t does the graph appear to be concave up?

(b) For what range of t does the graph appear to be concave down?

(c) Calculate the second derivative $g''(t)$, and use algebra to check your answers to parts (a) and (b).

Finally, we note the interplay of first and second derivatives. For the function $f(t) = t^3 - t$, there are two places where the first derivative changes sign, and they are important for locating the "peak" and the "valley" on the graph of f. But we found them both by solving $f'(t) = 0$. In this case it is pretty easy to tell which solution corresponds to the peak and which to the valley, because sketching the graph is easy. If the function were not so easily graphed (and if we did not have a computer or graphing calculator at hand), how would we know which was which? Well, the peak occurs where the graph is concave down (negative value of the second derivative) and the valley where the graph is concave up (positive value of the second derivative). So the *sign* of the second derivative is sufficient to distinguish between peaks and valleys.

Peaks go with negative second derivative and valleys with positive second derivative. That's important, because peaks and valleys on graphs of functions will often be the *highest* and *lowest* values of those functions — "best" and "worst" of something that the function measures. And if you are going to *optimize* the performance of something, it makes a difference whether you are finding the *best* it can do or the *worst*.

Terminology: Zeros, Critical Values, and Inflection Points

We have seen that we get a lot of information about the graph of a function f by finding the values of t at which $f(t)$ or $f'(t)$ or $f''(t)$ is zero. We summarize the common terminology associated with these values of t:

- The **"roots"** (or solutions) of $f(t) = 0$ tell us where the function *might* change from positive to negative values or *vice versa*; more specifically, they tell us where the graph either crosses or touches the horizontal axis. These numbers are also called **"zeros"**[1] of the function f. (See Figure 7.)

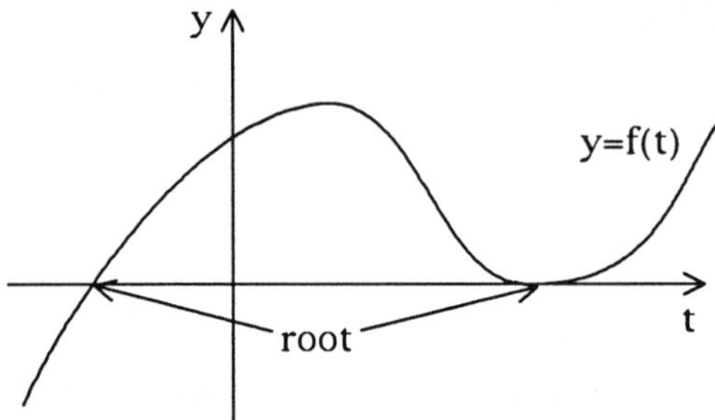

Figure 7. Crossing and touching the horizontal axis.

- The zeros of f' [roots of $f'(t) = 0$] tell us where the graph *might* change from increasing to decreasing or *vice versa*; more specifically, they tell us where the graph of f *might* have a peak or valley. These "leveling off" places are called **"turning points."** The formal terminology for a peak is **"local maximum point,"** and for a valley, **"local minimum point."** (See

---

[1]See Note 13 in Section 3.4 on the technical distinction between roots and zeros.

Figure 8.) Collectively, local maximum and minimum points are called "local extreme points."

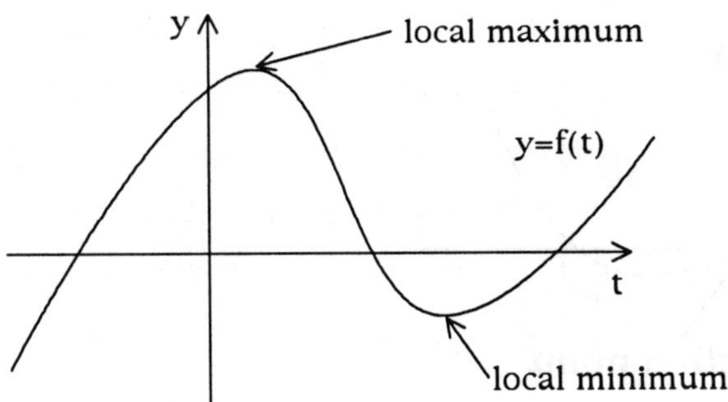

Figure 8. Turning points.

The values of the independent variable for which the first derivative is zero are called **"critical values"** or **"critical numbers."** Often, these values are called "critical points," a misnomer, as they are values of the independent variable and not points on the graph. Not all critical values correspond to local maximum points or local minimum points. For example, $f(t) = t^3$ has a zero derivative at the value $t = 0$, but this is neither a local maximum nor a local minimum. (See Figure 9.) However, if the graph of a locally linear function f does have a local maximum or a local minimum at $t = t_0$, then $f'(t_0) = 0$.

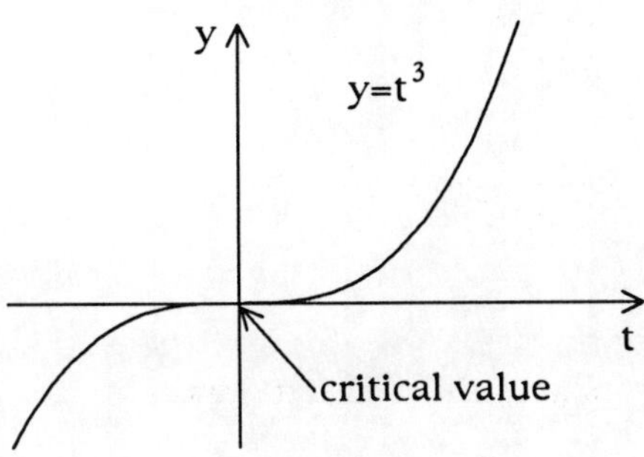

Figure 9. A critical value that does not correspond to a local extreme.

• The zeros of f'' [roots of $f''(t) = 0$] tell us where the graph of f *might* change from concave down to concave up or *vice versa*. When such a change actually occurs, the corresponding point on the graph is called an "inflection point." The origin is an inflection point of the graph of $y = t^3$ (Figure 9), and it is also an inflection point of the graph in Figure 10.

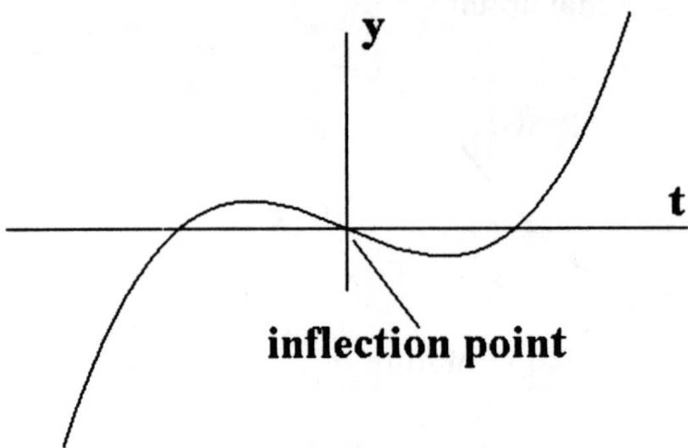

Figure 10. An inflection point that is not at a critical value.

Section Summary

Describe "concave up" in terms of the appearance of the graph.

Describe "concave up" in terms of the second derivative.

Describe "concave down" in terms of the appearance of the graph.

Describe "concave down" in terms of the second derivative.

Describe a "local maximum" on a graph.

Describe a "local maximum" in terms of the first and second derivatives.

Describe a "local minimum" on a graph.

Describe a "local minimum" in terms of the first and second derivatives.

Exercises

Calculate the second derivative of each of the following functions:

7. $f(t) = e^{3t}$ **8.** $g(t) = e^{-3t}$ **9.** $h(t) = e^t - t^3$

10. $u(t) = t + 1$ **11.** $v(t) = t^2 + t + 1$ **12.** $w(t) = t^3 + t^2 + t + 1$

13. *Figure 11 is a graph of the polynomial function* $f(t) = t^4 - 18t^2 - 10t + 39$.

(a) *Approximately where is* $f(t) = 0$?

(b) *Approximately where is* $f'(t) = 0$? *Mark those points on the graph of* f
in Figure 11.

(c) *On what intervals is* f' *positive?*

(d) *On what intervals is* f' *negative?*

(e) *Sketch a graph of* f' *in the space provided.*

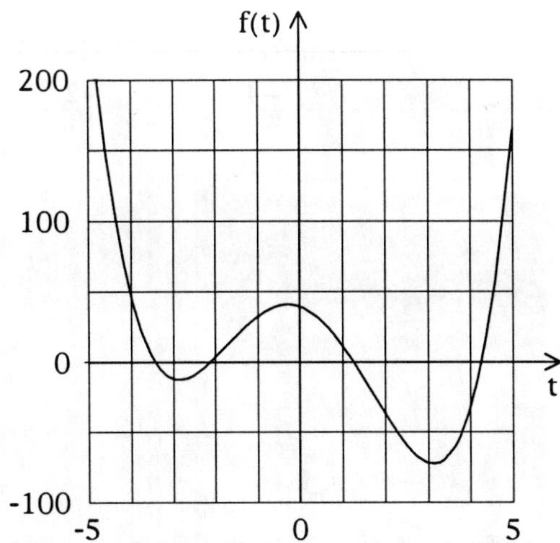

Figure 11. $f(t) = t^4 - 18\,t^2 - 10\,t + 39$.

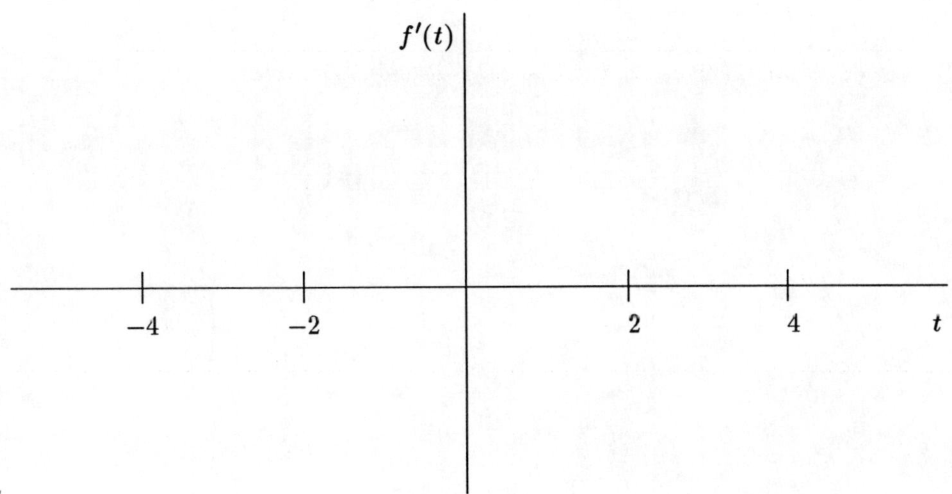

14. (a) For the function in Figure 11, $f(t) = t^4 - 18t^2 - 10t + 39$, where (approximately) is $f''(t) = 0$? (If you wish, you can determine algebraically exactly where f'' is zero.)

(b) Mark the points in Figure 11 where $f''(t) = 0$.

(c) On what intervals is $f''(t)$ positive?

(d) On what intervals is $f''(t)$ negative?

(e) Sketch a graph of $f''(t)$ in the space below.

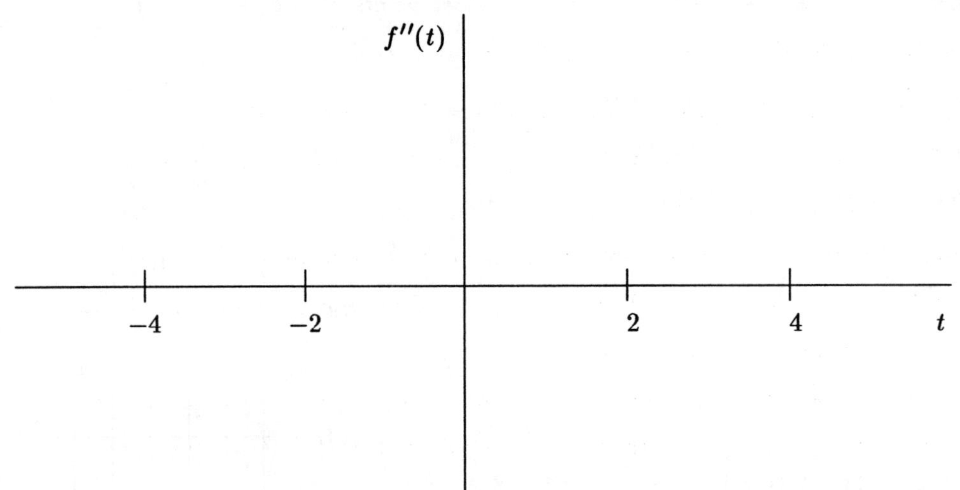

4.2 Solving Nonlinear Equations by Linearization: Newton's Method

We have seen that the analysis of the graph of a function leads naturally to the problem of finding roots. In most of the exercises in the previous section, we gave you functions for which solution of the equations $f(t) = 0$, $f'(t) = 0$, and $f''(t) = 0$ turned out to be an easy exercise in algebra — even though several of these equations were **nonlinear**.[2] The real world is seldom so generous. The fact is, most equations of interest are nonlinear, and nonlinearity is what makes solving equations hard.

We encountered this problem of finding roots (solving equations) once before (in Chapter 3), as we sought an appropriate k for fitting census data to the immigration model. We solved the problem there by using computer graphics to "zoom in" on a small segment of a graph at the point at which it crossed the horizontal axis. However, that method is somewhat tedious and absolutely requires fairly sophisticated technology. It also ignores calculus, which we will now put to work to develop a more efficient way to find roots — a method that is often used on computers, but that is not totally dependent on them.

Newton's Method

In contrast to the problem of solving nonlinear equations, solving *linear* equations is easy! And that gives us a clue as to where to look for another method for solving equations. The essence of differential calculus is to provide us the tool for finding directions along curves or, equivalently, finding tangent lines to graphs. Close to a point of tangency, a function looks a lot like its tangent line, and the function describing the tangent line is linear.

---

[2]An equation is "nonlinear" if it cannot be put in the "linear" form $mt + b = 0$, where t is the variable and m and b are constants. Similarly, a function $f(t)$ is "nonlinear" if the rule determining its values cannot be put in the "linear" form $f(t) = mt + b$. The graph of a linear function is a straight line; the graph of a nonlinear function is something other than a straight line.

So here's the plan: We start "close" to a solution, say at $t = t_0$. (How we find that starting point depends on the circumstances, but usually a crude graph or table of values will suffice.) We compute the tangent line to the graph at the point $(t_0, f(t_0))$ as a function of t, say $g(t)$. We solve $g(t) = 0$ to find the (only) number t_1 at which the tangent line crosses the horizontal axis. That number won't necessarily be the solution of $f(t) = 0$ that we seek, but it will usually be *much closer* than t_0 was (see Figure 12). Then we start over, replacing t_0 by t_1. The key to the success of this idea is the "much closer" that we achieve at each step, which usually makes it unnecessary to do more than a few steps.

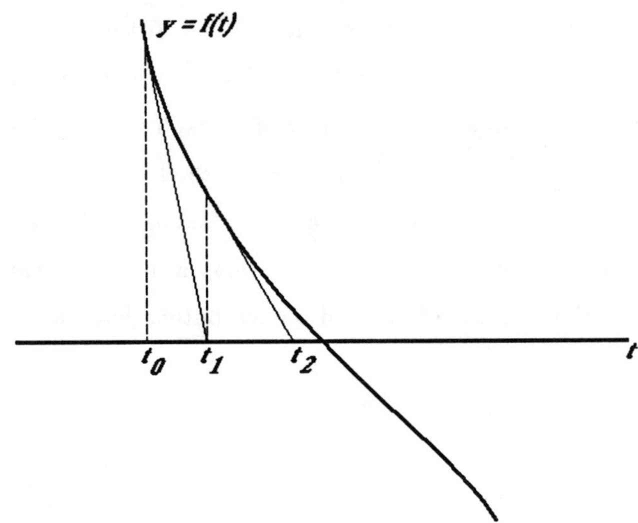

Figure 12. First two steps in using tangent crossings to find a solution of $f(t) = 0$.

Let's do the calculation. The slope of the tangent line to the graph of f at $t = t_0$ is $f'(t_0)$. The point $(t_1, 0)$ is also on the tangent line, because we *chose* t_1 to be the point where the tangent line crosses the t-axis. Thus, the slope is also "rise over run" or $\dfrac{0 - y_0}{t_1 - t_0}$. Now we set our two expressions for slope equal and solve for t_1. (Note that we don't actually need a formula for the tangent line as a function; equating the slope expressions gives us an equation that involves

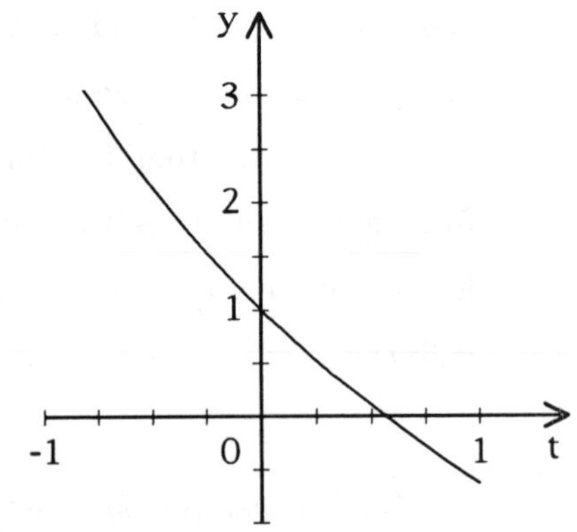

Figure 13. $y = e^{-t} - t$.

both t_1 and t_0, and that's enough to finish the calculation.)

Here is the solution:

$$f'(t_0) \;=\; \frac{-y_0}{t_1 - t_0}\,,$$

so

$$-\,y_0 \;=\; f'(t_0)\,(t_1 - t_0)\,,$$

so

$$t_1 - t_0 \;=\; \frac{-\,y_0}{f'(t_0)}\,,$$

and therefore

$$t_1 \;=\; t_0 - \frac{y_0}{f'(t_0)}\,. \tag{1}$$

Equation (1) tells us how to get from t_0 to t_1: Subtract the ratio of the function value at t_0 to the slope at t_0. (Look again at Figure 12, and explain the calculation to yourself in these terms: The run is the rise divided by the slope. Be sure you understand where the "$-$" comes from.)

Now we get t_2 from t_1 in exactly the same way. Then t_3 from t_2, and so on. In fact, what we have here is a *recursive* formula for generating the numbers t_1, t_2, t_3, ... from the starting point t_0:

$$t_{n+1} \;=\; t_n - \frac{f(t_n)}{f'(t_n)}\,. \tag{2}$$

Formula (2) is generally known as "Newton's Method" for solving equations. Yes, the same Newton who was responsible for the laws of motion, gravitation, and cooling, and for many of the most significant developments in calculus itself.[3]

Example: $f(t) = e^{-t} - t$.

(See Figure 13.)

Choose $t_0 = 0.1$.

Then $y_0 = f(t_0) \approx 0.804837$.

$f'(t) = -e^{-t} - 1$, so

$f'(t_0) \approx -1.904837$.

$t_1 = 0.1 - \dfrac{-1.904837}{0.804837} \approx 0.522523$

$t_2 \approx 0.566778$

$t_3 \approx 0.567143$

$f(t_3) \approx 4.55 \; 10^{-7}$

Another Example of the Use of Newton's Method

Let's see how Newton's Method works by finding the left-most solution of $f(t) = 0$ for the function $f(t) = t^4 - 18\,t^2 - 10\,t + 39$ (Figure 11). From the graph we see that the solution is between -4 and -3.

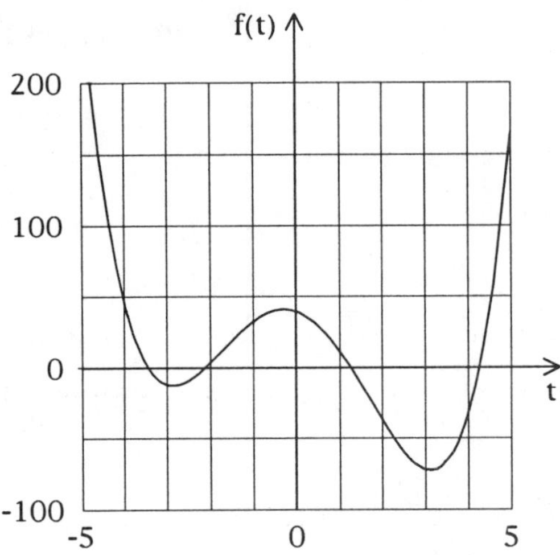

Figure 11R. $f(t) = t^4 - 18\,t^2 - 10\,t + 39$.

We could use either -4 or -3 as a starting point; let's use -4. We have $f'(t) = 4t^3 - 16t - 10$. We tabulate our work with a calculator (check the calculation of t_1 with your own calculator):

n	t_n	$f(t_n)$	$f'(t_n)$	$f(t_n)/f'(t_n)$
0	$-4.$	$47.$	-122	-0.385246
1	-3.614754	10.683527	-68.796815	-0.155291
2	-3.459463	1.402967	-51.069143	-0.027472
3	-3.431991	0.040318	-48.144002	-0.000837
4	-3.431154	0.000058	-48.055859	-0.000001
5	-3.431153	0.000058	-48.055859	-0.0000002

This is as far as we can go with six decimal place accuracy; the *change* in t in the last column is too small to affect the sixth place, and that is why neither f nor f' changed in the last step. We can safely conclude that the solution we seek is $t = -3.43115$ to five-place accuracy; the next digit remains in doubt because of our rounding.

---

[3]This formula is actually a variant of what Newton proposed for solving equations. The variant is due to his contemporary, Joseph Raphson, and is more properly known as the Newton-Raphson method. Newton is only known to have solved one equation with his method, a cubic polynomial equation.

Notice how rapidly the changes in the last column got small. Compare this to the zooming method we used earlier, with which we gained exactly one decimal place accuracy at each step — with the aid of a computer. The problems are not strictly comparable, but there we started much closer to the final answer and had only four-place accuracy. That only hints at how much more effective Newton's method can be — a matter we will take up in a lab related to this chapter. Of course, none of this discussion is necessary if your calculator understands symbolic entry of functions and has a $\boxed{\text{Solve}}$ key. In that case, consider this an explanation of what that $\boxed{\text{Solve}}$ key is doing.

You might be curious about why we didn't do a head-to-head comparison of Newton's method with zooming by solving equation (27) from Chapter 3 again:

$$(0.03929\ x + 0.25)\ e^{x/2}\ -\ 0.25\ -\ 0.17069\ x\ =\ 0.$$

We still have a problem with application of Newton's method to that equation: We don't have a formula yet for the necessary derivative. That's the subject of the next section.

Section Summary

How does Newton's Method for finding roots differ from the Zooming-In Method?

Exercises

1. Use Newton's Method to find the other roots of $f(t) = 0$ for the function $f(t) = t^4 - 18\ t^2 - 10\ t + 39$.

2. Use Newton's Method to find **all three** solutions of the equation $3x^2 = e^x$.

3. Let $f(t) = t^3 + 8t^2 + t - 6$.

    (a) Calculate the values of f at $t = -10$, $t = -5$, $t = 0$, and $t = 2$.

    (b) Why do your calculations in part (a) show that f has at least three zeros?

    (c) Why does f not have more than three zeros?

    (d) Use Newton's Method to find all zeros of f.

4. Let $f(t) = e^t + e^{-t} - 2t^2$.

    (a) Find the four zeros of f.

    (b) Find the three zeros of f'.

    (c) Find the smallest value of f and the two values of t where f has its smallest value.

    (d) Find the largest value of f between $t = -1$ and $t = 1$.

    (e) Sketch the graph of f.

5. Use Newton's Method to find the root of $t^2 - e^{-3t} = 0$.

4.3 The Product Rule

For your convenience, we repeat the problem posed in Chapter 3: We want to find a positive number x such that

$$(0.03929\, x \,+\, 0.25)\, e^{x/2} - 0.25 - 0.17069\, x \,=\, 0. \qquad (3)$$

As we did when we introduced the zooming method, we name the function defined by the left-hand side of the equation:

$$f(x) \,=\, (0.03929\, x \,+\, 0.25)\, e^{x/2} - 0.25 - 0.17069\, x. \qquad (4)$$

Then our problem is to find a positive root of $f(x) = 0$. Except for the name of the independent variable, this is just like the problem for which we derived Newton's Method in the previous section. Thus, we should find a solution with a small number of applications of the same recursion as in equation (2):

$$x_{n+1} \,=\, x_n \,-\, \frac{f(x_n)}{f'(x_n)}\,, \qquad (5)$$

where x_0 is a starting guess. But, before we can apply (5) even once, we have to know how to compute $f'(x_n)$.

We see from (4) that $f(x)$ is a sum of three terms, and we know what to do with sums when we want to find a derivative. Also, we already know how to differentiate two of the three terms. (Which two? What are their derivatives?) But one term is a *product* of a linear function and an exponential function. We know how to differentiate each factor separately, but what do we do with the product?

To see that the problem is a real one, let's practice with some products of functions for which we know the derivative of each factor *and* the product.

Exercise 1. (a) Suppose $g(x) = x$, and $h(x)$ also equals x. Let $f(x) = g(x) \cdot h(x)$; thus, $f(x) = x^2$. What are the derivatives of $f(x)$, $g(x)$, and $h(x)$?

(b) Suppose $g(x) = x$, and $h(x) = x^2$. Let $f(x) = g(x) \cdot h(x)$; thus, $f(x) = x^3$. What are the derivatives of $f(x)$, $g(x)$, and $h(x)$?

(c) Suppose $g(x) = e^{2x}$, and $h(x) = e^{3x}$. Let $f(x) = g(x) \cdot h(x)$. What is $f(x)$ as an explicit function of x? What are the derivatives of $f(x)$, $g(x)$, and $h(x)$?

(d) See if you can give a formula describing $f'(x)$ in terms of $g'(x)$ and $h'(x)$ that is valid for each of parts (a), (b), and (c).

A General Rule for Differentiating Products

It should be clear from Exercise 1 that $f'(x)$ is *not* $g'(x) \cdot h'(x)$ for any of the simple product functions just examined, so one "obvious" candidate for a "product formula" is *wrong*. Since we can't guess the answer, we are forced to use the only tool available for finding new derivative formulas: direct calculation of the difference quotient.

To that end, suppose $f(x) = g(x) \cdot h(x)$, where g and h are *any* functions that have derivatives (whether we already know them or not). We attempt to find $f'(x)$ by forming a difference quotient for a difference Δx that we will eventually allow to approach zero. That is, we will calculate an average rate of change over smaller and smaller intervals, in order to find the instantaneous rate of change.

Forming the difference quotient in the obvious way, we have

$$\frac{f(x + \Delta x) - f(x)}{\Delta x} = \frac{g(x + \Delta x) \cdot h(x + \Delta x) - g(x) \cdot h(x)}{\Delta x} . \qquad (6)$$

Now we can see what the problem is: The last expression doesn't break up in any obvious way into difference quotients (average rates of change) for g or h. However, we have not exhausted all the possibilities for different ways to write down this expression, and this is a good time to consider a new one.

We start by introducing dependent variables for the functions f, g, and h, say, $w = f(x)$, $u = g(x)$, and $v = h(x)$. Then we compute the increments in u and v that correspond to an increment Δx in x. For example, $\Delta u = g(x + \Delta x) - g(x) = g(x + \Delta x) - u$, so $g(x + \Delta x) = u + \Delta u$. Similarly, $h(x + \Delta x) = v + \Delta v$. Now we can write Δw [which is an abbreviation for $f(x + \Delta x) - f(x)$] in another form:

$$\Delta w \;=\; (u + \Delta u)\cdot(v + \Delta v) - u\cdot v$$

$$= \; u\cdot v + u\cdot(\Delta v) + (\Delta u)\cdot v + (\Delta u)\cdot(\Delta v) - u\cdot v$$

$$= \; u\cdot(\Delta v) + (\Delta u)\cdot v + (\Delta u)\cdot(\Delta v).$$

From this form we can see how to divide through by Δx and rewrite

$$\frac{\Delta w}{\Delta x} = \frac{f(x + \Delta x) - f(x)}{\Delta x}$$

in a way that relates average change in w to average changes in u and v:

$$\frac{\Delta w}{\Delta x} \;=\; u\cdot\frac{\Delta v}{\Delta x} + \frac{\Delta u}{\Delta x}\cdot v + (\Delta u)\cdot\frac{\Delta v}{\Delta x}\,. \tag{7}$$

We are ready to find limiting values as $\Delta x \to 0$. We assumed that g and h have derivatives (i.e., their average rates of change approach instantaneous rates of change as Δx approaches zero), so $\frac{\Delta u}{\Delta x}$ approaches $\frac{du}{dx}$ and $\frac{\Delta v}{\Delta x}$ approaches $\frac{dv}{dx}$. We leave it to you (in Exercise 7) to explain why Δu approaches 0 as Δx does. Now we see from (7) that $\frac{\Delta w}{\Delta x}$ approaches $u\cdot\frac{dv}{dx} + v\cdot\frac{du}{dx} + 0$ as Δx approaches 0, so

$$\frac{dw}{dx} \;=\; u\cdot\frac{dv}{dx} \;+\; v\cdot\frac{du}{dx}\,. \tag{8}$$

This result is called the **"Product Rule"** for derivatives; in terms of the original functions f, g, and h, we can write the rule in the following way:

$$\text{If } f(x) = g(x)\cdot h(x), \text{ then } f'(x) \;=\; g(x)\cdot h'(x) + h(x)\cdot g'(x). \tag{9}$$

In words, this says: "The derivative of a product is the first factor times the derivative of the second plus the second factor times the derivative of the first."

Exercise 2. Find the derivative of the function defined in equation (4):

$$f(x) \;=\; (0.03929\,x \;+\; 0.25)\,e^{x/2} - 0.25 - 0.17069\,x. \tag{5R}$$

This requires careful application of the rules for differentiating sums, products, constants, linear functions, and exponential functions of the form e^{kx}; take it one step at a time, and don't rush. In order to complete our calculation for Newton's method, the derivative must be right.

An Application of the Product Rule

Now we are ready to find the positive root of $f(x) = 0$, where

$$f(x) = (0.03929\ x + 0.25)\ e^{x/2} - 0.25 - 0.17069\ x,$$

by applying the Newton iteration to f. Both $f(x)$ and $f'(x)$ are complicated functions. Therefore, this could be a very tedious calculation[4] if we have to do more than a few steps of Newton's method. We start, as we did for the zooming example, with the graphical observation that the root lies between $x = 0.1$ and $x = 0.2$. (If we didn't have a graph, we could evaluate f at these two numbers and see that it changes sign somewhere between them — check it out.) We can use either of these numbers as x_0; suppose we set $x_0 = 0.2$. Then we calculate the entries of the following table (check the calculation of x_1, using your result from Exercise 2):

n	x_n	$f(x_n)$	$f'(x_n)$	$f(x_n)/f'(x_n)$
0	0.2	0.000839	0.152207	0.0551328
1	0.144867	0.000172	0.009001	0.1915060
2	0.125717	0.000020	0.006889	0.0029428
3	0.122774	0.0000005	0.006566	0.0000723
4	0.122701	0.0000000003	0.006560	0.00000004

Compare this with our earlier calculation (in section 3.4) to see a direct comparison between zooming and Newton's method. This table was created without using a computer, and it gives the solution to greater accuracy; we can see from the last computed "correction" that there will be no further change in the first six decimal places in the x column, so x_4 agrees with the exact solution to the six places shown.

---

[4]With a conventional calculator — on the other hand, with a symbolic calculator or a computer, it's a piece of cake.

Section Summary

Write out the Product Rule, and explain how to use it to calculate $\dfrac{d}{dt}\, t^2 e^{3t}$.

Exercises

Use the Product Rule to calculate the derivatives of the following functions:

3. $t^2 e^t$ **4.** $(t^3 + t)\, e^{-3t}$ **5.** $(t - 3)\, e^{2t}$

6. Verify that the product rule gives the right answer for the derivative of each of the products $f(x) = g(x) \cdot h(x)$ in Exercise 1:

(a) $g(x) = x$ and $h(x) = x$,

(b) $g(x) = x$ and $h(x) = x^2$,

(c) $g(x) = e^{2x}$ and $h(x) = e^{3x}$.

7. In the derivation of the Product Rule, explain why Δu approaches 0 whenever Δx does. (Think about the difference quotient that approximates the derivative; what would happen to the difference quotient if Δu did not approach 0 as Δx did?)

4.4 Reflection, Refraction, and the Chain Rule

It was early afternoon in late March; the bright sun mocked the cold wind blowing a few stray clouds across the sky. Marcus stuffed his hands in his pockets and hunched over against the wind. He was worried about his friend Tony. Tony had come down with mono at the beginning of the Spring Semester. He was so sick, and then so weak, that he had to withdraw from school for the semester. Tony still had to spend most of his time sitting in the apartment he shared with his older brother, Mario. Mario was in law school and rarely home.

Tony was by nature an active person. He had organized a team in the intramural basketball league. Marcus smiled to himself — they had done pretty well. Made it to the second round of the playoffs. Then they ran up against a team of off-season football players — and bounced off. They couldn't rebound anything against those guys. Still, they had kept it close. Tony started draining those long jump shots. When they finally started guarding him close, Ray was open in the corner.

As soon as the game was over and the season finished, Tony was talking about softball. Now inactivity was making Tony depressed. Just sitting around watching television was not Tony's thing. Marcus tried to think of something that would interest Tony, something to get him active, if only with his mind — maybe jigsaw puzzles. Marcus turned in at the apartment building, glanced at the forlorn clump of yellow jonquils that had been foolish enough to bloom in teeth of the winter wind, and pushed the buzzer for Tony's apartment.

As Marcus entered the room, he was surprised to see the TV off and a strange array of books, papers, and other stuff spread across a card table. Looking more closely, he saw one of those fast food soft drink cups full of water with a pencil sticking out. Marcus tossed his jacket on the couch and walked over to see what Tony was up to.

"Look at that cup. What do you see?" Tony asked with a surprising intensity.

"Well, it's a cup full of water with a pencil in it." Marcus responded.

"Right!" Tony said with enthusiasm. "Look at the point where the pencil comes out of the water."

Marcus looked instead at his friend's face. Maybe Tony had been even sicker than he thought. Still, it was good to see him interested in something. "OK," Marcus said, as he looked down at the cup.

"So, what do you see?" pressed Tony.

"The water and the pencil," Marcus answered slowly.

"Do you see how it seems to bend as it comes out of the water?"

Marcus looked again. "Yeah, it does seem to bend."

"Why does it do that?" asked Tony. "We know the pencil doesn't bend; it's the light that's bending as it comes out of the cup. How does the light decide how much to bend?"

"How do I know," responded Marcus, "that's the way the world works."

"Marcus, you have the soul of a CPA. There has to be a better explanation. I've been poking around in these books Mario brought. There's this French guy, Fermat, that had this explanation. It's a real neat idea. He figures that light travels at different speeds in different things — one speed in the air, another in water, another in glass and so on."

"Yeah, I remember Mrs. Brown telling us about that in high school physics," mused Marcus.

Tony pushed on, "But the neat part is that the light travels from one point to another over the path that takes the least time. It's all in here."

Tony picked up a book and pointed at a section with a picture of a weird bald-headed guy looking at a glass with a pencil in it.

Optimization in Nature[5]

Do the following experiment: Look at a mirror, and focus on a small object that you see in the mirror. You see that object because there are light rays traveling from it to the mirror and then to your eye. Light rays normally travel in straight lines, but they can be bent — for example, by reflection in mirrors. Now visualize three points in space: the object, your eye, and the *image* of the object in the mirror (see Figure 14). The ray from the object to its image in the mirror makes an angle with the mirror that we label α in Figure 14. The ray from the image to your eye makes an angle we call β. As you are looking in the mirror, estimate the sizes of α and β. Vary your position in the room with respect to the object, and, for each new position, estimate α and β again. How do you think α and β are related?

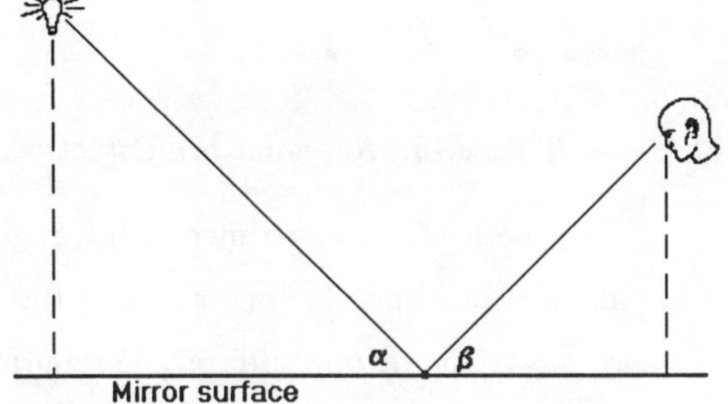

Figure 14. Reflection of a light ray in a mirror.

___

[5]The rest of Section 4.5 is based in part on "Somewhere Within the Rainbow" by Steven Janke, in *Applications of Calculus*, edited by Philip Straffin, MAA Notes, 1993, and in part on *Five Applications of Max-Min Theory from Calculus* by W. Thurmon Whitley, UMAP Module 341, COMAP, 1979.

Now try this experiment. Fill a glass with water, and place a pencil in the water at an angle (not vertical). Notice the apparent "bend" in the pencil at the surface of the water? You *know* the water is not actually bending the pencil, so it must be bending the light rays from the pencil to your eye. More precisely, a light ray from a point on the pencil (say, the bottom end) travels through the water along a straight line to the surface, and then it changes direction to travel along another straight line to your eye. We illustrate this situation in the left half of Figure 15 with the actual and apparent positions of the pencil; the heavy line is the route taken by a light ray from the end of the pencil to your eye. This bending phenomenon is called "refraction," and it occurs whenever a light ray passes from one medium to another in which the speed of light is different. In the right half of Figure 15 we have labeled the angles the light ray makes with the vertical α (in water) and β (in air). The question is much harder for refraction than for reflection (don't worry if you can't answer it yet): How do you think the angles α and β in Figure 15 are related?

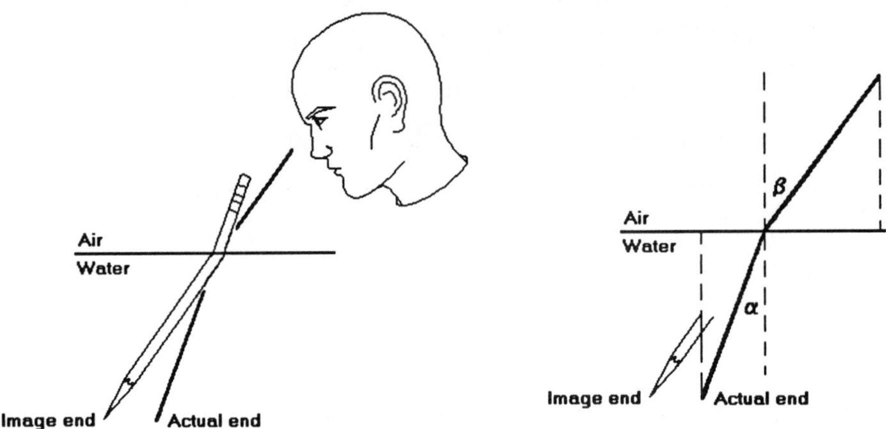

Figure 15. Apparent bending of pencil in water due to refraction.

In both of our experiments, the angle we have labeled α is called the "angle of incidence." For reflection, this means the angle at which the light ray meets the mirror surface. For refraction, it means the *complement* of the angle at which the light ray meets the water-air surface.[6] The angles we have labeled β are called, respectively, "angle of reflection" and "angle of refraction." In the case of reflection, you probably have a pretty good idea from our simple experiment how α and β are related, even if you have never thought about it before. In the case of refraction, it is unlikely that you can even guess the relationship — unless you remember it from a physics course.

Experiments with light rays, confirmed many times over a period of at least 300 years, suggest that light obeys a simple and plausible optimization law called "Fermat's Principle":[7] *Light follows a path that minimizes total travel time.* This "optimization" principle enables us to actually calculate the bending of light, whether by mirrors, water-air interfaces, lenses, prisms, or rainbows. All we have to do is write down a formula for total travel time as a function of something, and find the value of "something" that gives the minimum value of the function. As we have seen earlier in the chapter, one way to do that is to calculate a derivative and find its zeros — the "critical values" of the total-time function.

The Chain Rule and the Square Root Function

Marcus put down the book. Tony explained, "The details of this use a lot of math, calculus stuff. I need to be able to differentiate things like

$$\frac{d}{dx}\sqrt{p^2 + x^2}.$$

This seems to depend on the Chain Rule. You know, we were working on that when I got sick. I never did really understand that. How are you on the Chain Rule?"

"Man, I came over here ready to talk about the Pistons-Lakers game, the concert last weekend, the news that Ray has been dumped by his girl, and you want to talk about the Chain Rule!"

Tony's focus of attention wavered. "Ray got dumped, did he?" Then he returned to the books on the table. "Look, tell me how to to do the differentiation. Then we'll get something to eat and you can tell me about Ray. I don't get much chance to get help with this stuff. All Mario wants to talk about are precedents and briefs."

---

[6]A physics book would use the complementary angles (that is, angles with the perpendicular to the surface) in both cases. For the reflection experiment, we thought you would find it easier to estimate the angles with the surface. Whether you use angles with the surface or their complements doesn't really matter; we can arrive at the same conclusions either way.

[7]First formulated in 1657 by Pierre de Fermat (1601?-1665), a remarkable mathematician and physicist who made important contributions to both subjects. For example, he preceded Descartes in inventing analytic ("Cartesian") geometry.

Marcus pulled over a pad of paper. "OK, so what is p?"

Tony frowned, "I don't know, just some number."

"Well, does it depend on x?"

Tony looked at the book awhile, then said, "I don't think so. There are various cases of p, but once you pick it, it stays the same for the rest of the calculation."

"All right, it's a constant. So, first, you have to know how to differentiate the square root function" He wrote:

$$\frac{d}{dx}\sqrt{x} = \frac{1}{2\sqrt{x}} \, .$$

"Where does that come from?" asked Tony.

Marcus dragged the Calculus Reader out from under a stack of other books and paged through. "Here, read this while I find something to munch on. You got any chips?"

The Derivative of the Square Root Function

Let's write $y = \sqrt{x}$. The difference quotient that approximates $\dfrac{dy}{dx}$ is

$$\frac{\Delta y}{\Delta x} = \frac{\sqrt{x + \Delta x} - \sqrt{x}}{\Delta x} \, . \tag{10}$$

It's not obvious how to simplify this expression (in particular, how to find a factor of Δx in the numerator). We need to use Problem Solving Strategy 6 from Section 3.6 — algebraic manipulation to get a more useful form — but the most useful manipulation might not occur to you immediately.

A difference of square roots is the source of the difficulty; we could make that go away if we *rationalize the numerator*.[8] It's the same idea you learned in high school for rationalizing a denominator: Multiply and divide by the *sum* of the square roots. Since we are multiplying and dividing by the same quantity, we are not changing the value of the expression. Let's see what happens.

---

[8]This is not the first time this idea has come up in this course. Look back at Section 1.7, where we used rationalizing the numerator for a very different purpose.

<u>Algebraic step</u>   <u>Reason</u>

$$\frac{\Delta y}{\Delta x} = \frac{\sqrt{x+\Delta x} - \sqrt{x}}{\Delta x} \cdot \frac{\sqrt{x+\Delta x} + \sqrt{x}}{\sqrt{x+\Delta x} + \sqrt{x}}$$

$$= \frac{(\sqrt{x+\Delta x})^2 - (\sqrt{x})^2}{\Delta x\,(\sqrt{x+\Delta x} + \sqrt{x})}$$

We multiplied the two fractions.

$$= \frac{(x+\Delta x) - x}{\Delta x\,(\sqrt{x+\Delta x} + \sqrt{x})}$$

$$= \frac{1}{\sqrt{x+\Delta x} + \sqrt{x}}\,.$$

Exercise 1. *Fill in reasons for the other steps. Then finish the calculation by determining the limiting value as* Δx *approaches* 0. *Write your conclusion here:*

$$\frac{d}{dx}\,\sqrt{x} = \underline{\hspace{6cm}}\,. \tag{11}$$

Exercise 2. *(a) Calculate* $\frac{d}{dx}\,x\sqrt{x}$ *by using the Product Rule and formula (11).*

(b) Calculate the same derivative by using a difference quotient and a limiting value, as in the calculation preceding Exercise 1.

(c) Compare the results of parts (a) and (b), both in terms of algebraic equivalence of the two answers and in terms of labor required to get the answers.

Differentiating a Complicated Function

Tony looked up at Marcus, who had found a Coke and a bag of chips. Marcus was staring at the pencil in the cup. "OK Marcus, I see how to differentiate the square root function. Why didn't you say it was just the power rule with the power equal to $\frac{1}{2}$? *Now what do I do with something like*

$$\frac{d}{dx}\,\sqrt{p^2 + x^2}\ ?"$$

Marcus finished chewing. "You want a drink? I just helped myself."

"No, that's fine. Just tell me about the derivative before we get off the subject."

Marcus pulled the pad over. "When you want to differentiate a complicated function like

$$\frac{d}{dx}\sqrt{p^2 + x^2},$$

you use the Chain Rule. First you take the derivative of the square root function,

$$\frac{d}{du}\sqrt{u} = \frac{1}{2\sqrt{u}}.$$

Now replace the u with $p^2 + x^2$ and multiply by the derivative of u:

$$\frac{d}{dx}\sqrt{p^2 + x^2} \;=\; \frac{1}{2\sqrt{p^2 + x^2}}\frac{du}{dx}$$

$$=\; \frac{1}{2\sqrt{p^2 + x^2}}\,2x$$

$$=\; \frac{x}{\sqrt{p^2 + x^2}}\,.\text{''}$$

"Wait a minute," objected Tony, "where did this u come from? And how did you know to work on the square root first?"

"The u just represents the $p^2 + x^2$. You work from the outside in. The square root is the last operation you do to calculate the function, so you differentiate it first." Marcus picked up the Calculus Reader again. "Here, read this section and I'll come by tomorrow and we'll talk about it. ... You got any Cheese Whiz in here?"

Differentiating a Function-of-a-Function: The Chain Rule

Now that we have a formula for differentiating the square root function, we need to extend that to a formula for $\frac{d}{dx}\sqrt{u}$, where u is itself a function of x. Actually, "square root" has little to do with this step; this is a problem that we will encounter over and over in many different forms. If we write $y = \sqrt{u}$, then our problem takes this form: We have y as a function of u, and we have u as a function of x. Therefore, y is a function of x also. How do we find $\frac{dy}{dx}$ when we know $\frac{dy}{du}$ and $\frac{du}{dx}$?

Let's relate that question to something we have done before (Problem Solving Strategy 4, Section 3.6). In Chapter 2 we worked out a formula for the derivative of the rescaled function when the scale of the independent variable is changed. (Think speed in miles per hour vs. speed in miles per minute. The two rates differ by a factor of the number of minutes in an hour, i.e., the *rate* at which minutes increase when hours increase.) In the present context, that would apply if, say, u were a simple scale change,

$u = kx$. Then we would know already that $\frac{du}{dx} = k$, and

$$\frac{dy}{dx} \;=\; k\,\frac{dy}{du} \;=\; \frac{dy}{du}\,\frac{du}{dx}\,. \tag{12}$$

In words, this says that the instantaneous rate of change of y as a function of x is the rate of change of y as a function of u times the rate of change of u with respect to x. We know this is true if u is of the form $u = kx$; in fact, this equation "works," regardless of the relationship between u and x. Let's see why.

Suppose we fix a number x at which we want to know $\frac{dy}{dx}$, and we compute an approximating difference quotient for a small increment Δx. We write simply "u" for the value of u at x and $u + \Delta u$ for the value at $x + \Delta x$. That is, Δu is the corresponding increment in the intermediate variable. Similarly, we write y for the value of the "outer variable" at x and $y + \Delta y$ for the value at $x + \Delta x$, so Δy is the corresponding increment in the outer variable. Then $\frac{dy}{dx}$ is approximated by $\frac{\Delta y}{\Delta x}$, and simple algebra tells us that

$$\frac{\Delta y}{\Delta x} \;=\; \frac{\Delta y}{\Delta u}\,\frac{\Delta u}{\Delta x}\,. \tag{13}$$

We may not know yet what du is — or why it appears to "cancel" in equation (12) — but Δu is an ordinary numerical quantity, so, whenever it is not zero, it is subject to the algebraic cancellation law.

The two factors on the right in equation (13) approximate, respectively, the rate of change of y with respect to u and the rate of change of u with respect to x. Furthermore, the approximations to instantaneous rates of change all get better as the increment in x shrinks to zero, so when we take limiting values of all three quotients, we find

$$\frac{dy}{dx} \;=\; \frac{dy}{du}\,\frac{du}{dx}\,, \tag{14}$$

which is the same as equation (12), but *without the restriction that* u *should be a constant multiple of* x.

Simple as it is, equation (14) is perhaps the most important formula of differential calculus, because so many other formulas and calculations depend on it. Important results have names; the name of this one is the **"Chain Rule."** It is called that because it tells us how to differentiate "chains" of functions, i.e., how to find $\dfrac{dy}{dx}$ when y is a function of u and u is a function of x.

In particular, we can now calculate the derivative of $\sqrt{p^2 + x^2}$, where p is a constant. We set $u = p^2 + x^2$ and $y = \sqrt{u} = \sqrt{p^2 + x^2}$. Then $\dfrac{du}{dx} = 2x$, and $\dfrac{dy}{du} = \dfrac{1}{2\sqrt{u}}$, so

$$\frac{dy}{dx} \;=\; \frac{dy}{du}\frac{du}{dx} \;=\; \frac{1}{2\sqrt{u}}\,2x \;=\; \frac{x}{\sqrt{p^2 + x^2}}\,. \tag{15}$$

Exercise 3. If $y = \sqrt{e^x + x}$, find $\dfrac{dy}{dx}$.

Working with the Chain Rule

The next day, as Marcus walked down the street toward Tony's apartment building, the sunshine had given way to a heavy wet snow. It was the kind of snow that soaked into everything, and Marcus struggled to keep his paper bag from tearing apart. He had rye bread, pastrami, dill pickles, Dijon mustard, and potato salad. He didn't think Tony was eating too well. If only he had told the bag boy he wanted a plastic bag!

The scene in the apartment was much the same as the day before. However, the cup with the pencil had been moved to the kitchen counter. Marcus noticed a damp splotch on the papers on one corner of the card table. "How's it going, Tony? I got some stuff for lunch. You hungry?"

"In a minute," Tony replied, "look and see if I did this problem right."

"OK, let me see." said Marcus. "Exercise 3. What did you do?" He pulled the pad over:

$$y \;=\; \sqrt{e^x + x}$$

$$\frac{dy}{dx} \;=\; \frac{1}{2\sqrt{e^x + x}}\,\frac{d}{dx}\,(e^x + x)$$

$$=\; \frac{1}{2\sqrt{e^x + x}}\,(e^x + 1)$$

$$=\; \frac{e^x + 1}{2\sqrt{e^x + x}}\,.$$

"Looks fine to me. I think you've got it. Let's eat."

As Marcus made the sandwiches, Tony sipped on a Coke. "You know there is one advantage to being out of school for a semester. I get to work on what I am interested in when I am interested in it. If I want to read a novel all day, I can do that. If I want to study about light all day, I can do that. During the regular semester you have to work on things because they are due, not because you want to."

Marcus slid the sandwich across the counter and replied, "That's great, but what are you going to live on?"

Tony sighed, "Oh, I'll go to work for my Uncle Julius as soon as I get my strength back. But this afternoon I'm going to understand how Fermat's Principle works."

"Whatever turns you on," said Marcus, as he spooned out more potato salad.

Analysis of Reflection

In Figure 16, we have added some labels to the points and line segments in Figure 14 as a first step in analyzing the question of how α and β are related. We label the object, the viewer's eye, and the image of the object in the mirror by P, Q, and R, respectively. For a fixed position of the mirror, the object being viewed, and the eye, the quantities that are *constant* are the distance p from object to mirror, the distance q from eye to mirror, and the distance D between the projections of object and eye on the mirror. *In principle*, until we know the relationship between α and β, the point R could be anywhere along the mirror. We label its distance from the left edge of the figure by x, and we imagine that x can take any value from 0 to D. Our question, then, is what value of x produces the minimal travel time for a ray from P to R to Q? When we can answer that question, we will know how α and β are related.

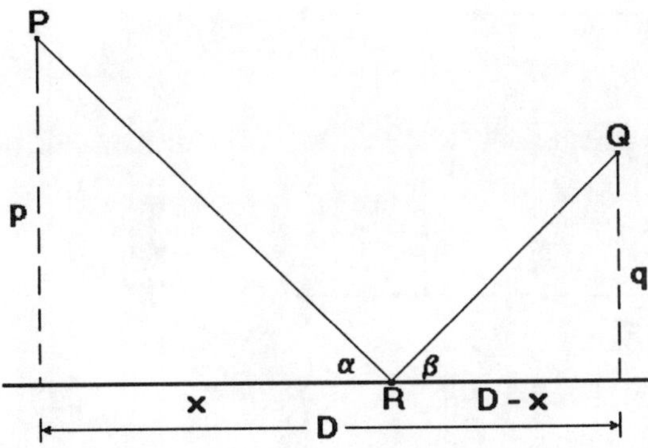

Figure 16. Reflection of light ray in a mirror.

Exercise 4. *(a) Figure 16 contains two triangles. We have explained the labeling of three of the six sides; explain the fourth distance label, $D - x$, in the figure.*
(b) Calculate the lengths of the two remaining sides, and label them accordingly.

Since the speed of light in air is constant, minimizing the time-of-travel for the light ray is equivalent to minimizing the *distance* traveled, i.e., the sum of the distances from P to R and from R to Q. Both distances depend on x, of course, so their sum does also.

Exercise 5. Show that, if we write $L(x)$ for the "length of path" function, then

$$L(x) \;=\; \sqrt{p^2 + x^2} + \sqrt{q^2 + (D - x)^2}\,, \qquad (16)$$

for any x between 0 and D.

In Figure 17 we show a graph of the function $L(x)$ defined by (16) for typical values of p, q, and D, along with a graph of its derivative. The graphs confirm that $L(x)$ has a unique minimum, that the minimum occurs somewhere between $x = 0$ and $x = D$, and that $L'(x) = 0$ at the minimum point. Thus, all we have to do is calculate $L'(x)$, set the result equal to zero, and solve for x.

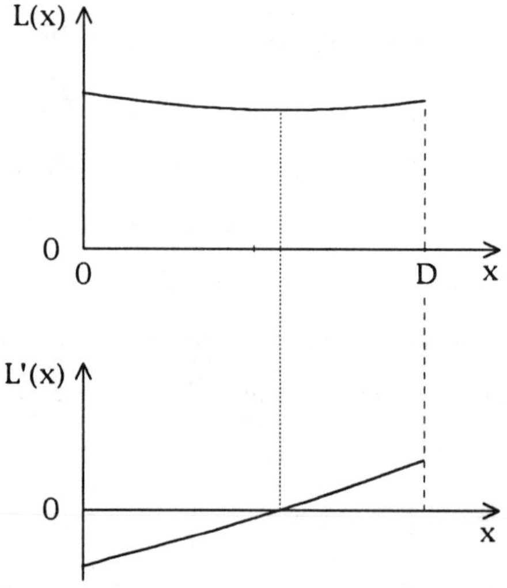

Figure 17. $L(x)$ and its derivative.

Figure 17 also has a message (several messages, actually) about the limitations of computers for solving the type of problem we are studying now. First, in order to draw the graph of L, we had to specify values of the constants p, q, and D. If we made different choices, would the graph look the same? We want to know the relationship between α and β *regardless* of the values of the constants. Presuming that the answer is "yes" — that Figure 17 really is "typical" — we would have a problem finding the leveling-off point on the graph of L by zooming in on it. That portion of the graph is already very flat, and when we zoom in we will see nothing but a horizontal line long before the scale is fine enough to tell us anything about the *lowest* point. We could zoom in instead on the graph of L' to find where it crosses the x-axis — but only if we know a *formula* for L', or at least some way to compute its values. Similarly, we could apply Newton's method to the equation $L'(x) = 0$, but for that we need to be able to compute values of *both* L' and its derivative, L''. Thus, even with a computer at hand, we still have to come to grips with the problem of calculating L'.

With our new tools in hand, we are ready to find the minimum value of the length function defined by

$$L(x) \;=\; \sqrt{p^2 + x^2} + \sqrt{q^2 + (D - x)^2}\,, \qquad\qquad \text{(16R)}$$

for any x between 0 and D. We have just calculated the derivative of the first term, and the second is similar — but with a small added complication. What's the derivative of $(D - x)^2$? (Whatever it is, adding q^2 won't change the derivative, because q is constant.) Well, $(D - x)^2$ is just another "function of a function" to which we can apply the Chain Rule. Specifically, if we write $v = w^2$ and $w = D - x$, then $v = (D - x)^2$. The Chain Rule tells us

$$\frac{dv}{dx} \;=\; \frac{dv}{dw}\frac{dw}{dx} \;=\; 2w(-1) \;=\; -2(D - x).$$

And, since the adding the constant q^2 does not change the derivative, we find

$$\frac{d}{dx}\left[q^2 + (D - x)^2\right] \;=\; -2(D - x).$$

Now, if we write $t = q^2 + (D-x)^2$ and $z = \sqrt{t}$, then

$$\frac{dz}{dx} = \frac{dz}{dt}\frac{dt}{dx} = \frac{1}{2\sqrt{t}}\left[-2(D-x)\right] = \frac{x-D}{\sqrt{q^2+(D-x)^2}}.$$

Finally, using the result from (16), we find that

$$L'(x) = \frac{dy}{dx} + \frac{dz}{dx} = \frac{x}{\sqrt{p^2+x^2}} + \frac{x-D}{\sqrt{q^2+(D-x)^2}}. \tag{17}$$

Well, it was quite a struggle, but we have our derivative! Notice what happened here: From two very simple formulas — the Power Rule and the Chain Rule — the first very specific, and the second very general, we differentiated a *very* complicated formula, one we could never have dealt with directly by using difference quotients.

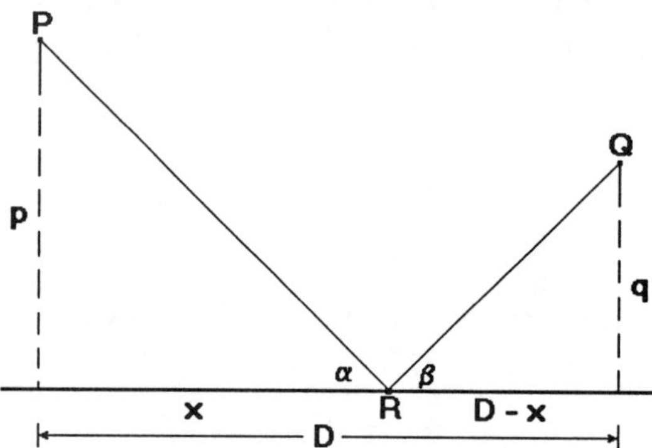

Figure 16R. Reflection of light ray in a mirror.

Now what did we learn about reflection? Well, from equation (17), $L'(x) = 0$ is equivalent to

$$\frac{x}{\sqrt{p^2+x^2}} = \frac{D-x}{\sqrt{q^2+(D-x)^2}}. \tag{18}$$

That still looks like a substantial algebraic challenge, but let's go back to Problem Solving Strategy 3, keeping our goal in focus. We wanted to determine the relationship between α and β, and the two quotients in equation (18) have a direct connection with those two angles. In Figure 16, we see that each numerator in (18) is the length of an "adjacent side" for one of the angles, and each denominator is the length of the corresponding hypotenuse.   Thus, (18) reduces to a very simple statement about the angles:

$$\cos \alpha \;=\; \cos \beta. \tag{19}$$

And, as both angles are between $0°$ and $90°$, the only way they can have the same cosine is to be the *same angle*.   Conclusion: *The angle of incidence equals the angle of reflection.*

That conclusion probably comes as no surprise to you; indeed, you probably figured it out from our mirror experiment.   The real reason we worked so hard to get this from Fermat's Principle is that we have now assembled all the tools for you to answer the harder question about refraction.

Analysis of Refraction

To help you get started, we repeat the refraction picture (Figure 15) with some added labels that use the same notation as in Figure 16:

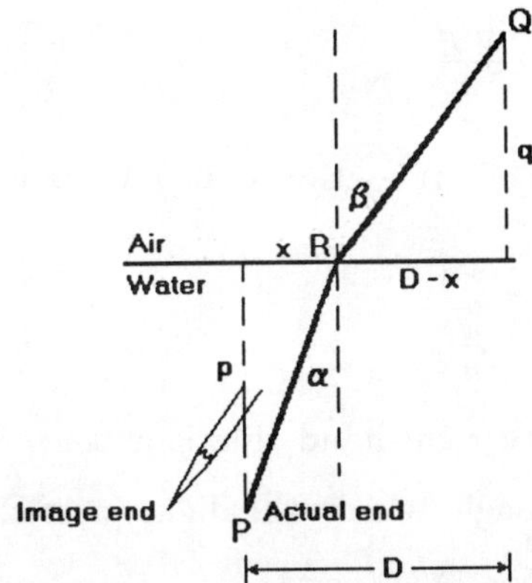

Figure 18. Angle of incidence and angle of refraction.

What's different about refraction is that there isn't just one "speed of light," but rather two — one speed in air and one in water.   In general, light has a different speed for each medium through which it travels.

The symbol c is commonly used for the speed of light. Suppose we write c_a for the speed of light in air and c_w for the speed of light in water. Similarly, we can write $T_a = T_a(x)$ for the travel time of the light ray through the air and $T_w = T_w(x)$ for the travel time through water. [The travel times are functions of x because they depend on where the ray passes from water to air on its way from the pencil tip to the eye.] Then, because the speed is constant in each medium, $c_w T_w$ is the distance traveled through water, and $c_a T_a$ is the distance traveled through air.

Exercise 6. *Show that the total travel time,* $T(x)$, *is given by*

$$T(x) \ = \ \frac{\sqrt{p^2 + x^2}}{c_w} + \frac{\sqrt{q^2 + (D-x)^2}}{c_a} \ , \tag{20}$$

for any x *between* 0 *and* D.

Since $T(x)$ is similar in form to $L(x)$, its graph is not very different from the one shown in Figure 17. Minimizing the total travel time for a refracted ray is essentially the same problem as minimizing the travel time (or distance) for a reflected ray. The formula has two added symbols, but $\frac{1}{c_w}$ and $\frac{1}{c_a}$ are constant factors in their respective terms.

Exercise 7. *(a) Calculate* $T'(x)$.

(b) Use the result to show that $T(x)$ *is minimized when*

$$\frac{\sin \alpha}{c_w} \ = \ \frac{\sin \beta}{c_a} \ . \tag{21}$$

One way to state the conclusion about refraction is to rewrite equation (21) in the form

$$\frac{\sin \alpha}{\sin \beta} \ = \ \frac{c_w}{c_a} \ . \tag{22}$$

The significance of this form is that the right-hand side is a *constant* that depends only on the two media through which the light passes. The classical form of the statement in equation (22) is:

Snell's Law.[9] *The ratio of the sine of the angle of incidence to the sine of the angle of refraction is a constant.*

---

[9]After Willebrord Snell, who discovered this relationship empirically in 1621, about two generations before Fermat stated his minimization principle.

We conclude this section with Thurmon Whitley's "Concluding Remarks" in *Five Applications of Max-Min Theory* (see Note 5):

> "Notice that the choice of air or water is not crucial to the derivation of Snell's Law. In fact, any two media through which light travels at a constant rate could have been used, with similar results.
>
> "You should note also that we derived Snell's Law without explicitly finding a critical value for $T(x)$. The actual solution of the equation $T'(x) = 0$ [see Exercise 7] would involve a cumbersome fourth degree polynomial. In addition, it is not at all important to have an explicit expression for a critical value. The ability to obtain useful results knowing only the existence of certain numbers (without knowing their values) is a phenomenon which occurs frequently in applied mathematics."

Section Summary

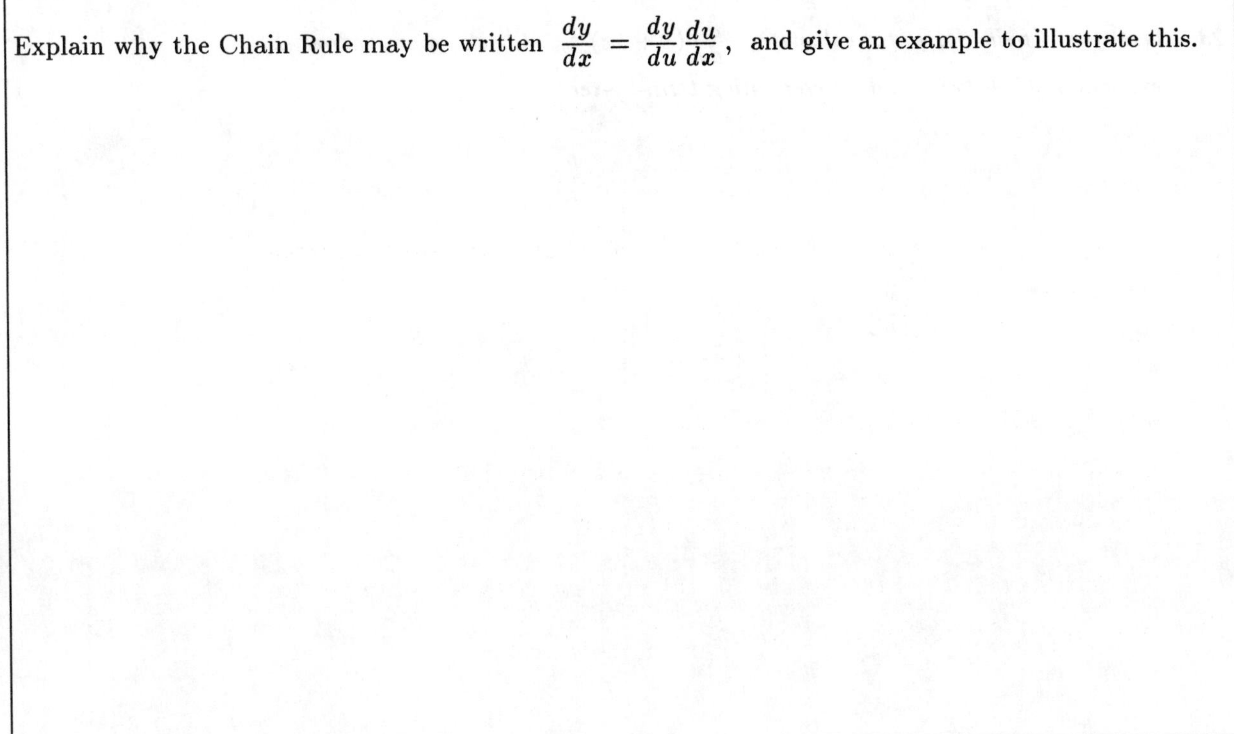

Explain why the Chain Rule may be written $\dfrac{dy}{dx} = \dfrac{dy}{du}\dfrac{du}{dx}$, and give an example to illustrate this.

Exercises

Use the Chain Rule and the rule for differentiating the square root function to calculate the derivative of each of the following functions:

8. $\sqrt{2t+3}$ 9. $\sqrt{2t^2+3}$ 10. $\sqrt{t^3-t}$

Use the Chain Rule and the Product Rule to calculate the derivative of each of the following functions:

11. e^{4t-6} 12. e^{t^2} 13. e^{t+4}

14. $t\,e^{4t-6}$ 15. $t\,e^{t^2}$ 16. $t\,e^{t+4}$

17. $t^2\,e^{4t-6}$ 18. $t^2\,e^{t^2}$ 19. $t^2\,e^{t+4}$

20. $\sqrt{t}\,e^{4t-6}$ 21. $\sqrt{t}\,e^{t^2}$ 22. $\sqrt{t}\,e^{t+4}$

23. Use the Chain Rule to calculate $\dfrac{d}{dx}(1+3x)^4$. Check your answer by expanding $(1+3x)^4$ and differentiating term-by-term.

4.5 Derivatives of Functions Defined Implicitly

Consider the ellipse (see Figure 19) with the equation

$$\frac{x^2}{4}+\frac{y^2}{9} = 1.$$

Suppose we want to know the slope of the tangent line at the point (x_0, y_0), where $x_0 = 1$ and $y_0 = 3\sqrt{3}/2 \approx 2.598$.

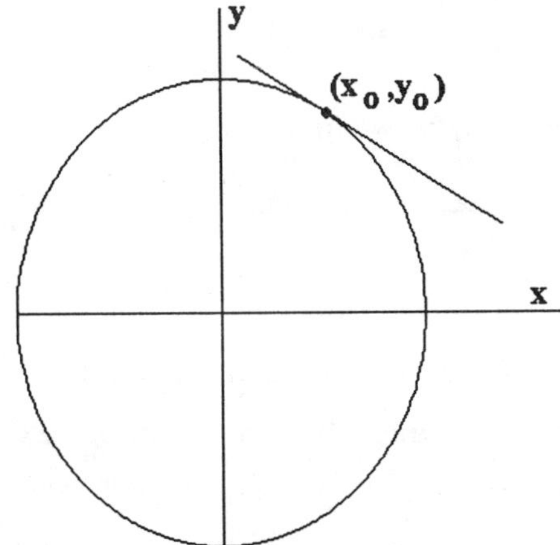

Figure 19. Graph of $\dfrac{x^2}{4}+\dfrac{y^2}{9}=1$.

One approach would be to find an explicit description $y = f(x)$ for the top half of the ellipse, differentiate that function, and evaluate the derivative at $x = 1$. You will carry out that computation in an Exercise 1.

However, before you do that, let's consider another approach. Whatever the functional relation between y and x, for each value of x we know that

$$\frac{x^2}{4}+\frac{y^2}{9} = 1. \tag{23}$$

So the function of x on the left in (23), obtained by squaring y, dividing by 9, and adding $\frac{x^2}{4}$, must be the same function as the one on the right side of (23), i.e., the function on the left must be the constant function 1.

Now we differentiate both the left-hand and the right-hand functions in (23). On the left, we use the Sum Rule, the Power Rule, and the Chain Rule to obtain

$$\frac{d}{dx}\left(\frac{x^2}{4}+\frac{y^2}{9}\right) = \frac{d}{dx}\frac{x^2}{4}+\frac{d}{dx}\frac{y^2}{9} = \frac{x}{2}+\frac{2}{9}\,y\,\frac{dy}{dx}.$$

On the right, the derivative of the constant 1 is 0. Since the two derivatives are equal, we may write

$$\frac{x}{2}+\frac{2}{9}\,y\,\frac{dy}{dx} = 0.$$

Solving for $\dfrac{dy}{dx},$ we obtain

$$\frac{dy}{dx} = \frac{-\frac{x}{2}}{\frac{2}{9}\,y} = -\frac{9x}{4y}. \tag{24}$$

In particular, at $x_0 = 1$ and $y_0 = 3\sqrt{3}/2,$ we find

$$\frac{dy}{dx} = -\frac{9}{6\sqrt{3}} \approx -0.866.$$

The ellipse in Figure 19 is the combined graphs of *two* functions of the form $y = y(x),$ one for the top half of the ellipse and one for the bottom half. Both of those functions are "**defined implicitly**" by equation (23):

$$\frac{x^2}{4}+\frac{y^2}{9} = 1. \tag{23R}$$

Our technique for calculating $\dfrac{dy}{dx}$ directly from the implicit definition gave us formula (24), which applies to *both* of the functions defined implicitly by equation (23).

Exercise 1. (a) Show that, for the top half of the ellipse,

$$y = 3\sqrt{1-\frac{x^2}{4}}.$$

(b) Calculate $\dfrac{dy}{dx}$ explicitly, substitute $x = 1,$ and show that your answer agrees with the one we obtained by our differentiation of the implicit definition.

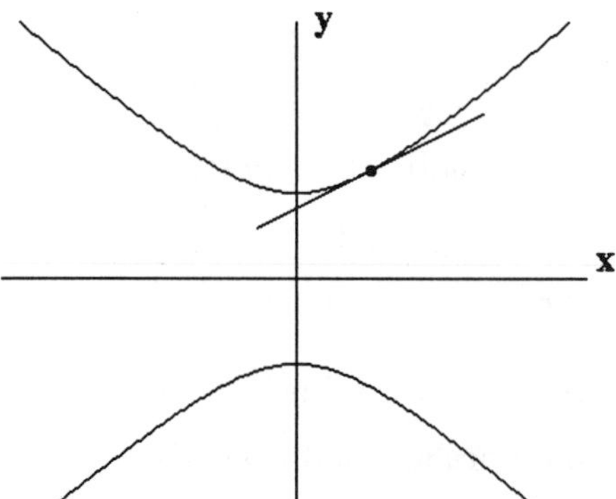

Figure 20. Graph of $y^2 - x^2 = 1.$

Exercise 2. Consider the graph of the equation $y^2 - x^2 = 1$, shown in Figure 20.

(a) Use the method of implicit differentiation to find the slope of the tangent line to the curve at $x = 1$ and $y = \sqrt{2}$.

(b) Use the explicit method of Exercise 1 to find the slope of the tangent line to the curve $(1, \sqrt{2})$.

The Derivative of the Natural Logarithm

We have formulas for the derivatives of all but one of the functions we have encountered so far — all except the natural logarithm. We will find a formula for this derivative in this section. Before we do, let's return to the definition of the derivative and look at difference quotients for $\ln(x)$.

In Figure 21 we graph the "difference quotient function"

$$y = \frac{\ln(x + 0.001) - \ln(x)}{0.001},$$

which should closely approximate the derivative of $\ln(x)$. Have you seen this function somewhere before?

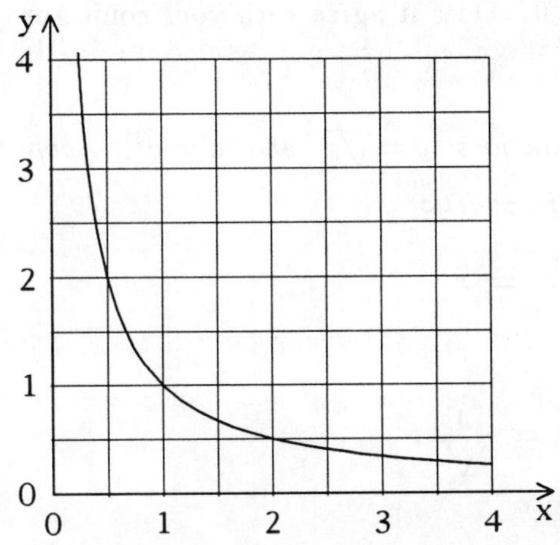

Figure 21. Graph of the difference quotient for $\ln(x)$.

Exercise 3. (a) Use the graph to estimate the difference quotient for $x = 1/4$, $x = 1/2$, $x = 1$, $x = 2$, and $x = 4$.

(b) On the basis of your estimates in (a), make a conjecture about the derivative of $\ln(x)$. (You will be able to check your conjecture soon; we will derive the formula for the derivative in the next paragraph.)

Now let's find a formula for the derivative. The implicit differentiation approach turns out to be just what we need. If $y = \ln(x)$, then, by the definition of the natural logarithm,

$$e^y = x.$$

If we differentiate both sides of this equation with respect to x, we obtain

$$e^y \frac{dy}{dx} = 1,$$

which leads to

$$\frac{dy}{dx} = \frac{1}{e^y}.$$

But, as $e^y = x$, we find

$$\frac{dy}{dx} = \frac{1}{x}.$$

What we have shown is that

$$\frac{d}{dx} \ln(x) = \frac{1}{x}. \tag{25}$$

Since the natural logarithm is only defined for positive numbers x, formula (25) is correct for all $x > 0$. Does it agree with your conjecture in Exercise 3?

Exercise 4. *Consider the pair of inverse functions $y = \sqrt{x}$ and $x = y^2$. Apply the implicit differentiation technique to the equation*

$$y^2 = x$$

to obtain the formula

$$\frac{dy}{dx} = \frac{1}{2\sqrt{x}}.$$

Derivatives of Inverse Functions

We have just seen two instances of the relationship between the derivatives of inverse functions — for the pair $y = \ln(x)$, $x = e^y$ and for the pair $y = \sqrt{x}$, $x = y^2$. Since we will use this relationship again to determine other derivatives of inverses of known functions, it is worth our while to look at derivatives of inverse functions in general.

Suppose f and g are an inverse pair of functions. So, if $y = f(x)$, then $x = g(y)$. When we apply one after the other, we get

$$g\big(f(x)\big) \; = \; x. \tag{26}$$

When we apply the Chain Rule to (26), we obtain

$$g'\big(f(x)\big)f'(x) \; = \; 1$$

or

$$f'(x) \; = \; \frac{1}{g'\big(f(x)\big)}. \tag{27}$$

In Leibniz notation, formula (27) becomes

$$\frac{dy}{dx} \; = \; \frac{1}{\dfrac{dx}{dy}}. \tag{28}$$

Thus, we see another instance (like the Chain Rule, for example) in which differentials appear to behave just like numbers. In words, formula (28) says that the derivatives of inverse functions are *reciprocals* of each other.

As an example of the use of formula (28), we repeat our argument for $y = \ln(x)$ and $x = e^y$:

$$\frac{dy}{dx} \; = \; \frac{1}{\dfrac{dx}{dy}} \; = \; \frac{1}{e^y} \; = \; \frac{1}{x}.$$

Section Summary

Describe the procedure for calculating the derivative of a function defined implicitly by an equation.

What is the formula for the derivative of the natural logarithm?

How are the derivatives of inverse functions related?

Exercises

5. (a) *Differentiate implicitly to find the slope of the tangent line to the curve* $x = y^2$ *at* $x = 1$ *and* $y = 1$.

(b) *Find the slope of the tangent line to the curve* $x = y^2$ *at* $(1,1)$ *by calculating* $\frac{dx}{dy}$ *and using formula (28).*

(b) *Find the slope of the tangent line to the curve* $x = y^2$ *at* $(1,1)$ *by first solving for* y *as an explicit function of* x.

Calculate the derivatives of the following functions:

6. $\ln(x + 2)$

7. $\ln(x^2 + 2)$

8. $x^2 + 2 \ln x$

9. $x \ln x$

10. $x^2 \ln(x + 2)$

11. $x \ln(x^2 + 2)$

4.6 The General Power Rule

We need one more general rule for differentiation. We know how to differentiate power functions such as $f(x) = x^2$ and $g(x) = x^9$; the derivatives are $f'(x) = 2x$ and $g'(x) = 9x^8$. We can summarize what we know about derivatives of power functions:

$$\frac{d}{dx}\, x^n = nx^{n-1}. \tag{29}$$

Here, n represents 0 or 1 or 2, or, in general, any nonnegative integer. The case of $n = \frac{1}{2}$ also fits this formula:

$$\frac{d}{dx}\, \sqrt{x} = \frac{d}{dx}\, x^{1/2} = \frac{1}{2}\, x^{\frac{1}{2}-1} = \frac{1}{2}\, x^{-1/2} = \frac{1}{2}\, \frac{1}{\sqrt{x}} = \frac{1}{2\sqrt{x}}.$$

What about the derivative of $y = x^{7/3}$? Does it fit the form of the Power Rule (29)? To answer this, we can use again the approach of differentiating implicitly. In this case, we know that

$$y^3 = x^7.$$

Now we can differentiate both sides with respect to x — because both exponents are positive integers:

$$3y^2\, \frac{dy}{dx} = 7x^6.$$

When we solve for $\dfrac{dy}{dx}$, we get

$$\frac{dy}{dx} = \frac{7x^6}{3y^2}.$$

If we substitute $y = x^{7/3}$, we get

$$\frac{dy}{dx} = \frac{7x^6}{3\left(x^{7/3}\right)^2}$$

or

$$\frac{dy}{dx} = \frac{7}{3}\, x^{6-\frac{14}{3}} = \frac{7}{3}\, x^{4/3}.$$

Note that if we set $n = \frac{7}{3}$, this does fits the pattern we have already seen:

$$\frac{d}{dx}\, x^n = nx^{n-1}. \tag{29R}$$

The Power Rule (29) holds for any rational power n. Indeed, as we will see in Section 4.8, it holds for any power function, whether n is rational or not.[10]

We end this section with an example that combines the Power Rule and the Chain Rule, much as we did in Section 4.4 when we differentiated the length and time functions for reflection and refraction. Suppose we want to differentiate the function $(e^{3x} + 5x)^{3/2}$. Using both the Power Rule and the Chain Rule we find

$$\frac{d}{dx}(e^{3x} + 5x)^{3/2} = \frac{3}{2}(e^{3x} + 5x)^{1/2} \frac{d}{dx}(e^{3x} + 5x).$$

After we finish the calculation of the second factor, we have

$$\frac{d}{dx}(e^{3x} + 5x)^{3/2} = \frac{3}{2}(e^{3x} + 5x)^{1/2}(3e^{3x} + 5).$$

Section Summary

> Write out the Power Rule for differentiation, and give an example of the use of this rule for which the power is not a positive integer.

Exercises

Calculate the derivatives of the following functions:

1. $x^{5/2}$

2. $x^{-5/2}$

3. $x^{5/2} + x^2$

4. $\sqrt{x^2 + 3}$

5. $(x^2 + 3)^{5/2}$

6. $(\ln x)^2$

7. $x^{4/3} \ln x$

8. $(\ln x + x)^{-1/3}$

9. $x(p^2 + x^2)^{-1/2}$

10. *Use implicit differentiation to show that the Power Rule holds for $\frac{d}{dx} x^{5/3}$.*

---

[10] It is not immediately clear what an irrational power means. For example, how do you compute 2^{π}? We discuss this in Section 4.8 also.

4.7. Differentials and Leibniz Notation

It's time to answer the question, "What's a differential?" Perhaps you have forgotten the question. It came up in Chapter 2 when we introduced the notation $\frac{dy}{dt}$ for "the instantaneous rate of change of y with respect to t." The "numerator" dy and the "denominator" dt were *called* "differentials" in order to justify the standard name "differential equation" for an equation containing a *derivative* of the unknown function. In fact, the portion of calculus that pertains to the study of derivatives is universally known as "differential calculus," so we really have an obligation to tell you why.

The concept of *differential* and the "quotient" notation for the derivative are part of the contribution to the development of calculus made by Leibniz[11] — a small but very important part. As we will see from time to time, notation can have *power*. Good notation is a powerful tool enabling us to develop concepts and to solve problems; bad notation can seriously hinder our efforts to do either.

We have already seen an example of the use of Leibniz notation in the development of a concept: the Chain Rule. The notation immediately suggests both the correct statement of the "rule" and a means for showing that it is indeed correct. Our other notation for derivatives, the "prime" notation that traces its heritage to Newton, would not have served us nearly so well.

To illustrate the power of the Leibniz notation for problem solving, we provide a small "problem" you can solve now — with some effort — with nothing but elementary geometry, and perhaps a calculator. Later, we will show how the problem can be solved with a quick pencil-and-paper calculation that uses Leibniz notation.

---

[11]Gottfried Wilhelm Leibniz (1646–1716), German philosopher and mathematician. He and Isaac Newton (1642–1727) independently assembled the key ideas of calculus into a coherent theory, Newton several years earlier, but without publishing until considerably later. This resulted in a bitter controversy over priority between the two men of genius and their followers. Today they are accorded equal status as "codiscoverers" of calculus.

Exercise 1. Estimate the total volume of the Earth's crust. You may assume the Earth is a sphere with a radius of 4000 miles, and the average depth of the crust is 20 miles. Write your answer here: ___________

As you may have guessed by now, the concepts of "differential" and "derivative" are closely related, but *they are not the same thing.* To explain the difference, we once again call on the fundamental concept for this course:

$$\text{slope} \ = \ \frac{\text{rise}}{\text{run}} \ .$$

That equation is meaningful, of course, only if there *is* a run. That is, it describes the calculation of average rate of change (slope), but it does not directly explain the calculation of instantaneous rate of change (derivative), except through an "approaching" process in which the run shrinks to zero.

Zooming In — Again

In Figures 22, 23, 24, and 25, we recall the zooming-in process by which we observed what happens when the run shrinks to zero: At most points on the graphs of most functions, a sufficiently small segment of the curve "looks like" a straight line, and the slope of that line is what we called the instantaneous slope of the function, i.e., the derivative. The figures here are like our zooming figures in Chapter 2, with these exceptions: (a) The point at which we are doing the zooming is at the left edge of each picture, not in the center. (b) We have added to each figure a dotted line *tangent* to the graph at the left edge of the figure.

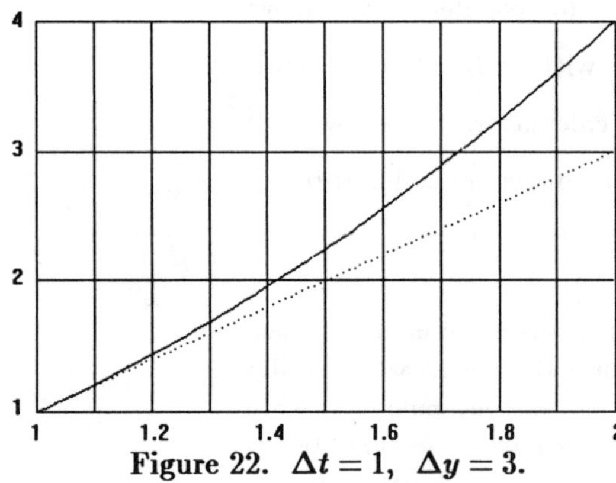

Figure 22. $\Delta t = 1, \ \ \Delta y = 3.$

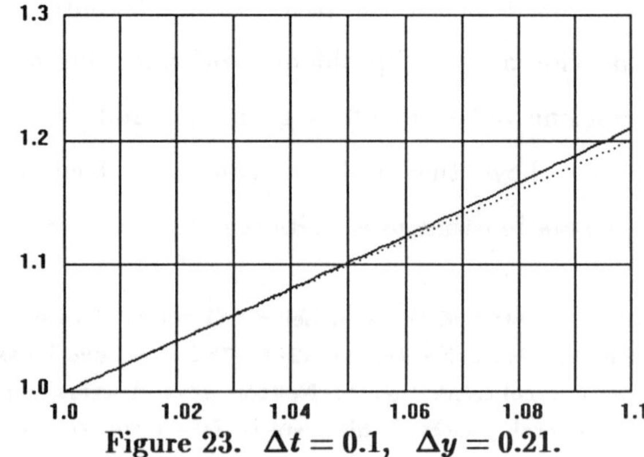

Figure 23. $\Delta t = 0.1, \ \ \Delta y = 0.21.$

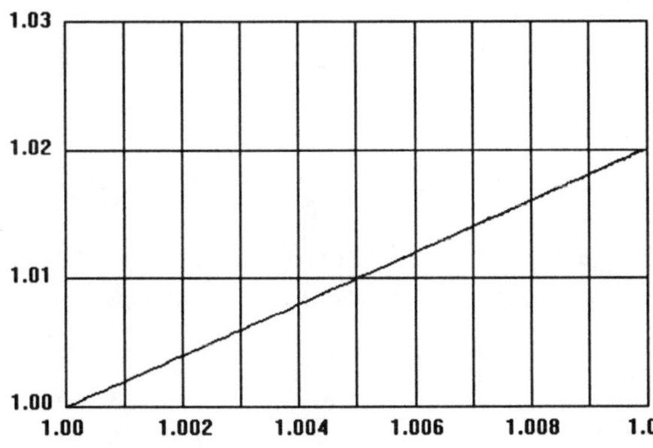

Figure 24. $\Delta t = 0.01$, $\Delta y = 0.02$.

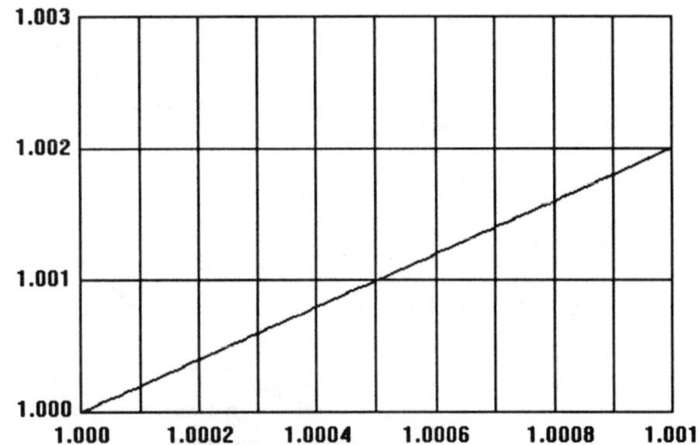

Figure 25. $\Delta t = 0.001$, $\Delta y = 0.002$.

Observe what happens as we zoom in: The curve eventually appears to coincide with the tangent line. Indeed, the line whose slope we called the derivative *is* the tangent line at the point at which we are calculating instantaneous rate of change. Conclusion: At any particular point on the graph of a "smooth" function $y = y(t)$, the derivative $\dfrac{dy}{dt}$ is the slope of the tangent line at that point.

As a slope, $\dfrac{dy}{dt}$ is also a "rise over run." What rise, and what run? Since the tangent line is a *line*, we can take any run we like and calculate the corresponding rise; the ratio of rise to run is constant. Suppose we *name* the run dt (even though it already has the name Δt) and the rise dy. Then we will have identified two quantities, dy and dt, whose ratio is indeed the derivative. That's it! The differential of t, calculated at a particular point on the graph of $y = y(t)$, is a (possibly small) *change* in t, and the differential of y is a *corresponding change* in y — not on the graph of $y = y(t)$, but on the tangent line at the point in question.[12]

---

[12]There is a technical problem with the way we have just described "differential," in that the word now has *two* definitions, depending on whether we are talking about independent or dependent variables. The same variable sometimes plays both roles in a single context (for example, the "intermediate" variable u that we introduced in our discussion of the Chain Rule). How are we to know that these possibly different meanings are actually the same? In Problem 9 at the end of the Chapter, we suggest a way to resolve this difficulty.

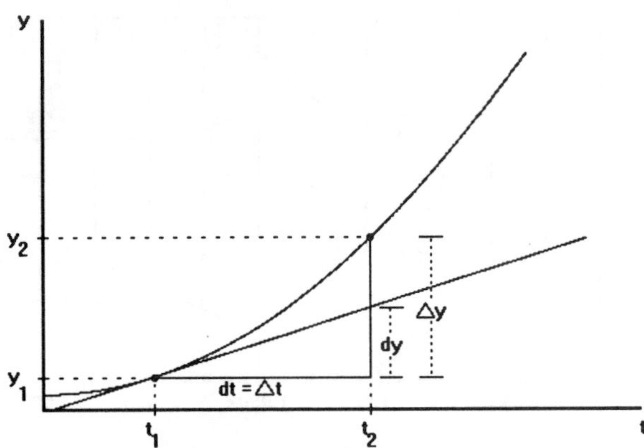

Figure 26. Rise to the curve (Δy) and rise to the tangent line (dy) for a given run Δt (or dt).

We summarize the previous paragraph in Figure 26, in which we deliberately exaggerate the size of the change in t to show clearly the distinction between the curve and its tangent line. Observe the *two different* "rises" in the figure:

Δy is the rise *to the curve* after a run of Δt.

dy is the rise *to the tangent line* after a run of dt.

When we take dt and Δt to be the *same* run, Δy and dy are clearly *different*, and it follows that the slopes $\dfrac{\Delta y}{\Delta t}$ (between two points on the curve) and $\dfrac{dy}{dt}$ (between two points on the tangent line) are also *different*.

However, we know the figure is an exaggeration; our zoom-in figures show what happens when we consider a very small run: The rise to the curve and the rise to the tangent line are *approximately equal*. In one sense, we already knew that. We introduced the derivative as the limiting value of difference quotients (slopes between points on the curve) when the run shrinks to zero. Thus, for a very small run, the slope between points is approximately the slope *at a point*, i.e., the derivative. If the slopes are approximately equal, the rises must be also. In another sense, we have a *new* idea in the approximate equality of the two rises:

$$dy \approx \Delta y. \tag{30}$$

This says that we can approximate an actual change in y values — a rise on the curve — by something that may be easier to compute, a rise on a straight line. Indeed, for the line, if we know the slope and the run, then we know the rise.

An Application of the Differential

To illustrate the point just stated, we now solve Exercise 1 by using a differential. Recall that we wanted to estimate the volume of the Earth's crust, knowing that the radius of the Earth is about 4000 miles and the thickness of the crust is about 20 miles. You can do this with the formula for volume of a sphere: $V = \frac{4}{3}\pi r^3$. All you have to do is calculate $V(4000) - V(3980)$, a ΔV! But, because 20 is "small" relative to 4000, we can estimate the actual rise in V by a differential, dV. The calculation by hand is a little simpler if we take the run in the negative direction, i.e., from 4000 to 3980. (What is the run?) We need the rate of change of V with respect to r, but that's easy: $\frac{dV}{dr} = 4\pi r^2$. Thus, $dV = 4\pi r^2 dr$ (rise equals slope times run). With $r = 4000$ and $dr = -20$, we get $dV = 4\pi(4000)^2(-20) = -128 \times 10^7 \times \pi$. (Why is the answer negative?) The only point at which we might need some help from a calculator is to estimate 128π; it's about 400.[13] Thus, we estimate the Earth's crust to have a volume of 4×10^9 (four billion) cubic miles. How does this compare with your estimate? Could you have done yours without a calculator?

Section Summary

Describe in your own words what the differential of a dependent variable is. Illustrate this with the formula for the area A of a square of side length s: $A = s^2$.

---

[13]We said we could do this problem with only pencil and paper, and we can. 128 is a power of 2 (which power?), which suggests approximating π by a fraction with a power of 2 in the denominator. 3.14 is close to 3.125, which is 25/8. Thus, 128π is approximately $(16 \times 8) \times 25/8 = 16 \times 25 = 400$.

Exercises

2. (a) *In the course of the calculation just completed, we also computed the area of the surface of the Earth. Where?*

    (b) *How is the volume of the crust related to the area of the surface? Is this reasonable? Why or why not?*

3. *Suppose we were able to determine that the (average) radius of the Earth is actually 3959 miles, plus or minus one mile.*

    (a) *Calculate the area of the Earth's surface.*

    (b) *How far off might this calculation be if our radius is actually off by one mile?*

    (c) *Find some region (city, state, country, continent) whose area is approximately equal to the possible error.*

    (d) *Calculate the percentage error. Is this error small or large?*

4.8 Derivation of the General Power Rule

Why Obtain General Formulas?

In this section we discuss more carefully the derivation of the Power Rule for rational powers; we also discuss (somewhat less carefully) its extension to irrational powers. You may reasonably wonder why we want to worry about this. After all, the Power Rule seems so obvious once one has seen the first few cases. Why should we work hard to achieve all these "extensions"? Are we making mountains out of molehills? Are we attacking this problem, to quote George Mallory's wonderful reason for climbing Everest, "Because it is there"? Certainly not — indeed, that will not be our justification for anything in this course. Here are some reasons to keep in mind as we work through the calculations.

First, effort expended at obtaining general formulas is well spent, because it makes it unnecessary to repeat essentially the same effort over and over for similar, but slightly different, cases that may arise later. *Example:* We worked hard to obtain the solution of exponential growth problems in general. Then we saw that the same concepts could be used to solve problems of radioactive decay, compound interest, cooling bodies, objects falling in a resisting medium, and simple electrical circuits. A direct attack on any one of these problems would have required repeating all the work invested in the first one, if we had not made the connection.

Second, the use of the same or similar algebraic notation, such as t^r, for many different types of objects hides a long history of discovery about the behavior of these objects with respect to various algebraic and limiting operations. *Example:* In order to make sense of the operation symbolized by $\sqrt{}$ when the object being operated on was negative, it was necessary to invent a whole new number system. Because the numbers in that system (complex numbers) behaved in most respects like real numbers, it was reasonable to retain the same algebraic notations and extend their meanings. This doesn't always happen; sometimes new systems require completely new notations, operations, and conceptual frameworks.

Third, and closely related to the second reason, it makes no sense to accept without question that objects assigned the same notation will behave in the same ways, when those objects may be quite different from each other. We obtained the first few cases of the Power Rule for positive integer powers from difference quotient approximations by simple algebraic manipulations, either factoring or expanding binomials, depending on the notation used. But those techniques did not help us find the derivative of the $\frac{1}{2}$ power, for example, and there is no reason to think that *any* algebraic technique would work for the π power, whatever that is. These various power functions are really very different objects; the fact that they behave the same way with respect to differentiation (as we shall see) is quite remarkable.

Fourth, differentiating power functions is not an abstract game. We have already seen (in the analyses of reflection and refraction) that non-integral power functions can model real physical behavior. So let's see why

$$\frac{d}{dt}\, t^r \; = \; r\, t^{r-1} \tag{31}$$

for *any* positive rational number r.

Here is the program: We will first show that $\frac{d}{dt}\, t^r \; = \; r\, t^{r-1}$ for any *positive rational*[14] number r. Then we will extend the Power Rule to *negative* rational powers and finally to *irrational* powers, such as $\sqrt{2}$ and π. Along the way we will discover more of the power of key concepts developed already, especially the Chain Rule.

The Power Rule for Rational Exponents

Suppose we write $r = \frac{p}{q}$, where p and q are positive integers. (In Section 4.6, we worked out the case with $p = 7$ and $q = 4$, and you worked out the case with $p = 5$ and $q = 3$ in Exercise 10 of that section.) We introduce a name for the r-th power function, say, $u = t^r$.

---

[14]A number is "rational" if it can be written as the *ratio* of two integers. That is, r is rational if $r = \frac{p}{q}$ where p and q are integers. Among the rational numbers are all the numbers you think of fractions (proper or improper), all terminating or repeating decimal numbers, and the integers themselves (because q may be 1). A real number that is not rational is called "irrational."

Then u is also equal to $t^{p/q}$. If we raise both sides of this equation to the q-th power, we have

$$u^q = t^p. \tag{32}$$

On the right in equation (32) we have a function of t whose derivative we know from the Power Rule, since p is a positive integer. On the left, we have the *same* function of t (because of the equality) written in a different form, namely, as the q-th power of some other function u. (We know exactly what function u is, but that's not important at the moment.)

From the positive integer version of the Power Rule, we know that $\frac{d}{du}u^q = qu^{q-1}$. Thus, the Chain Rule (with u^q as a function of u and u as a function of t) tells us that

$$\frac{d}{dt}u^q = \frac{d}{du}u^q\frac{du}{dt} = qu^{q-1}\frac{du}{dt}.$$

This gives us the derivative of the left-hand side of equation (32), which must equal the derivative of the right-hand side, so

$$q\,u^{q-1}\frac{du}{dt} = p\,t^{p-1}. \tag{33}$$

Now u is the function whose derivative we wanted to know, so we solve this equation for $\frac{du}{dt}$ and do a little algebra to simplify. (Write a reason for each step in the calculation.)

Step	Reason
$\dfrac{du}{dt} = \dfrac{p\,t^{p-1}}{q\,u^{q-1}}$	We divided both sides of (33) by $q\,u^{q-1}$.
$= \dfrac{p}{q}\dfrac{t^{p-1}}{u^{q-1}}$	
$= \dfrac{p}{q}\dfrac{t^{p-1}}{\left[t^{p/q}\right]^{(q-1)}}$	
$= \dfrac{p}{q}\dfrac{t^{p-1}}{t^{(q-1)p/q}}$	
$= \dfrac{p}{q}\dfrac{t^{p-1}}{t^{p-p/q}}$	
$= \dfrac{p}{q}\,t^{p-1-p+p/q}$	

$$= \frac{p}{q}\, t^{-1+p/q}$$

$$= r\, t^{r-1}.$$

Exercise 1. Use a similar technique to show that the Power Rule gives the correct derivative for the reciprocal function, i.e.,

$$\frac{d}{dt}\frac{1}{t} = -t^{-2}.$$

If you write $u = t^{-1}$, then u satisfies the equation $u\,t = 1$. Carefully apply the Product Rule to differentiate both sides of this equation, and solve for $\frac{du}{dt}$.

Exercise 2. If you calculate the derivative of the reciprocal function directly from a difference quotient, you have to simplify

$$\frac{\Delta u}{\Delta t} = \frac{\dfrac{1}{t+\Delta t} - \dfrac{1}{t}}{\Delta t}.$$

Do the necessary simplification, and show that this leads to the same result as in Exercise 1.

Exercise 3. Combine the previous results — the Power Rule for positive rational powers and the Power Rule for the -1 power — to derive the Power Rule for negative rational powers. Notation is important for making sense of this task. Suppose you want to differentiate t^r where r is a negative rational number. Then $r = -s$, where s is positive (in fact, s is the absolute value of r), so we know the Power Rule for t^s. If we write $u = t^r = t^{-s}$, then we can take reciprocals on both sides and write $u^{-1} = t^s$. Now differentiate to find $\frac{du}{dt}$, and then express the answer in terms of t and r.

Power Functions with Irrational Exponents

We turn now to the question of what might be meant by t^{π} or $t^{\sqrt{2}}$ or $t^{\sqrt{\pi}}$ or $t^{-\pi/4}$. When we can answer that question, we will also see that we can calculate the derivatives of such functions with rules that are mostly known to us already. Even though the techniques will turn out to be different from the algebraic calculations we have done already, the answer will be the *same* — the Power Rule.

We can "sneak up" on t^π by considering a sequence of rational approximations to π, namely, the successive decimal approximations to π:

$$3, \ 3.1, \ 3.14, \ 3.141, \ 3.1415, \ldots .$$

Each number in this list is *rational*, i.e., is the ratio of two integers. Thus, at least for positive values of t, it makes sense to use each of these numbers as an exponent on t. For example, $t^{3.14}$ means the same thing as $t^{314/100}$, which means the 100-th root of the 314-th power of t.

Exercise 4. *(a) Write an "operational" definition of* $t^{3.1415}$ *in terms of integer powers and roots.*

(b) Why did we restrict attention to positive *values of* t *for doing these operations?*

Now that we understand what is meant by an exponent that is a decimal fraction, we consider the graphs of

$$t^3, \ t^{3.1}, \ t^{3.14}, \ t^{3.141}, \ \text{and} \ t^{3.1415};$$

see Figure 27, in which the first four of these functions are graphed with dotted lines and the last with a solid line. Because each exponent is slightly larger than the one before, the order on the graph is "bottom to top." In fact, on the interval shown in Figure 27, only the first two of these power functions are visually distinct from the last.

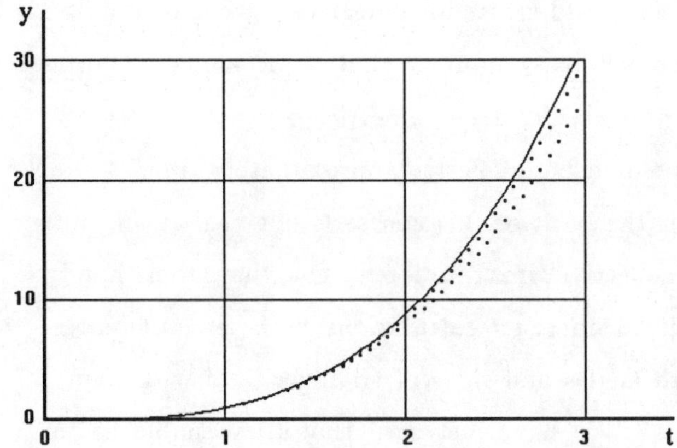

Figure 27. **Successive power functions approximating** t^π.

The power functions (the five shown and all the later ones determined by longer decimal approximations to π) *are* distinct, but the differences between them quickly get smaller and smaller, which suggests that there is a "limiting function" that the power functions are getting closer to as the powers get closer to π. That limiting function is what we call t^π. If we asked the computer to add the graph of t^π to Figure 27, we would see no change in the figure; the limiting function is visually indistinguishable from the power function shown already as a solid line.

Exercise 5. Write out formulas for the derivatives of each of the following functions:

$$t^3, \ t^{3.1}, \ t^{3.14}, \ t^{3.141}, \ and \ t^{3.1415}.$$

You know how to do this because each of the exponents is a rational number. Now use Figure 27 to explain in your own words why $\frac{d}{dt}\, t^\pi = \pi t^{\pi-1}$.

Symbols

We close this section with another observation about the power and convenience of abstract symbols. Every explicit calculation you will ever do with numbers — whether with pencil and paper, calculator, or computer — will be done with *rational numbers*. From the point of view of practical calculation, any other numbers are fictional. In particular, we can never use an exact value of π, or 2^π, or even $2^{3.1415}$ in an arithmetic calculation. But think of the extra work we would create for ourselves if we had to write out a rational approximation for every number that came along. Worse than that, all our familiar algebraic, trigonometric, and other "rules" (including those of calculus) would be, at best, "approximately true." The "real number system" that is the basis of this course is not *real* at all, but merely a figment of our collective imagination. On the other hand, irrational numbers enable us to do *exact* calculations with circumferences and diameters of circles, with angles and sides of triangles, and with many other important relationships. We have just seen that they enable us to talk about power functions without worrying about whether base, exponent, or value is rational — and the key derivative formula remains the same for all exponents, rational or not.

Summary

Much of our effort in this chapter has been devoted to developing the body of computational formulas that make "calculus" a CALCULUS, i.e., a system of calculation. Of course, most of us don't calculate just for the fun of it — we calculate in order to *solve problems*. Thus, our development of calculational tools has been embedded in the problem-solving process for coming to grips with, and solving, meaningful problems.

We now have developed all of the general rules for differentiation. These include the **"Constant Multiple Rule"**, the **"Sum Rule"**, the **"General Power Rule"**, the **"Product Rule"**, and the **Chain Rule"**. The last three of these were new in this chapter.

We introduced second (and higher) derivatives and considered the relationship between the graph of a function and the values of its derivatives. That led us to discover the importance of points at which the first derivative is zero (**"critical points"**) and at which the second derivative is zero (**"inflection points"** — usually). This brought us *back* to a problem encountered in the previous chapter, that of finding numeric solutions of equations in one variable. We saw, by putting the "local linearity" property to work, that we had an elegant way to quickly "sneak up on" a solution — **Newton's Method:**

$$t_{n+1} \ = \ t_n \ - \ \frac{f(t_n)}{f'(t_n)}, \quad \text{for} \quad n = 0, \ 1, \ 2, \ \ldots \ .$$

Compared to the zooming method of the previous chapter, Newton's Method is generally much faster for a given accuracy. However, the price for that speed is that a derivative has to be computed — and that brought us back to the need for more tools for calculating derivatives.

In particular, in order to apply Newton's Method to the root-finding task embedded in the homicide problem in Chapter 3, we had to differentiate a product of two functions whose derivatives we already knew. That led to the **"Product Rule."**

Next we turned to the physical properties of light rays to give a mathematical analysis of a natural "law" that takes the form of an optimization principle — Fermat's observation that light rays follow a path that minimizes travel time. Notice how this differs from the Newtonian "law" of motion on which we based our discussion of the falling body. Newton's laws of motion are statements about derivatives (e.g., acceleration) from which we obtain information about the function whose derivative is given (e.g., velocity). That requires "undoing" the derivative information, i.e., solving a differential equation. Fermat's law, on the other hand, makes an assertion about a travel time function — not about a rate of change — and analysis of the minimization statement calls for calculating a derivative. In general, we expect calculating derivatives to be easier than solving differential equations, but Fermat's principle forced us to confront harder derivative calculations than we had seen before — indeed, to develop some new formulas.

In particular, analysis of the angle of reflection led us naturally to the most important derivative formula of all, the **"Chain Rule."** When combined with the formulas known already, plus a direct calculation of the derivative of the square root function, the chain rule provided the tool we needed for differentiating the travel time functions for both reflection and refraction.

We examined the meaning of **"differential"** as a free-standing object. We found that we could interpret the differential of a dependent variable to mean "rise" to a tangent line, as opposed to *difference* of a dependent variable, which is the rise to a point on the graph. For small values of the "run," the tangent line and the curve are close together, so "differential" and "difference" approximate each other. We used that approximation to estimate the volume of the Earth's crust — modeled as a *difference* of the volumes of two concentric spheres — and we saw that the corresponding *differential* was easier to calculate. In the next chapter, we will use this approximation in the other direction. In particular, we will see that we can generate approximate solutions to a *differential* equation by replacing the differentials by *differences.*

Derivative Formulas

Derivatives of exponential functions (Chapter 2):

$$\frac{d}{dt}\,e^t \;=\; e^t, \quad \text{where } e \text{ is the natural base, } 2.71828\ldots. \tag{34}$$

$$\frac{d}{dt}\,e^{kt} \;=\; k\,e^{kt}, \quad \text{for any constant } k. \tag{35}$$

$$\frac{d}{dt}\,b^t \;=\; (\ln b)\,b^t, \quad \text{for any constant exponential base } b. \tag{36}$$

Derivatives of polynomial functions (Chapter 2):

Use the Sum Rule to differentiate term-by-term and use the Constant Multiple Rule and Power Rule on each term:

$$\frac{d}{dt}\,ct^n \;=\; cnt^{n-1}, \quad \text{for any power } n \text{ and any constant } c. \tag{37}$$

Derivative of the natural logarithm:

$$\frac{d}{dt}\,\ln t = \frac{1}{t}. \tag{38}$$

Derivatives of power functions:

$$\text{Power Rule:} \quad \frac{d}{dt}\,t^r \;=\; r\,t^{r-1}, \quad \text{for any real constant } r. \tag{39}$$

Scaling formulas for independent and dependent variables:

If $T = kt$, where k is constant, then (from Chapter 2)

$$\frac{d}{dt}\,f(kt) \;=\; k\,\frac{d}{dT}\,f(T). \tag{40}$$

$$\textbf{CHAIN RULE:} \quad \frac{dy}{dt} \;=\; \frac{dy}{du}\frac{du}{dt}. \tag{41}$$

Constant Factor Rule (from Chapter 2):

$$\frac{d}{dt}\,Af(t) \;=\; A\,\frac{d}{dt}\,f(t), \quad \text{for any constant } A. \tag{42}$$

Formulas that apply to combinations of functions:

$$\text{Sum Rule:} \qquad \frac{d}{dt}\big[f(t)+g(t)\big] \;=\; \frac{d}{dt}\,f(t)+\frac{d}{dt}\,g(t) \tag{43}$$

$$\text{Product Rule:} \qquad \frac{d}{dt}\big[g(t)\cdot h(t)\big] \;=\; g(t)\cdot\frac{d}{dt}\,h(t)+h(t)\cdot\frac{d}{dt}\,g(t). \tag{44}$$

$$\text{Power Rule:} \qquad \frac{d}{dt}\,f(t)^{r} \;=\; r\,f(t)^{r-1}\frac{d}{dt}\,f(t), \tag{45}$$

for any real constant r.

$$\text{Exponential Rule:} \quad \frac{d}{dt}\,e^{f(t)} \;=\; e^{f(t)}\,\frac{d}{dt}\,f(t). \tag{46}$$

Practice with Calculations

*Exercises in this category include (a) calculations that can be done by machines and (b) practice on important topics from courses that precede calculus. You need to develop **facility** with both categories — not because such calculations are a central feature of the course, but because they should not frustrate you or keep you from concentrating on the more important parts of the course by occupying a lot of your time or by leading to lots of mistakes. Even though routine calculations can be done quickly and accurately by computer or calculator, you need to develop judgment about when to use a machine and when not to, and you need to know how to tell when you might have pressed the wrong button. These skills are acquired and sharpened by **practice**.*

You should expect to see exercises like these as some portion of your homework assignments, quizzes, and tests.

For each of the following functions, find $\dfrac{dy}{dx}$.

1. $y = xe^{-x}$

2. $y = e^{(x^2 + x)}$

3. $y = (2x^3 + 2x)^2$

4. $y = \ln\left(x^2 + 1\right)$

5. $y = 2^x\left(x^2 + 1\right)$

6. $y = v^2 + v$, where $v = \sqrt{u^2 + 1}$ and $u = x^3 - x$

For each of the following functions, find $\dfrac{dy}{dt}$.

7. $y = 7t^3 - 5t^2 + 13t - \sqrt{2}$

8. $y = e^{t^2}$

9. $y = \sqrt{t^3 + 1}$

Calculate each of the following derivatives:

10. $\dfrac{d}{dt}\left(6t^5 - 4t^3 + 7\right)$

11. $\dfrac{d}{dw}\dfrac{1}{w^4}$

12. $\dfrac{d}{dt}\, t^{\frac{3}{5}}$

13. $\dfrac{d}{dt}\left(\dfrac{3}{5}\right)^t$

14. $\dfrac{d}{dx}\, x\, e^x$

15. $\dfrac{d^2}{dx^2}\, x\, e^x$

16. $\dfrac{d}{dy}\sqrt{y + 1}$

17. $\dfrac{d}{dx}\sqrt{x^2 + 1}$

18. $\dfrac{d}{dx}(x \ln x + 1)^{3/2}$

19. $\dfrac{d}{dw}\left(7w^7 - 3w^3 + 8\right)$

20. $\dfrac{d}{dx}\sqrt{x^5 + x^2 + 1}$

21. $\dfrac{d}{dy}\,(y + 1)e^{-2y}$

22. $\dfrac{d}{dx}\ln\left(x^2 + 2\right)$

23. $\dfrac{d}{dt}\dfrac{(\ln t)^3}{t}$

24. $\dfrac{d}{dt}\ln 5t$

25. $\dfrac{d}{dx}\dfrac{1}{(x + 1)^2}$

Find the derivative of each of the following functions:

26. $f(x) = x^3 + 4x^2 - 2x - 7$

27. $g(t) = t - e^{-3t}$

Find the second derivative of each of these functions:

28. $f(x) = x^3 + 4x^2 - 2x - 7$

29. $g(t) = t - e^{-3t}$

30. Given $y = (1 + t^2)(t^3 - 3t^2 + 1)$, calculate $\dfrac{dy}{dt}$ two ways:

(a) using the Product Rule;

(b) multiplying the two factors before differentiating.

Verify that the two answers you get are equivalent.

31. If $f(x) = x$, find $f'(5)$.

32. If $z = y^3 + y + 2$ and $y = 3x + 1$, find $\dfrac{dz}{dx}$.

Suppose f and g are functions such that $f'(2) = 4$, $g'(2) = -3$, $f(2) = -1$, $g(2) = 1$, $f'(1) = 2$, and $g'(-1) = 5$. For each of the following functions, find the value of the derivative at $x = 2$:

33.[15] $s(x) = f(x) + g(x)$ **34.** $p(x) = f(x)g(x)$

35. $q(x) = f(x)/g(x)$ **36.** $h(x) = f(g(x))$

37. $k(x) = g(f(x))$

38. Suppose $y = f(u(t))$, $f'(1) = 2$, $f'(3) = 8$, $f'(5) = 13$, $u(1) = 3$, $u(3) = 20$, $u'(1) = 5$, and $u'(3) = 51$. Find $\dfrac{dy}{dt}$ at $t = 1$.

Find (approximately) a number x that solves each of the following equations:

39. $x = 5 \ln x$ **40.** $x^3 + 5x = 10$

---

[15]Exercises 32-36 are adapted from *Calculus Problems for a New Century*, edited by Robert Fraga, MAA Notes Number 28, 1993.

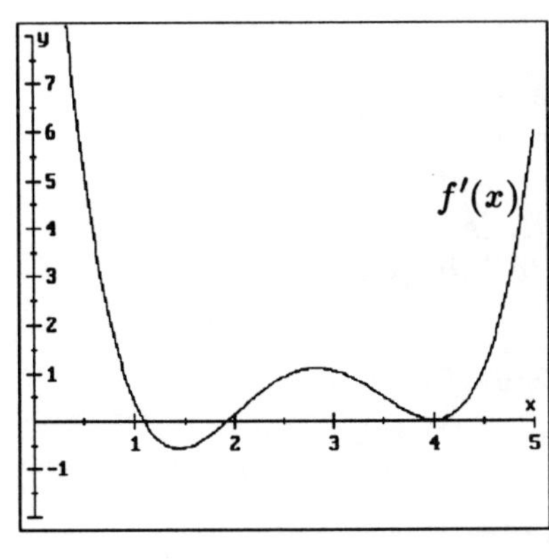

41. The figure at the left shows the graph of the *derivative* $f'(x)$ of an unknown function $f(x)$.

(a) From the figures $\boxed{A}$–$\boxed{I}$ below, choose the one that shows the graph of $f(x)$.

(b) From the figures $\boxed{A}$–$\boxed{I}$ below, choose the one that shows the graph of $f''(x)$.

Justify each of your choices in parts (a) and (b).

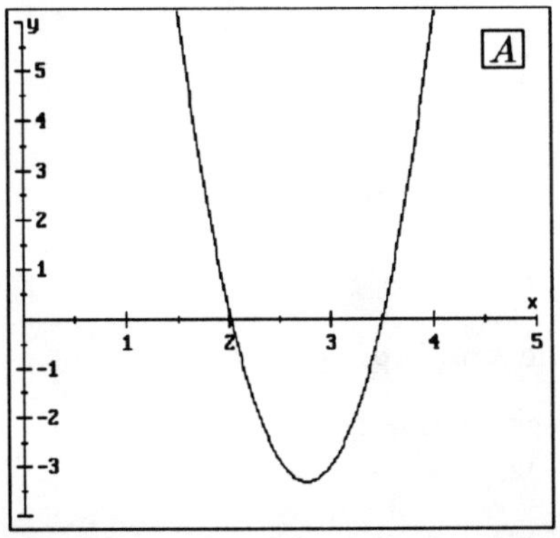

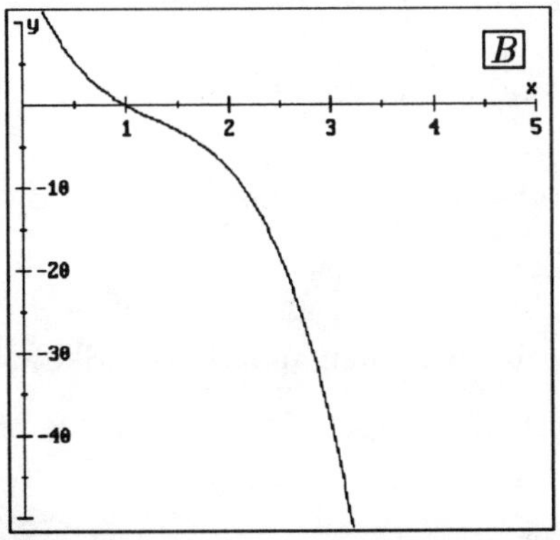

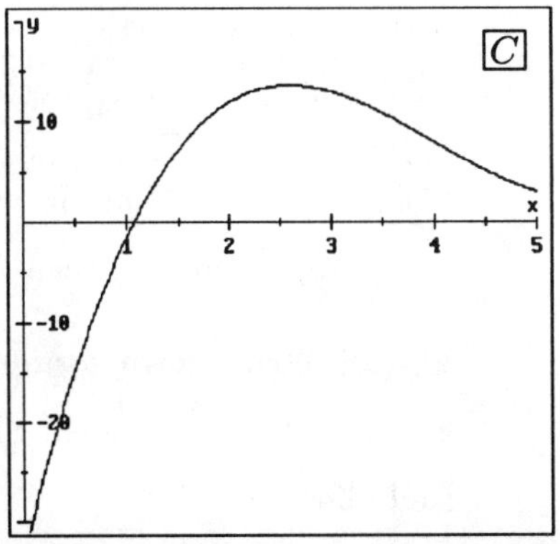

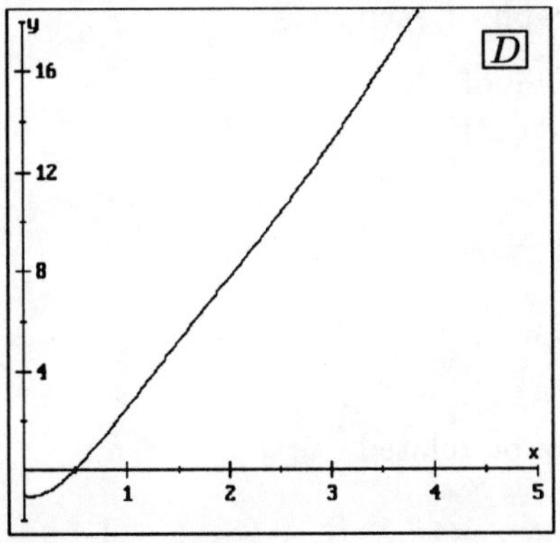

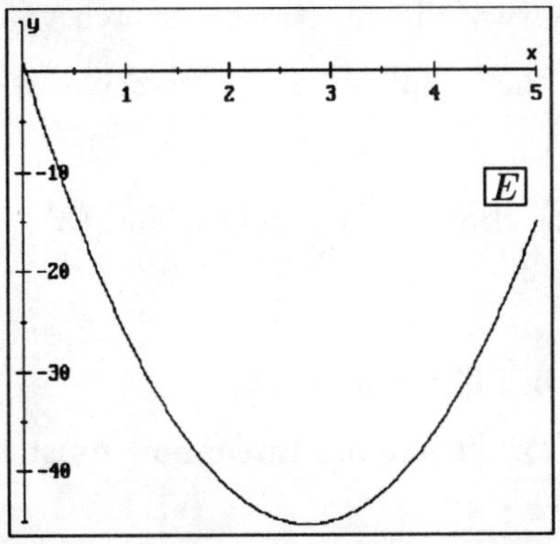

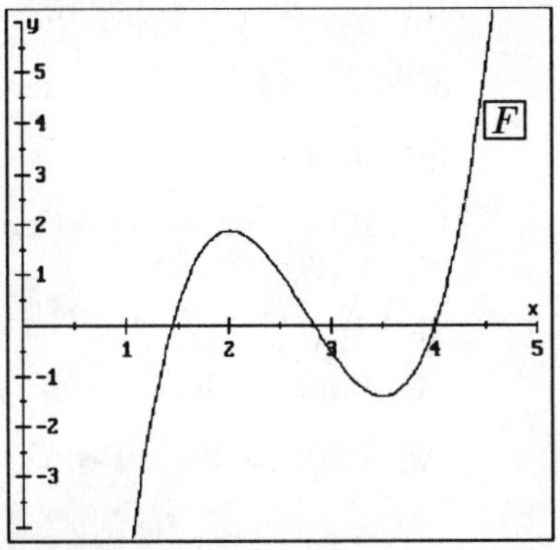

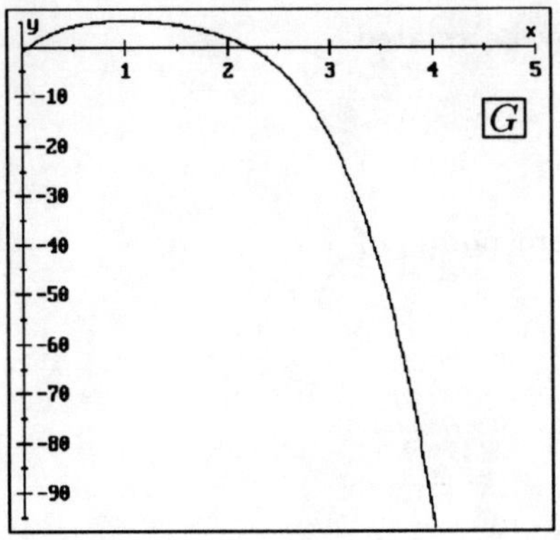

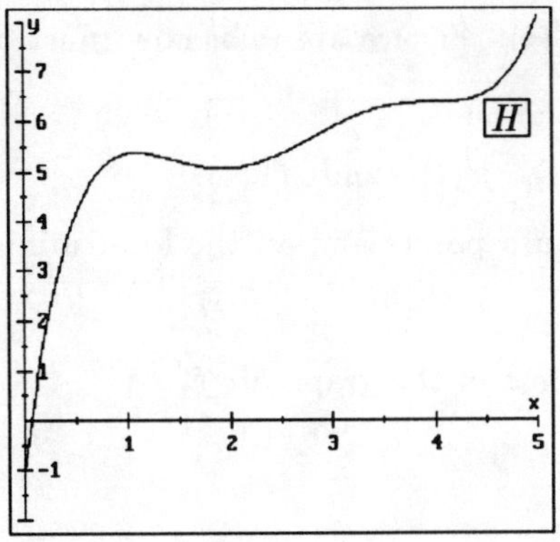

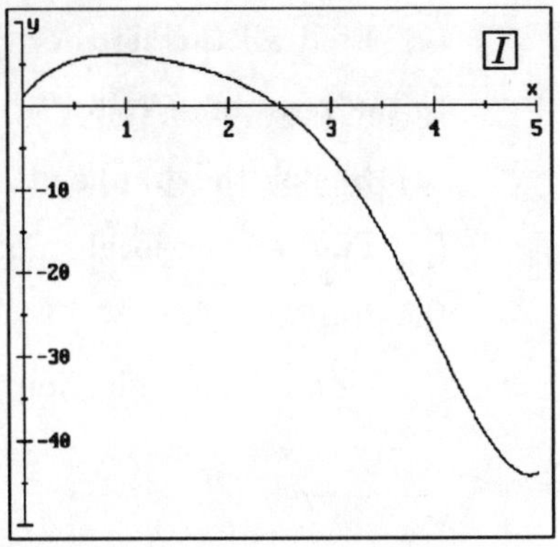

Conceptual Exercises

Exercises in this category use and/or further develop the concepts introduced in this and earlier chapters. The level of difficulty is similar to that of the exercises embedded in the text and at the ends of sections. While many of these exercises are presented in "realistic" contexts, they are not representative of real problems. Rather, these exercises are to help you get ready for tackling real problems.

You should expect to see exercises like these as a major portion of your homework assignments, quizzes, and tests.

1. Use the ideas developed in Section 4.1 (as many as apply) to sketch the graph of each of the following functions (see Calculations 26-29):

(a) $f(x) = x^3 + 4x^2 - 2x - 7$;

(b) $g(t) = t - e^{-3t}$.

Compare your hand-drawn graphs with graphs drawn by calculator or computer. (Do your own first!)

2. (a) Write down formulas for two even functions. Sketch the graph of each of these functions. Sketch the graph of the derivative of each of these functions.

(b) Write down formulas for two odd functions. Sketch the graph of each of these functions. Sketch the graph of the derivative of each of these functions.

(c) What conclusion(s) do you draw from parts (a) and (b)?

3. Let $f(t) = t^4 - 10t^2 + e^{2t}$.

(a) Find all the zeros of $f(t)$. (There are four.)

(b) Find all the zeros of $f'(t)$. [There are three; how must they be related to the zeros of $f(t)$?]

(c) Find all the zeros of $f''(t)$. [There are two; how must they be related to the zeros of $f'(t)$?]

(d) Sketch the graphs of $f(t)$, $f'(t)$, and $f''(t)$.

(e) Find all the local maximum points and all the local minimum points of the graph of f.

(f) Find all the inflection points of the graph of f.

4. (a) Calculate $\dfrac{dy}{dt}$ if $y = (1 + 2t)\, e^{-3t}$.

(b) Check your formula for the derivative at three different values of t by calculating $\dfrac{\Delta y}{\Delta t}$ for an appropriately small Δt.

5. (a) Find a function of t whose derivative is t.

(b) Find a function whose derivative is t^2.

(c) Find a function whose derivative is t^3.

(d) Find a function whose derivative is t^n, where n is any positive integer. Verify your result by showing that the derivative of your "answer" function really is t^n.

(e) Does your answer to part (d) work if $n = 0$? Why or why not?

6. In the previous exercise, we used the phrase "a function whose derivative is" several times. Such a function is called an **"antiderivative,"** and the process of finding it (i.e., of undoing differentiation) is called **"antidifferentiation."**

(a) Find three *different* antiderivatives of t^7.

(b) Formulate a general principle that enables you to find infinitely many antiderivatives of a given function, once you know one of them.

7. Find an antiderivative of the function e^{kx}, where k is any non-zero constant, and x is the independent variable. [Hint: Guess the form of such a function, differentiate it, and compare the derivative to e^{kx}. Then adjust your guess.]

8. Suppose you know antiderivatives for the functions $f(t)$ and $g(t)$. How would you find an antiderivative for $f(t) + g(t)$? Use a differentiation rule to justify your answer.

9. Suppose you know an antiderivative for the function $f(t)$. How would you find an antiderivative for $c \cdot f(t)$, where c is a constant? Use a differentiation rule to justify your answer.

10. Find an antiderivative of each of the following functions:

(a) $f(x) = x^3 + 4x^2 - 2x - 7$;

(b) $g(t) = t - e^{-3t}$.

11. Imagine the graph of a polynomial of degree 3 or higher. (If you have difficulty imagining one, draw one in the margin.) Imagine that this graph is a map of a (relatively flat) road. Imagine that you are riding along this road (in the positive x-direction) on a bicycle.

(a) Describe what you experience as you pass through a point at which the second derivative of the polynomial is zero.

(b) What are you doing along a stretch of the road on which the second derivative is positive?

(c) What are you doing along a stretch of the road on which the second derivative is negative?

12.[16] If $y = f(u)$ and $u = g(x)$, then $y = f(g(x))$; this last function is called the "**composite**" of f and g.[17] For example, if $f(u) = u^2$ and $g(x) = x^3 + x$, then $f(g(x)) = (x^3 + x)^2$. What can be said about the oddness or evenness of the composite of f and g if

(a) f and g are both odd;

(b) f and g are both even;

(c) f is odd and g is even;

(d) f is even and g is odd?

Give an example for each of the cases in parts (a) through (d); verify that the composite in each case has the oddness or evenness property you said it should have.

13. Water is the only common liquid whose greatest density occurs at a temperature above its freezing point. (This phenomenon favors the survival of aquatic life by preventing ice from forming at the bottoms of lakes.) According to the *Handbook of Chemistry and Physics*, a mass of water that occupies one liter at $0\,°C$ occupies a volume of $1 + aT + bT^2 + cT^3$ liters at $T\,°C$, where $0 \leq T \leq 30$, and where the coefficients are

$$a = -6.42 \times 10^{-5}, \quad b = 8.51 \times 10^{-6}, \quad \text{and} \quad c = -6.79 \times 10^{-8}.$$

Find the temperature between $0\,°C$ and $30\,°C$ at which the density of water is greatest.

---

[16]Adapted from *Calculus Problems for a New Century*, edited by Robert Fraga, MAA Notes Number 28, 1993.

[17]The Chain Rule tells us how to differentiate composite functions.

14. Suppose $z^2 = x^2 + y^2$, where x, y, and z are all functions of t. Find $\frac{dz}{dt}$ in terms of x, y, $\frac{dx}{dt}$, and $\frac{dy}{dt}$.

15. Suppose a can (a "right circular cylinder" to a mathematics student) is to be made to hold 63 in^3 of Blastola Cola. The material for the sides and bottom costs 0.12¢/in^2, and the material for the easy-open top costs 0.5¢/in^2.

(a) What should the height and radius of the can be to minimize the cost of the material?

(b) What is the minimum cost?

16. Show that $3x^2 - 3x + 1$ is never negative. (Graphing the function with your calculator is not enough. You have to justify your answer.)

17. In a recent (fictional) study, scientists formulated the following equation to represent the amount i of information retained from one day of studying:

$$i = k s c^{3/2},$$

where s is the number of hours slept, c is the number of hours spent cramming, and k is a proportionality constant. Assuming that the average freshman prepares for finals by cramming every hour that she or he is not sleeping, how much should a student sleep in the course of a day to maximize the knowledge retained? Justify that your answer yields a maximum.

18. If Newton's method is applied to $f(x) = 3x + 4$, and x_0 is chosen to be 15, what will x_{17} be?

19. Find the point between 0 and 4 where $f(x) = x^3 - 3x^2 + 9x + 5$ is the smallest. Explain your answer.

20. Find at least one solution to the equation $4x^3 - x^4 = 30$, or explain why no such solution exists.

21. For what point (or points) on the graph of $y = 3 - x^2$ does the tangent line go through the point $(2,0)$?

22. (a) Suppose you are using Newton's method to find a root of $f(x) = 0$ for the function f whose graph is shown in Figure 28. If your starting point is at $x_0 = 2.5$, label where (approximately) x_2 will be.

(b) Suppose you try again with Newton's method to find a root of $f(x) = 0$ for the same function f. If your starting point is at $x_0 = 5$, label where (approximately) x_2 will be.

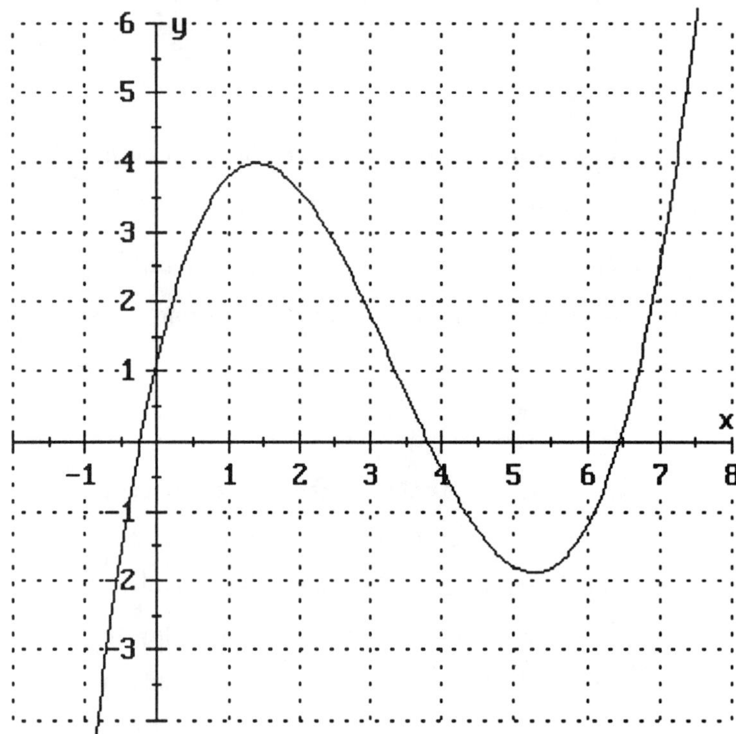

Figure 28. Start Newton's Method (a) at 2.5 and (b) at 5.

23.[18] A ship is moving directly away from shore at a speed of 15 knots. A person on the ship is walking toward the bow at a speed of 3 knots.

(a) How fast is this person moving away from shore?

(b) What differentiation rule does this problem illustrate?

---

[18]Adapted from *Calculus Problems for a New Century*, edited by Robert Fraga, MAA Notes Number 28, 1993.

24. The graphs of three functions appear in Figure 29. Identify which is the graph of f, which is the graph of f', and which is the graph of f''.

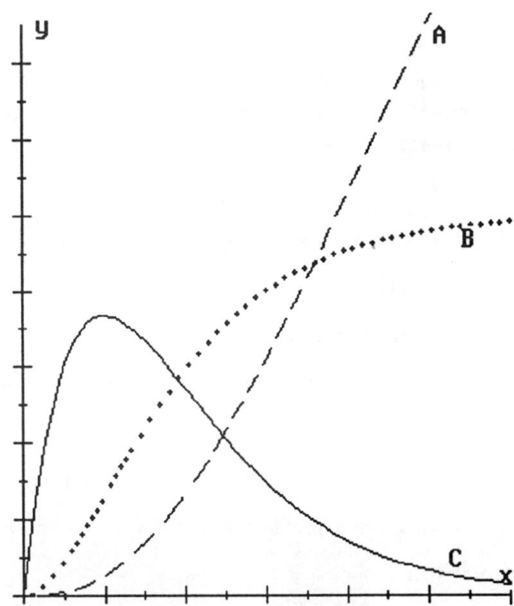

Figure 29. Identify the function f and its first two derivatives.

25. When you cough, your windpipe contracts. The velocity v at which air comes out depends on the radius r of your windpipe. If R is the normal (rest) radius of your windpipe, then (for each possible radius r), $v = a(R - r)r^2$, where a is a constant.

(a) What value of r maximizes the velocity?

(b) What is that maximum velocity?

(Both answers may involve either or both of the constants a and R.)

26. Sketch the graph of a function that is

(a) increasing at an increasing rate;

(b) increasing at a decreasing rate;

(c) decreasing at an increasing rate;

(d) decreasing at a decreasing rate.

Problems and Projects

The problems presented here are not intended for individual homework assignments or for tests. These problems should be attempted by groups of two to four students sharing ideas, whether in the classroom or elsewhere. Some of the more extensive problems are suitable for projects with time frames ranging from a class period to a week.

1. (a) Explain in geometric terms (i.e., by using symmetries and slopes) why the derivative of an odd function is an even function and why the derivative of an even function is an odd function. (See Exercises 2, 4, 5, and 6 in Section 4.1 and Conceptual Exercise 2 for examples that illustrate both of these statements.)

(b) Explain in algebraic terms (i.e., by using difference quotients) why the derivative of an odd function is an even function and why the derivative of an even function is an odd function.

(c) Explain in calculus terms (i.e., by using the Chain Rule) why the derivative of an odd function is an even function and why the derivative of an even function is an odd function.

2. The square root of a number A can be calculated by solving the equation $x^2 - A = 0$ numerically.

(a) Show that application of Newton's method to this equation leads to a simple "repeated averaging" scheme. What is being averaged at each step?

(b) Start with $x_0 = 1$, and compute $\sqrt{7}$ by this scheme. Compare your result with that obtained from the square root key on your calculator. Could your calculator actually be doing repeated averaging?

3. Figure 30 repeats Figure 1 in Chapter 2; it depicts Thomas Malthus' view of the eventual population crisis, with population growing exponentially and food supply growing linearly. Suppose we give Malthus the benefit of the doubt on both growth rates and try to determine when the food supply would become inadequate. We set $t = 0$ in 1800, and we take one unit on the food scale to mean just enough food to feed the entire population in that year. The unit of population is the population in 1800, i.e., $P(0) = 1$. We interpret the food function $F(t)$ to mean *potential* for producing food at a given time; if we were to zoom in on the graph, we might find that a reasonable initial value is $F(0) = 5$. Suppose the population grows at a rate of 2% per year (compounded continuously), and the slope of the food function is 0.1. In what year would the potential food supply be just adequate for the population? What would happen after that? In what year would the potential food surplus (relative to the population) be greatest? (Our numbers here are fictional — they should not be taken seriously.)

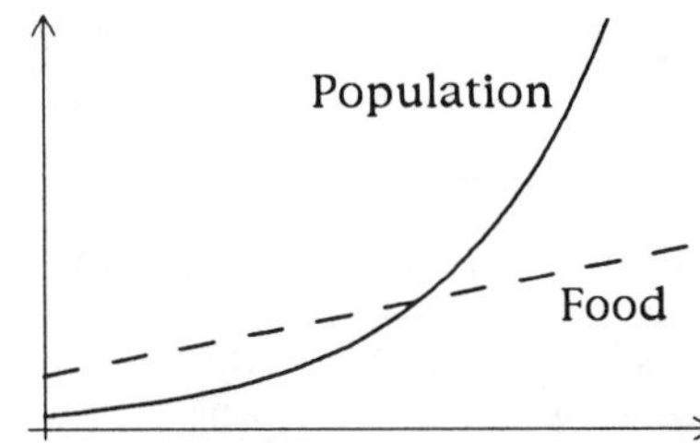

Figure 30. Malthus's view of population growth.

4. We show again Figure 17, which includes the graph (for a specific combination of p, q, and D, with $p > q$) of the length-of-path for a reflected ray as a function of where the ray would hit the mirror. (The derivative of the length-of-path function is also shown.)

(a) It is clear from this figure that $L(x)$ also has a *maximum* value on the interval $0 \le x \le D$ — and it *does not* occur at a point where $L'(x) = 0$. Where does it occur?

(b) We repeat here the equation defining $L(x)$:

$$L(x) = \sqrt{p^2 + x^2} + \sqrt{q^2 + (D-x)^2}\,, \qquad (16R)$$

for any x between 0 and D. Does your answer to part (a) depend on the relative sizes of p and q? If so, how?

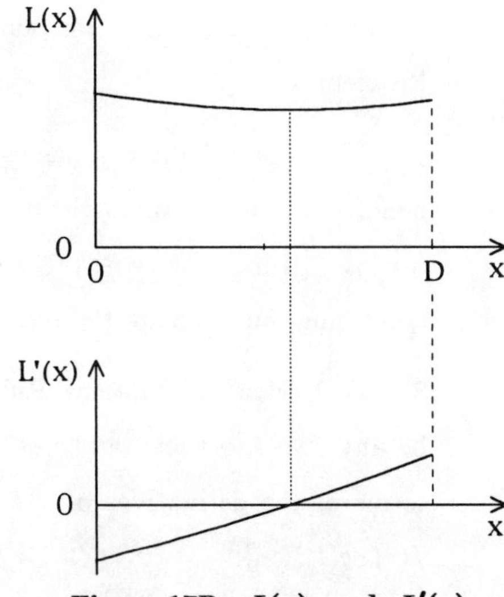

Figure 17R. $L(x)$ and $L'(x)$.

5. Here again is the formula for $L'(x)$:

$$L'(x) = \frac{x}{\sqrt{p^2 + x^2}} + \frac{x - D}{\sqrt{q^2 + (D - x)^2}}, \qquad (17R)$$

for any x between 0 and D.

(a) Find the second derivative of $L(x)$, and show that it is always positive.

(b) Explain why the unique solution of $L'(x) = 0$ must be the x-coordinate of a minimum point. [Hint: See Exercise 9 in Section 4.6. The p-term and the q-term of L' each produce two terms in the derivative; combine each pair by using a common denominator, and you should find that each numerator must be positive.]

6. Here again is the formula for the refractive time-of-travel function:

$$T(x) = \frac{\sqrt{p^2 + x^2}}{c_w} + \frac{\sqrt{q^2 + (D - x)^2}}{c_a}, \qquad (20R)$$

for any x between 0 and D. Show that the second derivative of T is always positive, and therefore that any solution of $T'(x) = 0$ is the x-coordinate of a minimum point. [Hint: See Exercise 7 in section 4.4 and Problem 5.]

7.[19] Suppose a light ray passes through a plate of glass, such as a window pane, entering at an angle α and leaving at an angle γ. Show that $\alpha = \gamma$. [Hint: Draw a picture that shows the path of the ray entering, traversing, and leaving the plate. Use Snell's Law.]

8. (a) Derive a "Quotient Rule" for differentiation. Let $f(t)$ and $g(t)$ be any two functions, and find a way to express the derivative of $\dfrac{f(t)}{g(t)}$ in terms of the derivatives of $f(t)$ and $g(t)$. [Hint: Write the quotient as $f(t) \cdot g(t)^{-1}$.]

(b) Use your rule to find the derivative of $\dfrac{t^2 + 1}{e^t}$.

(c) Write the function in part (b) as $(t^2 + 1)e^{-t}$, and confirm that you get the same answer for the derivative by using the Product Rule.

---

[19]Adapted from *Five Applications of Max-Min Theory from Calculus* by W. Thurmon Whitley, UMAP Module 341, COMAP, 1979.

9. In Note 12 (Section 4.7), we noted the problem of showing that "dt" means only one thing, whether t happens to be the independent or the dependent variable.

(a) First consider the case of the function $y = t$, that is, the function for which independent and dependent variables always have the same value. Sketch the graph of this function. Explain why $dy = dt$ in this case.

(b) Now consider a situation in which y is a function of u and u is a function of t. Thus, u is both an independent and a dependent variable. Explain why the Chain Rule requires that du means the same thing whether u is viewed as independent or dependent.

10. In an early laboratory, you studied the "logistic" growth equation $\frac{dP}{dt} = cP(M - P)$ as a model for growth of fruit flies. Among other things, you located the approximate size of the population at the time when it is growing most rapidly.

(a) Explain why the population must be growing most rapidly at a time at which the second derivative $\frac{d^2P}{dt^2}$ is zero.

(b) Differentiate both sides of the differential equation to find an expression for the second derivative. (The Product Rule works for differentiating the right-hand side, but you can make the computation easier if you rewrite the expression in another form first.) Be careful with your differentiation; you are differentiating with respect to t, not P, so the Chain Rule must come into play every time you run into the unknown function P.

(c) What does your expression for the second derivative tell you about the population size when the growth rate is maximal? How does this fit with your data from the laboratory?

11. In the text we gave meaning to the expression t^π as a function by treating it as a "limiting function" obtained from successive rational approximations to π. Since we knew the Power Rule applied to rational powers of t, we were able to show (Exercise 5 in Section 4.8) that the Power Rule also applies to this irrational-power function. With the aid of the Chain Rule and the rule for differentiating the natural logarithm function, you can show that the Power Rule applies to all power functions:

(a) Explain why t^r must be the same thing as $e^{r \ln t}$ for every pair of positive real numbers t and r, regardless of what's variable and what's constant.

(b) Now suppose t is the independent variable, and r is a constant. Explain why the function $y = t^r$ is the same as $y = e^u$, where $u = r \ln t$.

(c) Use the Chain Rule to find $\dfrac{dy}{dt}$, and simplify to show that the result is the same as the Power Rule.

12. Develop the following argument to establish the Product Rule by using the Chain Rule and without resorting to difference quotient calculations:

(a) Let $x = x(t)$ and $y = y(t)$ be any two functions. Expand the square $(x + y)^2$ to find an equal expression in terms of x^2, y^2, and xy.

(b) Solve your equation for xy.

(c) The other side of the equation now contains only squares of unknown functions, which you can differentiate with the aid of the Chain Rule. Thus, when you differentiate both sides, you should find an expression for $\dfrac{d}{dt}\, xy$. With any luck (skill?), it will turn out to be the Product Rule.

13.[20] When a drug is injected into the bloodstream, its concentration $C(t)$ at t minutes after injection is given by a formula of the form

$$C(t) \;=\; K\, \frac{e^{-bt} - e^{-at}}{a - b}\,,$$

where K, a, and b are positive constants, and $a > b$. When does the maximum concentration occur?

14.[20] A patient's "reaction" $R(x)$ to a drug dose of size x is given by a formula of the form

$$R(x) \;=\; A\, x^2\, (B - x),$$

where A and B are positive constants. The "sensitivity" of the patient's body to a dose of size x is defined to be $R'(x)$.

(a) What do you think the domain of x is? What is the physical meaning of the constant B? Of the constant A?

(b) For what value of x is R a maximum? (Your answer should contain the constant B.)

---

[20]Adapted from *Calculus Problems for a New Century*, edited by Robert Fraga, MAA Notes Number 28, 1993.

(c) What is the maximum value of R?

(d) For what value of x is the sensitivity a maximum?

(e) Why is it called "sensitivity"?

15.[20] A square plot of ground 100 feet on a side has corners labeled A, B, C, D clockwise. Pipe is to be laid in a straight line from A to a point P on the side BC and thence to C. (P could be one of the corners B or C.) The cost of laying the pipe is \$20 a foot if it goes through the lot (because it must be laid underground) and \$10 a foot if it is laid along one of the sides of the square. What is the most economical way to lay the pipe?

16.[20] A rectangular computer chip is made of ceramic material with circuitry placed in an interior rectangle. Two parallel edges of the circuitry are each 1 mm from the corresponding edge of the chip, and the other two edges are each 2 mm from the corresponding edges of the chip. How should a chip whose area is 200 square millimeters be designed in order to maximize the area of the circuitry it can accommodate?

17.[20] Lee and Dana are in a rowboat half a mile from the nearest point on a relatively straight shore line when Dana suddenly becomes ill. The closest place to find a telephone to call for help is at a seaside restaurant one mile from the nearest point on shore. Lee plans to row to a point on the shore and then jog to the restaurant. Lee can jog at 6 miles per hour and row at 1.5 miles per hour.

(a) How long would it take to get to the restaurant if Lee rows to the nearest point on shore?

(b) How long would it take if Lee rows directly to the restaurant?

(c) What point on the shore would make the trip to restaurant as quick as possible? How long would it take to get to the restaurant?

(d) How fast would Lee have to row so that the quickest route would be to row directly to the restaurant?

18.[20] A truck traveling on a flat interstate highway at a constant rate of 50 MPH gets 4 miles to the gallon. Fuel costs \$1.15 per gallon. For each mile per hour increase in speed, the truck loses a tenth of a mile per gallon in its mileage. Drivers get \$27.50 per hour in wages, and fixed costs for running the truck amount to \$12.33 per hour. What constant speed should a dispatcher require on a straight run through 260 miles of Kansas interstate to minimize the total cost of operating the truck?

19.[20] Your local pizza delivery uses square boxes that are made from rectangular pieces of corrugated cardboard, each 47 cm by 90 cm. The box is made by cutting out six small squares, three from each of the 90 cm edges, one square at each corner and one in the middle of the edge. The result is then folded into a box in the obvious fashion. (If it isn't obvious, order one with sausage and mushrooms to investigate.[21]) How does one design a box in this fashion to obtain the largest volume?

20. What is the cone of largest volume that can be formed by rotating a right triangle of fixed hypotenuse h around one of its legs? What is the volume of that cone?

21.[20] Take a circle of radius r, and cut out a sector subtended by an angle θ, producing a Pac-Man-like figure. Join the cut edges together (the upper and lower jaws of Pac-Man) to form a cone. What angle θ produces the largest volume of such a cone?

22.[20] What is the area of the largest triangle that can be formed in the first quadrant by the x-axis, the y-axis, and a tangent to the graph of $y = e^{-x}$?

23. American Flight 1003 is traveling from Minneapolis to New Orleans, and United Flight 366 is traveling from Los Angeles to New York. Both flights are at 33,000 feet, and the flight paths intersect over Ottumwa, Iowa. At 1:30 PM (Central time), the American flight is 32 nautical miles (horizontally) from Ottumwa and is approaching it on a heading of 171° at a rate of 405 knots. The United flight is 44 nautical miles from Ottumwa and is approaching it on a heading of 81° at a rate of 465 knots.

(a) At this instant, how fast is the distance between the planes decreasing?

---

[21]If you order from Domino's, you will get an octagonal box that is made from a single rectangular piece of cardboard. If you ignore the clever diagonal cuts, the basic structure of removing squares is still there, but you will have to use a little imagination to see it.

(b) How close will the planes come to each other? Will they violate the FAA's minimum separation requirement of 5 nautical miles? Will they collide if Air Traffic Control does not take action to separate them?

(c) What time will it be at the time of closest approach? Is there enough time for ATC to take appropriate action?

24. Figure 31 shows a direction field for $\dfrac{dy}{dt} = \dfrac{3}{10}\, y\,(3-y)$ for t ranging from -2 to 6 and y also ranging from -2 to 6. Figure 32 shows a direction field for $\dfrac{dy}{dt} = \dfrac{3}{10}\, t\,(3-t)$ with both t and y ranging from -2 to 5.

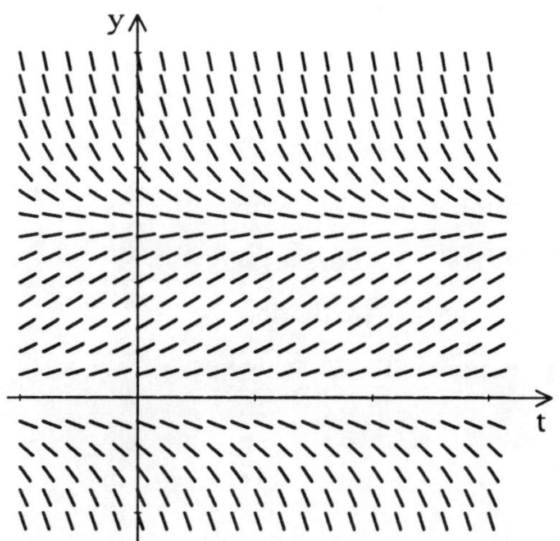

Figure 31. Direction field for $\dfrac{dy}{dt} = \dfrac{3}{10}\, y\,(3-y)$.

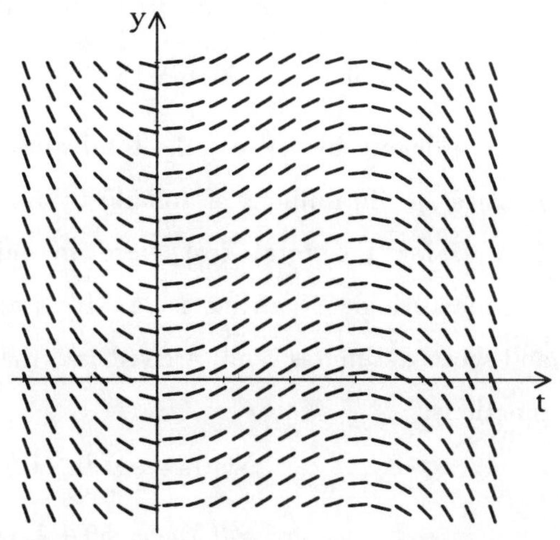

Figure 32. Direction field for $\dfrac{3}{10}\, t\,(3-t)$.

(a) On Figure 31, sketch the solution of $\dfrac{dy}{dt} = \dfrac{3}{10}\, y\,(3-y)$ for which $y(0) = 0.5$.

(b) On Figure 32, sketch the solution of $\dfrac{dy}{dt} = \dfrac{3}{10}\, t\,(3-t)$ for which $y(0) = 0.5$.

(c) Solve the initial value problem: $\dfrac{dy}{dt} = \dfrac{3}{10}\,(3t - t^2)$ with $y(0) = 0.5$.

(d) Your graph in part (b) should show that the solution in part (c) has a root near $t = 5$. Take that as a starting guess, and find the next approximation to the root from Newton's Method.

The next three problems are based on pages 6-10 of Whitley (see Note 18) who in turn based his discussion on "Pricing, Investment, and Games of Strategy" by Georges Brigham, in *Management Sciences, Models and Techniques*, C. W. Churchman and M. Verhulst (eds.), Pergamon Press, 1960. We quote the scenario from Whitley:

> "Several years ago, the Boeing Aircraft Company was faced with the problem of determining the selling price for a new model jet airliner. The basic problem was to find the price per aircraft which would maximize the company's profit. In this particular case, Boeing had one competitor, which had a similar plane. It was understood that the companies would charge the same price, since any price adjustment by one company would automatically be met by the other. Thus, the price would not affect the relative shares of the market. It could, however, have a significant impact on the total size of the market."

We denote by $N(p)$ the total number of airliners that would be sold at price p (in millions of dollars) by Boeing and its competitor. We write $C(X)$ for the total cost (also in millions of dollars) to Boeing of manufacturing X airliners. Analysts at Boeing made market predictions and cost estimates and arrived at the following expressions for these functions:

$$N(p) = -78p^2 + 655p - 1125 ; \tag{47}$$

$$C(X) = 50 + 1.5X + 8X^{3/4}. \tag{48}$$

25. (a) Find the range of prices p for which $N(p) \geq 0$.

(b) If P denotes the total profit (in millions of dollars) to Boeing, and h is the fraction (between 0 and 1) of the market to be won by Boeing, show that

$$P = phN(p) - C(hN(p)). \tag{49}$$

(c) Calculate $P'(p)$. [Hint: Use the Chain Rule.]

(d) Show that the critical price p that maximizes profit P must satisfy the condition

$$p + \frac{N(p)}{N'(p)} = C'(X), \quad \text{if } N'(p) \neq 0. \tag{50}$$

Also show that if $N'(p) = 0$ then $P'(p) \neq 0$ unless h is zero. Thus, if Boeing has any market share at all, then condition (50) must hold at the critical price.

26. (a) Calculate $N'(p)$ and $C'(X)$.

(b) Substitute into equation (50) from part (a), equation (47), and $X = hN(p)$ to find a condition on the critical price that (almost) involves only p. You can't quite get rid of h, but, in principle, one could set various values of market share and solve for the critical price. However, you may find that the equation to be solved is sufficiently complicated that you wouldn't even want to try solving it by Newton's Method. We'll try another tack first.

(c) Calculate $C'(X)$ for $X = 60$, 80, 100, and 120. You should find that C' doesn't change much over a wide range of production numbers. Why is that? [Hint: Calculate *its* rate of change, and explain why it is small when X is large.]

(d) The implication of (c) is that the right-hand side of (50) is almost constant. Select an "average" value to replace the right-hand side. Then clear the fractions to rewrite (50) as a quadratic equation in p. Now you can solve for the (approximately) optimal price with your calculator.

(e) The equation in part (d) has two solutions; explain why the profit-maximizing price is about five million dollars per plane, regardless of the number made or the market share.

27. (a) Find the maximum value of the total market function N defined by (47). What price produces the largest number of sales? How many planes will be sold at this price?

(b) How many planes can be sold by both companies at the optimal price you arrived at in the previous exercise?

(c) If Boeing makes 70 planes at the optimal price, what will their profit be? How many planes will the competitor make, and what will Boeing's market share be?

(d) (with help from a computer) For each Boeing production level X from 50 to 200 (in steps of 10), calculate $C'(X)$, solve (50) for the optimal price, and find the total production for both companies, the Boeing market share, and the Boeing profit. From your tabulated results, advise the company on their pricing and production strategies.

CHAPTER 5

Applications of Euler's Method

We begin this chapter by picking up where we left off in Chapter 4, looking at differentials in terms of "rise over run." In particular, we will see that the differential approximation to a difference of function values is also the heart of a numerical method for solving initial value problems — a method that appeared in an early lab under the title "Euler's Method." The rest of the chapter will be devoted to models (both discrete and continuous) of the behavior of epidemics and of the evolution of prices in the marketplace. In the first of these contexts, we will see that Euler's Method can be applied to initial value problems that have more than one dependent variable, a fact we will put to use in Chapter 6 as we develop a method for solving differential equations that involve the *second* derivative.

In most of the problems we have investigated up to now, we have been interested in a *quantitative* description of some phenomenon; we wanted to be able to say *what* the population would be at some time in the future, *where* the falling body was at a particular time, and *when* the murder was committed. At times we have also studied *qualitative* features of our models, for example, the approach to maximum current in the RL circuit model (Chapter 3, Problem 6), to ambient temperature in the cooling model (the homicide problem in Section 3.2), and to terminal velocity in the falling body model (Chapter 3, Problem 7). Many applications of the ideas of calculus are focused *primarily* on qualitative descriptions of phenomena. A frequent question is, "Are there reasonable assumptions that lead to behavior similar to that observed?"

Following our look at Euler's method, we begin a brief study of epidemics that is more qualitative that quantitative, in that the only numbers available are determined by data from *past* epidemics; the value of the model lies in its ability to predict qualitative behavior of future epidemics. We will use the 1968-69 Hong Kong flu epidemic as the context for our study, but we will see that the conclusions can be applied to many other epidemics. Among other things, we will be able to address such questions as, "Why is it that some virulent and highly contagious diseases (e.g., polio and smallpox) have been wiped out or controlled by inoculation and some less serious diseases (e.g., measles and rubella) haven't?"

We then turn to the study of price evolution in an economy in which prices adjust so that supply equals demand. Plausible simplifying assumptions, when incorporated into a differential equation model, lead to a problem we have already seen several times — and therefore know how to solve. Essentially the same assumptions, when incorporated into a *difference* equation model, lead to the need for new solution techniques and (under some circumstances) to very different behavior from that seen in the continuous model. That "very different behavior" will be the subject of a laboratory experience associated with this chapter. In particular, the lab will provide our first look at *unstable* behavior, a subject to which we will return in Chapter 8.

5.1 Euler's Method

Perhaps, while reading the last part of the previous chapter, you experienced that feeling expressed in the immortal words of Yogi Berra: "It was *déjà vu* all over again." We were discussing the meaning of "differential" for the first time. And yet, there was something familiar about approximating the change in the dependent variable (rise on the curve) by a simple product, slope (of a line) times run.

Think back to an early lab in which you studied a proposed model for constrained population growth. The proposal was that the growth rate for the population should be jointly proportional to the population itself and the room left for additional population. Thus, the proposal stated that the derivative — the instantaneous growth rate — of the unknown population function should have a certain form expressed in terms of the same unknown function. That's a differential equation. Since we also had a known starting population, we were attempting to solve an initial value problem. And, even though we never saw a formula for the solution function, we *did* solve the problem: We generated numerical and graphical representations of the population function over a long enough time frame to see the population rise to its maximum level. How did we do that?

Let's restate the problem in symbols so we can use the power of notation. We let t denote time and $P = P(t)$ denote the unknown population function. The growth rate proposal plus the known starting population, P_0, gives us the following initial value problem:

$$\frac{dP}{dt} = cP(M - P), \quad \text{with } P = P_0 \text{ at } t = 0, \tag{1}$$

where c and M are known constants.

Our strategy for generating approximate values of P was this: We selected a time step Δt, so we had a sequence of time values $t_0 = 0$, $t_1 = 0 + \Delta t$, $t_2 = t_1 + \Delta t = 2 \cdot \Delta t$, and so on. Starting from P_0, we wanted to compute $P_1 = P(t_1) = P_0 + \Delta P_0$, then $P_2 = P(t_2) = P_1 + \Delta P_1$, and so on. That is, to get from each P to the next one, we needed to compute a rise — to the *curve*. Because we had no way to do that exactly (and we still don't[1]), we computed the rise to the *tangent line* as slope times run:

$$dP = cP(M - P)dt \tag{2}$$

$$= cP(M - P)\Delta t,$$

since $dt = \Delta t$ for the independent variable. Then we found each "new P" as "old P plus dP" or "old P plus slope times run." In symbols, that gave us the recursive formula

$$P_{k+1} = P_k + cP_k(M - P_k)\Delta t, \quad k = 0, 1, 2, \dots. \tag{3}$$

That's **"Euler's Method"** for solving an initial value problem: Starting from a known point, we move to the next one by computing the *rise* as "slope times run," where the slope at the known point is given by the differential equation, and the run is chosen by the problem solver.

---

[1]But we *will* in Chapter 8.

In general, if the slope at every point is given by a differential equation of the form

$$\frac{dP}{dt} = f(P), \tag{4}$$

and a starting point P_0 is known, then Euler's Method computes values of an approximate solution function by the recursion

$$P_{k+1} = P_k + f(P_k)\Delta t, \quad k = 0, 1, 2, \dots . \tag{5}$$

If the right-hand side of the differential equation (4) is a function of t rather than of P, say, $g(t)$, then the Euler recursion takes the form

$$P_{k+1} = P_k + g(t_k)\Delta t, \quad k = 0, 1, 2, \dots . \tag{6}$$

The computation of the "new P" is always "old P plus slope times run," regardless of how the slope is calculated.

We give an example of the computations for the differential equation

$$\frac{dP}{dt} = P + 1$$

with the initial condition

$$P(0) = P_0 = 2.$$

To illustrate the process, we calculate the first three steps in the approximation with $\Delta t = 0.1$:

$$P_0 = 2,$$

$$P_1 = P_0 + (P_0 + 1)\Delta t = 2 + 2(0.1) = 2.2,$$

$$P_2 = P_1 + (P_1 + 1)\Delta t = 2.2 + (3.2)(0.1) = 2.52,$$

$$P_3 = P_2 + (P_2 + 1)\Delta t = 2.52 + (3.52)(0.1) = 2.872.$$

As long as the run is sufficiently small, Euler's Method succeeds because $dP \approx \Delta P$. Thus, the method can be viewed as repeated application of the differential approximation. In the next section we apply this approximation scheme to the study of epidemics.

Section Summary

Finish the following description of Euler's Method for the initial value problem

$$\frac{dP}{dt} = f(P), \text{ with } P(0) = P_0.$$

If we set $t_0 = 0$, $t_1 = \Delta t$, $t_2 = 2\Delta t$, $t_3 =$ __________ , ... ,

and

$P(t_1) = P_1 = P_0 + f(P_0)\Delta t,$

$P(t_2) = P_2 =$ __________________

$P(t_3) = P_3 =$ __________________   (and so on)

Exercises

1. Given the differential equation $\dfrac{dP}{dt} = P^2$ and the initial condition $P(0) = P_0 = 1$, use Euler's Method to generate P_1, P_2, and P_3 with a time step of $\Delta t = 0.1$.

2. Given the differential equation $\dfrac{dP}{dt} = P^2$ and the initial condition $P(0) = P_0 = 1$, use Euler's Method to generate P_1, P_2, and P_3 with a time step of $\Delta t = 0.01$.

3. Given the differential equation $\dfrac{dP}{dt} = t^2$ and the initial condition $P(0) = P_0 = 1$, use Euler's Method to generate P_1, P_2, and P_3 with a time step of $\Delta t = 0.1$.

4. Given the differential equation $\dfrac{dP}{dt} = t^2$ and the initial condition $P(0) = P_0 = 1$, use Euler's Method to generate P_1, P_2, and P_3 with a time step of $\Delta t = 0.01$.

5. Given the initial value problem $\dfrac{dP}{dt} = t^2 + 1$ with $P(0) = P_0 = 2$, use Euler's Method to generate P_1, P_2, and P_3 with a time step of $\Delta t = 0.1$.

6. Given the initial value problem $\dfrac{dP}{dt} = P^2 + 1$ with $P(0) = P_0 = 2$, use Euler's Method to generate P_1, P_2, and P_3 with a time step of $\Delta t = 0.1$.

5.2 Modeling Epidemics

Hong Kong Flu

During the winter of 1968-69, the United States was swept by a virulent new strain of influenza, named "Hong Kong flu" for its place of discovery. We will study and model the spread of the disease through a single urban population, that of New York City. The data available to us (Figure 1 and Table 1) consists of weekly totals of "observed excess pneumonia-influenza deaths," that is, the number of such deaths in excess of the average number to be expected from other sources.

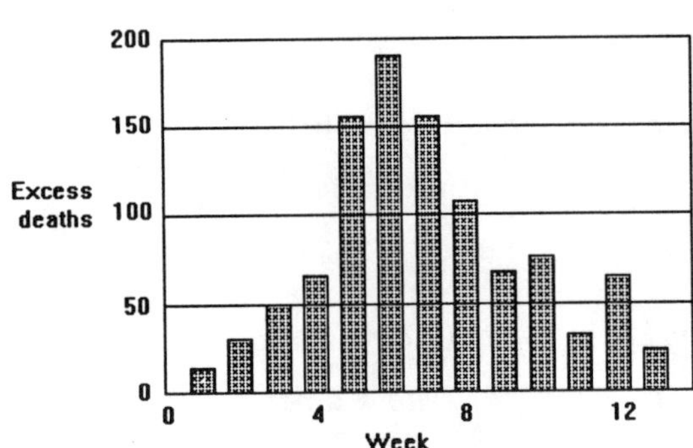

Figure 1. Flu-related deaths.

WEEK	EXCESS DEATHS
1	14
2	28
3	50
4	66
5	156
6	190
7	156
8	108
9	68
10	77
11	33
12	65
13	24

Table 1. Flu-related deaths.[2]

Relatively few flu sufferers died from the disease or its complications. However, we may reasonably assume that the number of excess deaths in a week was proportional to the number of new cases of flu in some earlier week, say, three weeks earlier. Thus, the figures in Table 1 reflect (proportionally) the rise and subsequent decline in the number of new cases of Hong Kong flu. We want to model the spread of such a disease so we can predict what might happen with similar epidemics in the future.

---

[2]Source: Centers for Disease Control, as quoted in "Mathematical Models for Urban-Suburban Spread of Disease" by Geoffrey M. Davis, a paper submitted for Graduation with Distinction from Duke University, 4/11/88.

In a flu epidemic, what is changing? What do we want to measure from the past or present or predict about the future? At any given time, we might want to know the number of people who are infected. We also need to know the number of individuals who have been infected and have recovered; these people now have an immunity to the disease.[3] If we ignore movement into and out of the infected area, then the remainder of the population is still susceptible to the disease. So at any time, the fixed total population (approximately 7,900,000 in the case of New York City) may be divided into three distinct groups: those who are *infected*, those who have *recovered*, and those who are still *susceptible*.

The Initial Value Problem

For a new disease, as Hong Kong flu was in 1968, the initial number of infected individuals is quite small, and everyone else is initially susceptible. Under the assumptions we have made, how do you think the number of susceptible individuals should vary with time? The number of recovered individuals? The number of infected individuals? On the axes provided here, sketch what you think the graph of each of these functions looks like.

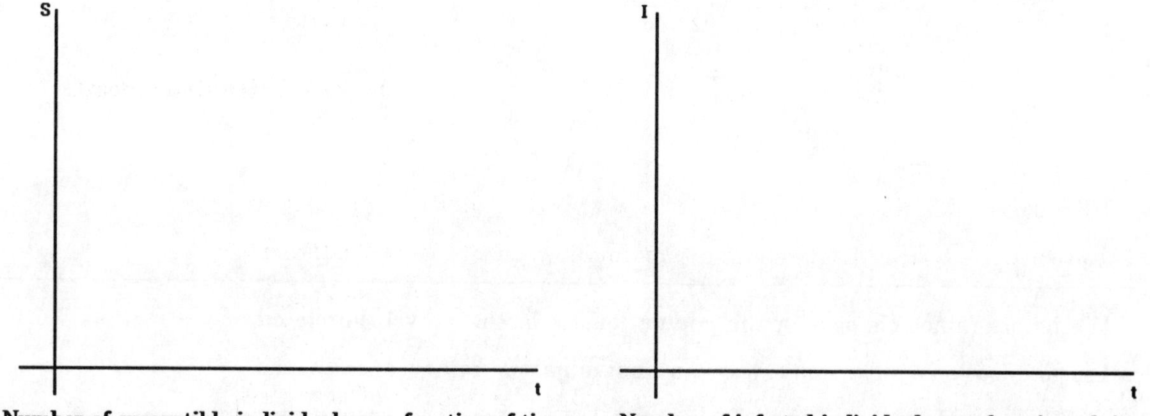

Number of susceptible individuals as a function of time.   Number of infected individuals as a function of time.

---

[3]We include in this group the relative handful who do not recover, but die; they too have immunity to future infection. In the jargon of epidemiology, the name for this entire group of people, living and dead, is "removals." They have been removed from both the susceptible and infected populations and can never return to either.

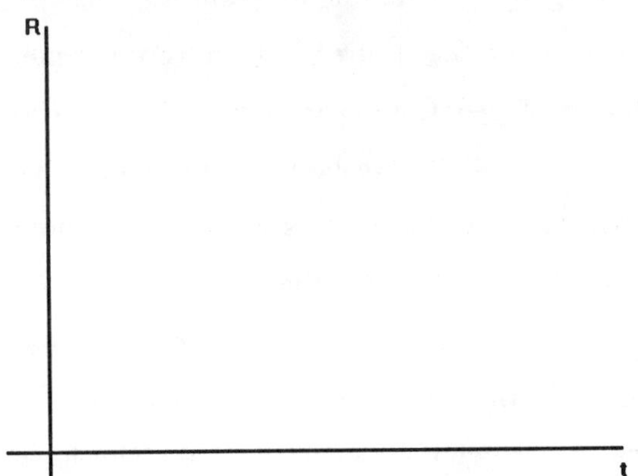

Number of recovered individuals as a function of time.

We designate the number of susceptible individuals as a function of time by $S(t)$, the number of recovered individuals by $R(t)$, and the number of infected individuals by $I(t)$. If we let N represent the total population (which we assumed to be constant), then for each time t we have

$$S(t) + R(t) + I(t) = N. \tag{7}$$

In the space below, list some factors that you think might govern the *rates of change* of S, I, and R. We will investigate a model that accounts for some, but probably not all, of the factors you list.

We ignore minor changes in the population by births, travel, unrelated deaths, and so on. The only way an individual is removed from the susceptible group is by becoming infected with the disease. We assume that the rate of change of S with respect to time depends on the number of individuals in the susceptible category, the number of individuals in the infected category, and the amount of contact there is between them. Suppose that each infected individual has a fixed number β of contacts per day that are sufficient to spread the disease. Not all of these contacts are with individuals susceptible to the disease; if we assume a homogeneous mixing of the population, the fraction of these contacts that are with susceptibles is $\dfrac{S(t)}{N}$. Thus, on the average, each infected individual generates $\beta\,\dfrac{S(t)}{N}$ new infected individuals per day. Accounting for the new cases of the disease generated by all the infected individuals, we have

$$\frac{dS}{dt} = -\frac{\beta}{N}\,I(t)\,S(t)\,. \tag{8}$$

Exercise 1. Explain carefully how each component of equation (8) follows from the text preceding. In particular, why is the factor of $I(t)$ present? Where did the minus sign come from?

The quotient $\dfrac{\beta}{N}$ appears so often in what follows, that we find it useful to introduce a single symbol for it; we let

$$\alpha = \frac{\beta}{N}\,.$$

Using this notation, we write equation (8) as

$$\frac{dS}{dt} = -\alpha\,I(t)\,S(t)\,. \tag{9}$$

Now we turn to the rate of change of $I(t)$, for which we need to consider both the movement of individuals from the susceptible group into the infected group (as we have just done) and the movement of individuals from the infected group to the recovered group. We assume that a fixed fraction λ of the infected group will recover during any given day. Thus,

$$\frac{dI}{dt} = \alpha\,I(t)\,S(t) - \lambda\,I(t)\,. \tag{10}$$

Exercise 2. Explain carefully how each component of equation (10) follows from the text preceding. In particular, why are there two terms? Why is it reasonable that the rate of flow from the infected population to the recovered population should depend only on $I(t)$? Where did the minus sign come from? Now explain why

$$\frac{dR}{dt} = \lambda\,I(t)\,. \tag{11}$$

Exercise 3. (a) Suppose $\lambda = \dfrac{1}{4}$. Explain why the average length of time that an individual remains infected is approximately four days.

(b) For a general λ, what is the average number of days that an individual remains infected?

Exercise 4. *From equations (9), (10), and (11), what is the sum $\dfrac{dS}{dt} + \dfrac{dI}{dt} + \dfrac{dR}{dt}$ as a function of t? How could you have predicted this result from equation (7)?*

We now have the three quantities of interest, $S(t)$, $I(t)$, and $R(t)$, governed by a system of three differential equations. Each equation has an initial condition, that is, a (presumably) known value at the start of the epidemic.

Exercise 5. *Explain why the following are reasonable initial values for the Hong Kong Flu epidemic in New York:*

$$S(0) = 7,900,000; \tag{12}$$

$$I(0) = 10; \tag{13}$$

$$R(0) = 0. \tag{14}$$

We must determine the constants β and λ experimentally, i.e., we select them so that the resulting functions fit the data on hand. For the New York flu epidemic, reasonable values turn out to be $\beta = 0.6$ and $\lambda = 0.34$. (This value of λ corresponds to an observation that the average flu sufferer remains infectious for about three days — see Exercise 3.) In a lab associated with this chapter, we will experiment with other values to see what effect they have on the solution functions.

Here is a summary of our problem: We want to find functions $S(t)$, $I(t)$, and $R(t)$ that satisfy the following differential equations and initial conditions:

$$\frac{dS}{dt} = -\alpha\, I(t)\, S(t), \qquad S(0) = 7{,}900{,}000, \qquad \text{(9R, 12R)}$$

$$\frac{dI}{dt} = \alpha\, I(t)\, S(t) - \lambda\, I(t), \quad I(0) = 10, \qquad \text{(10R, 13R)}$$

$$\frac{dR}{dt} = \lambda\, I(t), \qquad R(0) = 0. \qquad \text{(11R, 14R)}$$

This looks like *three* initial value problems, but it is really *only one*. That's because each of three differential equations involves at least one of the other unknown functions; we say the three differential equations are "linked."

The Euler's Method Approximation

As we did with initial value problems involving a single differential equation, we can approximate a solution by using Euler's Method. What is new here is that at each time step t_k we have three approximate quantities — S_k, I_k, and R_k — and we have to alter each one by adding *its* approximate rise. For example,

$$S_{k+1} = S_k - \alpha I_k S_k \Delta t.$$

Exercise 6. (a) Why is the approximate rise equal to $-\alpha I_k S_k \Delta t$? (In this case the "rise" is actually a drop.)

(b) Look back at the differential equations (10) and (11), and fill in the other two iteration formulas:

$$I_{k+1} =$$

$$R_{k+1} =$$

Exercise 7. If we take $\beta = 0.6$ and $\lambda = 0.34$, then $\alpha = \dfrac{\beta}{N} = 7.6 \times 10^{-8}$. With time measured in days, and $\Delta t = 1$ day, you can use your formulas (from Exercise 6) and ours to generate the following table. We have filled in the first three rows; you fill in the next three.

k	t_k	S_k	I_k	R_k
0	0	7,900,000	10	0
1	1	7,899,994	13	3
2	2	7,899,986	17	7
3	3			
4	4			
5	5			

It would be extremely tedious to extend this table to, say, 13 weeks if we did all the work with a calculator. In Figure 2, we show a computer-generated solution for each of the three population groups. Decide which curve is which, and label the curves accordingly. How do these curves compare with the ones you drew at the beginning of this section?

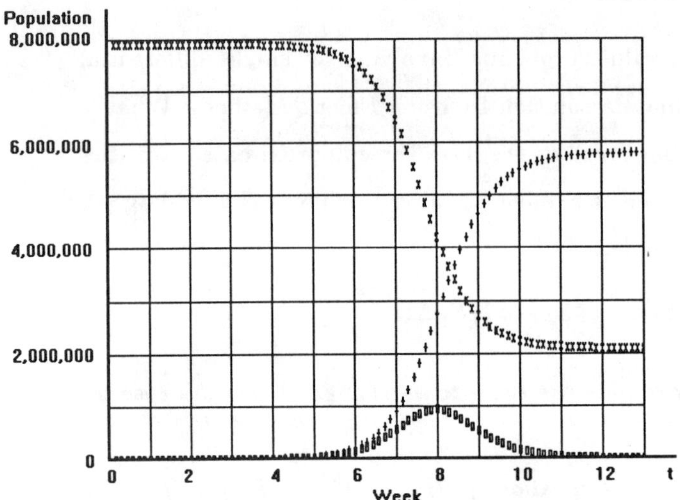

Figure 2. Susceptible, infected, and recovered populations
generated by the theoretical model with $\beta = 0.6$ and $\lambda = 0.34$.

*Exercise 8. Observe that, for some time at the beginning of the epidemic, the
susceptible population remains nearly constant.*

(a) Explain why this means $\frac{dI}{dt} \approx (\beta - \lambda)I(t)$ during that time.

*(b) From your observation in part (a), explain why the infected population
tends to grow exponentially at the start of the epidemic, provided β is greater
than λ.*

*(c) What happens if β is less than λ? Interpret your answer in terms of the
meanings of β and λ.*

Estimating and Using the Parameters[4]

In the previous section we took it for granted that the parameters β and
λ could be estimated somehow, and therefore it would be possible to
generate numerical solutions of the differential equations, as we did in
constructing Figure 2. (If β is known and the total population N is
known, then α is easily computed.) In Exercise 3 you explained why λ

---

[4]Based on Chapter 2 of *Calculus: The Language of Change* by K. D. Stroyan
(August, 1990), the emerging textbook for the University of Iowa calculus reform
project. Some of our information comes from "Herd Immunity," a student project
handout by A. L. Miller and K. D. Stroyan.

should be the reciprocal of the average number of days of infection; indeed, for many contagious diseases, the infection time is approximately the same for most infecteds and is known by observation. There is no *direct* way to observe β, but there is an indirect way, which we take up now.

Consider the ratio of β to λ:

$$\frac{\beta}{\lambda} \;=\; \beta \cdot \frac{1}{\lambda}$$

$$= \text{(the number of ``close'' contacts per day per infected)}$$

$$\times \text{(the number of days infected)}$$

$$= \text{the number of ``close'' contacts per infected individual.}$$

We call this ratio the "contact number," and we write $c = \dfrac{\beta}{\lambda}$. The contact number c is a combined characteristic of the population and of the disease; in similar populations, it measures the relative "contagiousness" of the disease, because it tells us indirectly how many of the contacts are "close enough" to actually spread the disease. We will now put our calculus to work to show that c can be estimated after the epidemic has run its course. Then β can be calculated as $c\lambda$.

Consider again equations (9) and (10), which we write in a form that displays β explicitly, since we need to find β before we can find α:

$$\frac{dS}{dt} \;=\; -\frac{\beta}{N}\, I(t)\, S(t)\; ; \qquad\qquad (8\mathrm{R})$$

$$\frac{dI}{dt} \;=\; \frac{\beta}{N}\, I(t)\, S(t) - \lambda\, I(t)\;. \qquad\qquad (15)$$

We observe about these two equations that the most complicated term in both would cancel and leave something simpler if we were to divide equation (15) by equation (8)[5] — provided we can figure out what it means to divide the derivatives on the left. But the Chain Rule solves that puzzle for us: The number of infecteds at any given time is implicitly a function of the number of susceptibles, so the rate of change $\dfrac{dI}{dS}$ makes sense, and the Chain Rule tells us that

---

[5]Of course, we could make an even stronger cancellation statement if we were to *add* the two equations, but that just leads to something we already know: equation (11). Because of equation (7), equation (11) is actually redundant; if you know S and I, you know R also.

$$\frac{dI}{dt} = \frac{dI}{dS}\frac{dS}{dt}.$$

Thus,

$$\frac{\frac{dI}{dt}}{\frac{dS}{dt}} = \frac{dI}{dS}.$$

Put another way, the Chain Rule says that, if it looks like differentials cancel, they really do. Now we can divide equation (15) by equation (8) and simplify:

Step	Reason

$$\frac{dI}{dS} = \frac{\frac{\beta}{N}\,I\,S - \lambda\,I}{-\frac{\beta}{N}\,I\,S}$$

$$= -1 + \frac{\lambda N}{\beta S}$$

$$= -1 + \frac{N}{cS} \qquad \text{We substitute } c = \frac{\beta}{\lambda}.$$

$$= -1 + \frac{1}{c(S/N)}.$$

(Write a reason next to each of the other steps of the calculation.)

The ratio S/N that turned up in the last step is just the *fraction* of the population that is susceptible, so we can simplify further by introducing variables, say, i and s, to stand for the infected and susceptible *fractions* of the population[6]: $i = I/N$ and $s = S/N$, so $\frac{di}{ds} = \frac{dI}{dS}$, and

$$\frac{di}{ds} = -1 + \frac{1}{cs}. \tag{16}$$

Equation (16) is a differential equation that determines (up to dependence on an initial condition) the infected fraction as a function of the susceptible

---

[6]This is another scaling of variables, this time of both independent and dependent variables. Since we are scaling both variables by the same factor, the rate of change of one with respect to the other does not change. We could have scaled all of the variables, S, I, and R, at the outset of our discussion, which would have been a natural thing to do if we had decided to focus on "Hong Kong flu" rather than "Hong Kong flu in New York." By making "fraction of the population" the variable for each group rather than "number in the group," the absolute size of the population becomes irrelevant. An added benefit that arises from this scaling is that β plays the role for which we found it necessary to create another parameter, α.

fraction. Three features of equation (16) are particularly worth noting:

(a) The only parameter that appears is c, the one we are trying to determine.

(b) The equation is *independent of time*; that is, whatever we learn about the relationship between i and s, it must be true for the entire duration of the epidemic.

(c) The right-hand side is an explicit function of s, the *independent* variable.

The conclusion we can draw from observation (c) is that $i = i(s)$ is a function whose derivative is $-1 + \frac{1}{cs}$. So, to solve equation (16), all we have to do is find a function that has this derivative. Of course, c will appear in the answer [observation (a)]. Since the relationship between i and s will be independent of time [observation (b)], we can write it down at *two different* times at which we know i and s; the equality of these two expressions will give us an equation we can solve for c.

So what's a function whose derivative is $-1 + \frac{1}{cs}$? Well, it has two terms, so we can take it a term at a time (see Conceptual Exercise 8 in Chapter 4). The first term, -1, is the derivative of $-s$. We can write the second term as $\frac{1}{c} \cdot \frac{1}{s}$, a constant times the *reciprocal* of s, so a function that has the right derivative will be the same constant times a function whose derivative is the reciprocal of s. We know that such a function is $\ln(s)$: $\frac{d}{ds} \ln s = \frac{1}{s}$.

Exercise 9. *(a) Show that $i = -s + \frac{1}{c} \ln s$ is a solution of (16).*

(b) Show that any solution of (16) must have the form

$$i = -s + \frac{1}{c} \ln s + k, \tag{17}$$

where k is a constant. (Recall that any two solutions of the same differential equation on a given interval must have a difference that is constant.)

The immediate implication we can draw from (17) is

$$i + s - \frac{1}{c} \ln s = k, \tag{18}$$

i.e., the expression on the left is *constant* for all values of time. Now we need to know the values of i and s at two different times in order to find from (18) an equation that involves c alone.

The times at which we are likely to be able to estimate i and s with some precision are just before the epidemic $(t = 0)$ and after it has run its course $(t = \text{"infinity"})$. If we write i_0 and s_0 for the starting fractions and i_∞ and s_∞ for the ending fractions (or the limiting values of i and s as t becomes large), then equation (18) tells us that

$$i_0 + s_0 - \tfrac{1}{c} \ln s_0 = i_\infty + s_\infty - \tfrac{1}{c} \ln s_\infty . \tag{19}$$

The *meaning* of "the epidemic has run its course" is that there are essentially no infecteds left in the population, i.e., $i_\infty = 0$ (see Figure 2). Furthermore, the typical epidemic starts with a very small number of infecteds being introduced into a largely susceptible population, so, relative to s_0, we have $i_0 \approx 0$ (again see Figure 2). Those observations simplify (19) to an approximate equality:

$$s_0 - \tfrac{1}{c} \ln s_0 \approx s_\infty - \tfrac{1}{c} \ln s_\infty . \tag{20}$$

Exercise 10. *Solve the approximate equation (20) for the contact number c to show that*

$$c \approx \frac{\ln s_\infty - \ln s_0}{s_\infty - s_0} . \tag{21}$$

Exercise 11. *(a) For the 1968-69 flu epidemic in New York City, what was s_0?*

(b) Use Figure 2 to estimate the number of recovered individuals (R_∞) and then the fraction of recovered individuals (r_∞) at the end of the epidemic. $(R_\infty$ is also the total number of people who got infected — a number that is known to public health officials if the reporting of cases is thorough.)

(c) Find s_∞ from r_∞. (The number of remaining susceptibles cannot be counted directly, so it must be computed from the known total number of cases.)

(d) Now finish the calculation of c from approximate equation (21), and then calculate β. Does your result agree with our use of $\beta = 0.6$ in Figure 2? (It should, because that figure is the only evidence you have about r_∞.)

How Effective Is Inoculation?

We turn our attention now to the questions raised at the beginning of the chapter regarding the possibility of wiping out or controlling contagious diseases by inoculation.[7] For a disease of the type we have been considering (one that confers future immunity on its sufferers), a population in which almost everyone has had the disease would itself protect those who have not had the disease: There would not be enough susceptibles left in the population to allow an epidemic to get under way. This sort of "group" protection is called "herd immunity." The idea of inoculation is to create herd immunity by stimulating the antibodies that create immunity in as many people as possible, *without* actually giving those people the disease. Thus, inoculation creates a direct path from the susceptible group to the "recovered" group[8] without passing through the infected group. And a large-scale inoculation program to head off an impending epidemic does this rapidly enough to artificially lower the "initial" susceptible population to a "safe" level — safe enough that if a trace level of infection enters the population, a few people may get sick, but no epidemic will develop.

At least, that's how it is supposed to work. In order to be sure it *will* work, public health officials need to know what fraction of the susceptible population must be inoculated. That's the question we address now.

In Exercise 8 you observed that the time derivative of I would be initially *negative* if β is less than λ — and, if I decreases with time, then clearly no epidemic develops. In the notation of this section, the

---

[7] Another possible strategy for controlling contagious disease is quarantine. If everyone who has a disease *knows* they have it and is voluntarily or forcibly isolated from the entire susceptible population, then there is no danger of the disease spreading. Unfortunately, most diseases have a period of transmissibility *before* symptoms are apparent, so quarantine (even with full cooperation, which is often impossible) is usually not a viable control strategy. Of the diseases mentioned so far in this chapter, the only one that must show a symptom to be transmissible is smallpox, and, until it was reduced to very low levels of occurrence by inoculation, it was never feasible to isolate all of its sufferers. Furthermore, quarantine alone is at best a temporary strategy, because it leaves large, susceptible populations unprotected from the next trace occurrence of infection that shows up.

[8] As we observed in Note 3, the technical term for the "R" group is "removals," a term that includes all ways of being removed from the susceptible and infected groups — recovery, death, quarantine, and inoculation.

condition "$\beta < \lambda$" is equivalent to "$c < 1$" because $c = \beta/\lambda$. However, Exercise 8 was based on a simplification of equation (10) that got the susceptible population out of the picture — and now we want to consider how the susceptible population itself contributes to the increase or decrease of the infected population. So let's go back to equation (15), which relates the rate of change of the number of infecteds to the *fraction* of susceptibles:

$$\boxed{\text{Step}} \qquad\qquad\qquad\qquad \boxed{\text{Reason}}$$

$$\frac{dI}{dt} \;=\; \frac{\beta}{N}\, I(t)\, S(t) - \lambda\, I(t) \qquad\qquad (15\text{R})$$

$$\;=\; \beta\, I(t)\, s(t) - \lambda\, I(t)$$

$$\;=\; \lambda\, [c\, s(t) - 1]\, I(t).$$

Exercise 12. Write a reason next to each step in the preceding calculation.

Thus, we see that, to assure a negative derivative for I, we *don't* have to have $c < 1$ (which is likely to be impossible); it's enough to have $c\, s(t) < 1$. In particular, to have I decreasing from time zero, it's enough to have $c\, s_0 < 1$, or $s_0 < \frac{1}{c}$.

Now we have a handle on how to determine an inoculation strategy: Determine the contact number c for a given disease from historical data, take its reciprocal, and reduce the susceptible fraction of the population below that level. "*The* contact number" is a little misleading, because, for a given epidemic, this number depends on both the disease and the population. But, if we can determine a worst-case or a likely-case value of c, then we can determine what we have to do to protect against that level of contagiousness. For example, in the United States from 1912 to 1928, the contact number for measles was 12.8. The reciprocal of this number is about 0.078; if we assume that c is still 12.8, then reducing the susceptible population to 7% of the total population of any given area should confer herd immunity on the entire population. This requires vaccination of *more* than 93% of the population, because the measles vaccine is only 95% effective; that is, 5% of those inoculated do not acquire immunity.

Exercise 13. *What fraction of a population in the United States must be vaccinated to assure herd immunity against measles?*

Our discussion of contact numbers, herd immunity, and other aspects of epidemics continues in Conceptual Exercises 2-8 and Problems 1-4 at the end of the chapter.

5.3 Modeling the Evolution of Prices in a Simple Economy[9]

Continuous Price Evolution

We begin our study of price evolution with some assumptions about the economic mechanism whereby prices shift to balance supply and demand. First we consider a very simple market economy in which there is just one "good" being produced and bought.   Moreover, we'll assume that transactions are being made at all times (i.e., continuously) and that buyers and sellers respond immediately to changes in the price. All of these are unrealistic assumptions; we'll look at adding more realistic (and more complicated) features later. For the moment, we want to see if we can give a reasonable mathematical description of this situation.

Consider the "demand" for the good. We assume that the demand D (in bushels, say, for a good such as peaches) is a function of the price p. Lacking any specific knowledge of demand functions, we make the simplest possible assumption, namely, that demand is a *linear* function of price:

$$D(p) = a + b\,p\;,\tag{22}$$

where $b < 0$.

Exercise 1. *(a) What is the significance of the fact that b is negative?*
(b) What does a represent? Should it be positive or negative?

We assume that the "supply" Q (also in bushels, in the case of peaches) has the same simple form:

$$Q(p) = c + d\,p\;,\tag{23}$$

where d is positive.[10]

Exercise 2. *(a) What is the significance of the fact that d is positive?*
(b) Should we expect c to be positive or negative? Does it depend on the commodity?

---

[9] Adapted from *An Introduction to Mathematical Models in Economic Dynamics*, by David Clements, Polygonal Publishing, 1984.

[10] We use Q for supply instead of S because (a) S stands for too many other things, and (b) Q is traditional in economics texts. Think of Q as standing for "quantity."

We want to find the "equilibrium price" p^*, the price at which $Q = D$. A glance at Figure 3 shows that there is such a price, namely, the p-coordinate of the point of intersection of the two lines.[11]

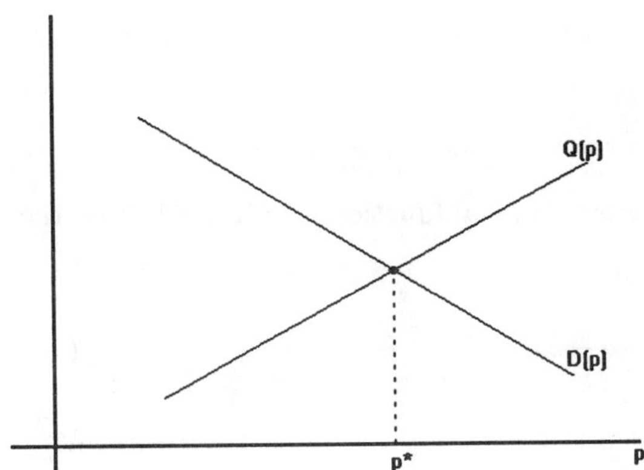

Figure 3. Model supply and demand functions and the equilibrium price.

As usual, we are interested in *change*: Given a starting price at which supply and demand *do not* balance, does the price moves toward equilibrium? If so, how? These questions require that we introduce time t as the independent variable and study how supply, demand, and price vary as functions of time. To do this, we have to make some assumption about the way the price reacts to an imbalance between supply and demand. We assume that the rate of change of price as a function of time, $\dfrac{dp}{dt}$, depends on the "excess demand,"

$$E(p) = D(p) - Q(p). \tag{24}$$

Specifically, we assume that $\dfrac{dp}{dt}$ is proportional to $E(p)$:

$$\frac{dp}{dt} = \gamma\, E(p)\,, \tag{25}$$

where γ is positive. Now we can combine the assumptions made so far to formulate a differential equation for the unknown price function p.

Exercise 3. (a) Why is γ positive?

(b) Give reasons for the remaining steps in the following calculation.

---

[11] In economics texts (including that of Clements referenced in Note 9), these figures are drawn with the independent variable, price, on the vertical axis.

Step	Reason

$$\frac{dp}{dt} = \gamma \left[D(p) - Q(p) \right]$$

We substituted from (24) for $E(p)$ in (25).

$$= \gamma \left[a + bp - (c + dp) \right]$$

$$= -\gamma(d - b)p + \gamma(a - c)$$

Thus, we can describe the price p, a function of time t, as the solution of an initial value problem:

$$\frac{dp}{dt} = -\gamma(d - b)p + \gamma(a - c), \tag{26}$$

$$p(0) = p_0, \tag{27}$$

where p_0 is the price at the initial time.

For example, if $a = 12$, $b = -1.5$, $c = -1$, $d = 1$, $\gamma = 0.17$, and $p_0 = 5$, then the differential equation (26) is

$$\frac{dp}{dt} = -0.17(1 + 1.5)p + 0.17(12 + 1) = -0.425p + 2.21,$$

so the initial value problem becomes

$$\frac{dp}{dt} = -0.425p + 2.21 \quad \text{with} \quad p(0) = 5. \tag{28}$$

Exercise 3. *(a) Show that the unique solution of the initial value problem (28) is*

$$p(t) = 5.2 - 0.2e^{-0.425t}. \tag{29}$$

(b) What is p^* in this case?

Exercise 4. *(a) Use (26) to give a description of p^* in terms of a, b, c, and d. (Hint: What is $\frac{dp}{dt}$ when $p = p^*$?)*
(b) Does your general description of p^* in part (a) agree with your calculation of the specific case in the preceding exercise?

We may simplify the *appearance* of equation (26) by introducing a single symbol, say α, for the constant $\gamma(d - b)$ and another single symbol, say β, for the constant $\gamma(a - c)$. Then the initial value problem becomes

$$\frac{dp}{dt} = -\alpha p + \beta \quad \text{with} \quad p(0) = p_0. \tag{30}$$

We have seen this problem many times now; in particular, in Chapter 3 you learned how to solve it by scaling the dependent variable. Presumably, that is what you did for the particular case in Exercise 3. Now we ask you to tackle the general case; as we have seen on many other occasions, the calculation is actually less messy with letters than with numbers as the coefficients.

Exercise 5. (a) *Solve the initial value problem stated in (30). [You may find it useful to begin by rewriting the right-hand side of the differential equation in the form $-\alpha\left(p - \frac{\beta}{\alpha}\right)$.]*

(b) *Substitute the definitions of α and β to express the solution function in terms of the constants a, b, c, d, γ, and p_0.*

(c) *Use the result of Exercise 4 to show that the solution function can be written in the form*

$$p(t) = (p_0 - p^*)\, e^{-\gamma(d-b)t} + p^* \, . \tag{31}$$

(d) *Check that the general solution in (31) agrees with your calculation of the particular solution in Exercise 3.*

(e) *What does the graph of the function defined by (31) look like? How does the graph change for different starting values p_0? In the space below, make a rough sketch showing several solutions starting from different initial points.*

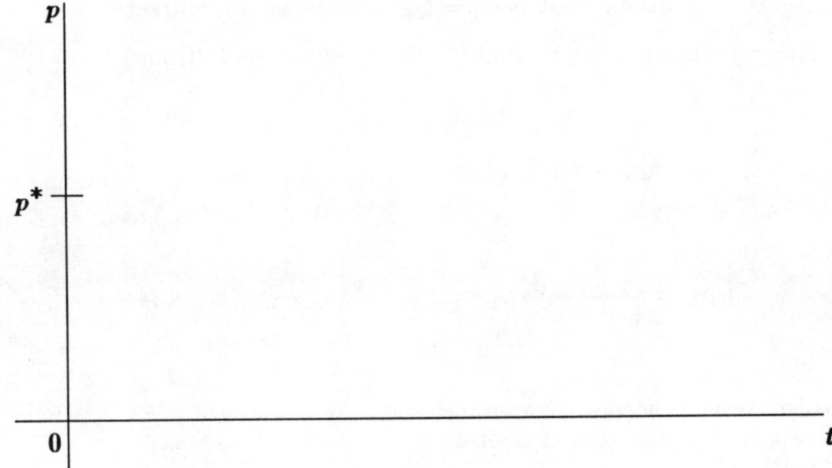

Discrete Price Evolution

You might object to the continuous model for prices described in the previous section because suppliers cannot react instantaneously to changes in prices. Indeed, for many commodities, goods are exchanged only at regular intervals, say, once a day or once a week at market. We now consider a model that incorporates discrete trading times but retains the other assumptions of the continuous model.

We assume that trading times occur at regular time intervals of length Δt.[12] Thus, the trading times are $t_0 = 0$, $t_1 = \Delta t$, $t_2 = 2 \cdot \Delta t$, ... We denote the price at trading time t_k by p_k; similarly, the demand and supply at this trading time are denoted D_k and Q_k, respectively. The linearity assumptions on the demand and supply [formerly (22) and (23)] take the form

$$D_k = a + b p_k \tag{32}$$

and

$$Q_k = c + d p_{k-1} . \tag{33}$$

Notice the "p_{k-1}" in the formula for Q_k: The suppliers base the amount to bring to the market at time t_k on the price at the *previous* market time, p_{k-1}.

Now we impose the condition that $D_k = Q_k$. This is a "market clearing" assumption; everything must be sold.[13] From equations (32) and (33), we have

$$a + b p_k = c + d p_{k-1};$$

hence,

$$p_k = \frac{c-a}{b} + \frac{d}{b} \cdot p_{k-1}. \tag{34}$$

---

[12]In this discussion, there is no reason to assume that Δt is "small." For example, in the case of a farm product such as peaches, the product might be brought to market once a week during a certain season.

[13]Think about peaches again, or any perishable good that cannot be taken back from the market and stored. If peaches are left over and thrown away, the "effective price" per bushel is the total revenue divided by the *total* number of bushels, not just the ones that were sold. Thus, while the suppliers may try to impose a higher price, the market imposes the effective price.

This recursive formula allows us to substitute a starting value, p_0, on the right to find p_1, then substitute p_1 to find p_2, and so on, through the entire price evolution, that is, for as many trading times as the market stays in business.

We have encountered recursive formulas for generating sequences of numbers on at least three other occasions: discrete "natural" growth at the end of Chapter 2, Newton's Method for solving equations in Chapter 4, and Euler's Method for solving initial value problems earlier in this chapter. Our discussion of price evolution is analogous to the natural growth discussion in Chapter 2 in the following way: There we had both discrete and continuous models (initial value problems, one with a difference equation and the other with a differential equation) based on the *same* assumption about biological growth (growth rate proportional to the population); here we have both discrete and continuous initial value problems to model price evolution that are based on the *same* assumptions about supply and demand. With biological growth, the discrete model with a very small time step closely approximated the continuous model, and the same could be said about the *solutions* of the two kinds on initial value problems (see the Summary in Chapter 2). Could we have a similar connection between the discrete and continuous price models? In particular, will the solutions of the difference equation (34) display the "decaying exponential" behavior[14] of the solutions

$$p(t) = (p_0 - p^*)\, e^{-\gamma(d - b)t} + p^* \qquad\qquad (31\text{R})$$

to the differential equation

$$\frac{dp}{dt} = -\gamma(d - b)p + \gamma(a - c)? \qquad\qquad (26\text{R})$$

---

[14]We won't belabor the point here, but we could draw the same analogy and ask the same questions about epidemic models. When we applied Euler's Method to the S-I-R differential equation model, we turned it into a discrete model governed by recursive (difference) equations — the equations you used in Exercise 7 of Section 5.2 to start generating a solution — the same equations we used to generate Figure 2. We could just as reasonably have started with the discussion of epidemics with the discrete model and then have derived equations (9)-(11) by letting Δt shrink to 0. We took it for granted that the Euler (discrete) approximation would indeed approximate the solution of the system of differential equations. However, we will see here that the answer to the question just asked isn't automatically "yes."

We can attempt to answer that question by actually solving (34), that is, by finding an explicit formula for p_k in terms of the starting price p_0 and the constants a, b, c, and d. This is a different problem from that of solving the discrete natural growth equation, so we will have to develop a somewhat different technique.

Geometric Sums

We start as we did with the continuous price evolution model by introducing single symbols for the complicated constants, say, $\alpha = \frac{c-a}{b}$ and $\beta = \frac{d}{b}$. Then (34) assumes the simpler form

$$p_k = \alpha + \beta p_{k-1}. \tag{35}$$

Now we write out the first few cases for $k = 1$, 2, and 3:

$$p_1 = \alpha + \beta p_0;$$

$$p_2 = \alpha + \beta p_1$$

$$= \alpha + \beta\,(\alpha + \beta p_0) = \alpha + \beta\alpha + \beta^2 p_0;$$

$$p_3 = \alpha + \beta p_2$$

$$= \alpha + \beta\,(\alpha + \beta\alpha + \beta^2 p_0) = \alpha + \beta\,\alpha + \beta^2\alpha + \beta^3 p_0.$$

We see a pattern emerging; in fact,

$$p_k = \alpha\,(1 + \beta + \beta^2 + \cdots + \beta^{k-1}) + \beta^k p_0 \tag{36}$$

for each positive integer k.

Equation (36) is a "solution" of the difference equation (35) in the sense that it expresses each price p_k in terms of the starting price p_0 and the constants α and β. However, it is not an "explicit solution" because of the "$\cdots$" (ellipsis) in the middle of the summation. Our next task is to evaluate the summation that appears in parentheses in (36).

For the time being, let us suppose the values of β and k are fixed, and let's give a name to the summation in (36), say,

$$S = 1 + \beta + \beta^2 + \cdots + \beta^{k-1}. \tag{37}$$

If we multiply both sides of this equation by β, and write the result in such a way that like powers of β match up with (37), we have

$$\beta S = \beta + \beta^2 + \cdots + \beta^{k-1} + \beta^k \tag{38}$$

Next, we subtract equation (38) from equation (37) to get

$$S - \beta S = (1 - \beta)\, S = 1 - \beta^k. \tag{39}$$

Finally, we solve the second equation in (39) for S to get an explicit expression involving β and k — but no ellipsis. (Here we must assume β is not equal to 1. Why do we need to assume that?) When we replace S by its definition from (37), we get

$$1 + \beta + \beta^2 + \cdots + \beta^{k-1} = \frac{1 - \beta^k}{1 - \beta}. \tag{40}$$

Exercise 6. (a) Check the result in formula (40). Start with $\beta = \frac{1}{2}$ and $k = 4$; do both sides give you the same number?

(b) Try $\beta = 2$ and $k = 5$.

(c) Choose two other combinations of β and k, and check that the formula is correct for your choices.

This is not the first time we have seen something growing in proportion to its size at the previous step; we have already associated the names "exponential" and "geometric" with such quantities. Indeed, the terms of the sum in equation (40) are values of the discrete exponential function with base β. What is new here, thrust on us by the need to solve a new discrete initial value problem, is the idea of *summing* such terms. The sum that is the left-hand side of (40) is called a **"geometric sum."**

Exercise 7. Your derivation of formula (40) required the assumption that $\beta \neq 1$. But the sum on the left in (40) clearly has a value when $\beta = 1$, a value that you can easily write down without an ellipsis.

(a) What is it when $k = 3$? When $k = 4$?

(b) What is $1^0 + 1^1 + 1^2 + \cdots + 1^{k-1}$ in general ? Write your general answer here: ____________

Exercise 8. (a) Substitute equation (40) into equation (36) to find an explicit solution for p_k in terms of α, β, and p_0.

(b) Now replace α and β by their definitions in terms of our original constants, a, b, c, and d, to get a solution of the difference equation (34). Simplify the result.

(c) Write your solution in the space below; you will need this formula for a laboratory experiment associated with this chapter.

A Comparison of the Continuous and Discrete Models

In Figures 4 and 5, we show the solution (31) to the continuous initial value problem (26)–(27) and the solution you just worked out to the discrete initial value problem (34), both with the numerical values you used in Exercise 3: $p_0 = 5$, $a = 12$, $b = -1.5$, $c = -1$, $d = 1$, and (for the continuous case) $\gamma = 0.17$. For this particular selection of constants, the two models lead to similar evolutions of prices toward the "equilibrium" price p^* at which supply and demand exactly balance. (Compare the prices on even-numbered trading days.) For any reasonable selection of constants, the continuous solution *always* has the same general shape that you see in Figure 4 (why?). However, the discrete model has some features (in addition to the odd-even oscillation seen here) that the continuous model does not exhibit; we will explore these features in our laboratory experiment.

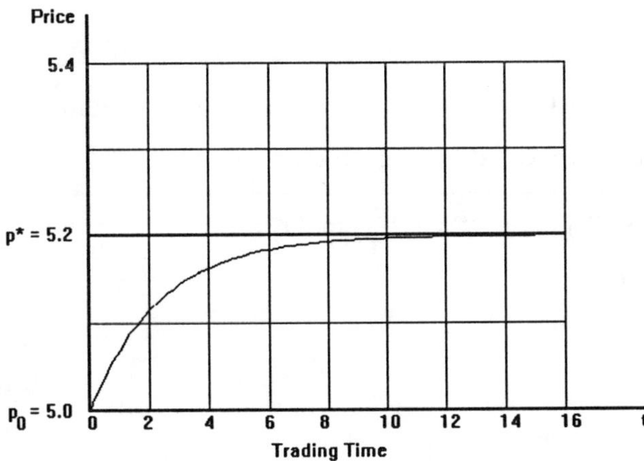

Figure 4. Solution of the continuous price model
for $a = 12$, $b = -1.5$, $c = -1$, $d = 1$, $\gamma = 0.17$.

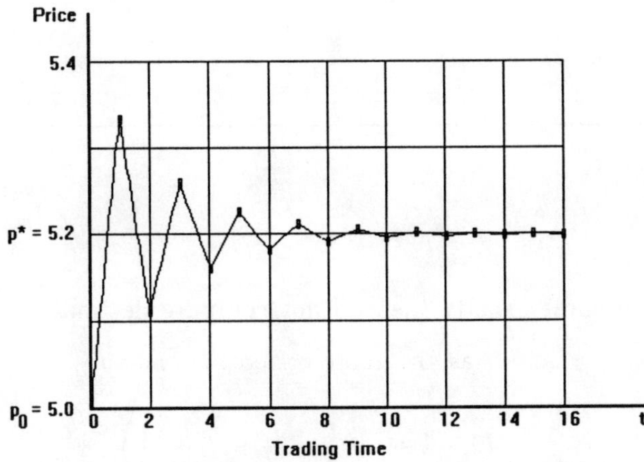

Figure 5. Solution of the discrete price model
for $a = 12$, $b = -1.5$, $c = -1$, $d = 1$.

Section Summary

The important messages about discrete price evolution will appear in the laboratory; in this summary we ask you to concentrate on the important new mathematical technique that appeared here: geometric sums.

Explain carefully why $1 + r + r^2 + r^3 + \ldots + r^n = \dfrac{1}{1-r}$ for $r \neq 1$.

Exercises

Calculate each of the following sums exactly (i.e., do not compute decimal approximations):

9. $1 + 3 + 9 + 27 + \ldots + 3^{10}$

10. $1 + \frac{1}{3} + \frac{1}{9} + \frac{1}{27} + \ldots + \left(\frac{1}{3}\right)^{10}$

11. $5 + 25 + 125 + \cdots + 5^9$ (Watch the first term!)

12. $3^6 + 3^7 + 3^8 + \ldots + 3^{12}$

13. $1 + \frac{1}{4} + \frac{1}{4^2} + \frac{1}{4^3} + \cdots + \frac{1}{4^8}$

SUMMARY

We began this chapter with a formalization of Euler's Method, which we had seen informally in an early laboratory. Euler's Method is a tool for generating numerical and graphical solutions to initial value problems; with a computer or graphing calculator at hand, we can "solve" an initial value problem, whether or not we know how to find a "formula" for the solution. At the same time, we must keep in mind that Euler's Method relies on a discrete approximation to the given differential equation — that is, it replaces the differential equation with a difference equation. Thus, we need to consider how small the step size should be for the approximation to be "adequate" in the sense that the generated discrete solution "tracks" along the unknown solution curve. In principle, we can always take a smaller step size if we need to, but in practice there are limits to our patience for waiting while computer (never mind a calculator!) does tens of thousands of calculations.

For a method as powerful as Euler's, the underlying idea is amazingly simple: At each step, we want to know a "rise," and we already know how to calculate "slope," so we multiply "slope" by "run" (step size). When we add the computed "rise" to the "current" function value, we get the "next" function value — approximately.

Of course, Euler's Method has nothing to do with either epidemics or economics — it's a *general tool*. But we demonstrated the power of this tool by applying it to the S-I-R epidemic model, which has (a) three interrelated dependent variables and (b) no possibility of a simple "formula" solution for any of the variables as functions of time.[15] The effectiveness of Euler's Method is shown in the simultaneous solutions for all three functions shown in Figure 2. In particular, the "infected population" graph has roughly the same shape as the (presumably proportional) excess death data shown in Figure 1.

---

[15]It *is* possible to find formulas relating the dependent variables to each other; see Exercise 9 in Section 5.2 and Problem 4.

Among other details picked up along the way, we saw (Exercise 8 in Section 5.2) that an epidemic for which the S-I-R model is appropriate must have roughly exponential growth of infecteds in its early stages. We saw earlier (Chapter 2, Problem 16; see also Conceptual Exercise 7 in this chapter) that the early growth of the AIDS epidemic in this country followed a cubic pattern, so it definitely does not fit the S-I-R model. Given what we know about how difficult it is to transmit the HIV virus, it is not surprising that the "total mixing" assumptions of the S-I-R model do not fit the AIDS epidemic.

After using the discrete Euler approximation to generate graphical solutions, we turned to the question of estimating the parameters in the model by using observable data. For that purpose, we found the Chain Rule to be an essential tool for finding a time-independent relationship between the susceptible and infected populations — by solving a differential equation for i (the infected fraction) as a function of s (the susceptible fraction). When we substituted observable "starting" and "ending" data, we found we could estimate the "contact number" c for the epidemic. Knowing the average time $\frac{1}{\lambda}$ that each infected remained contagious, we could calculate the "mixing" coefficient β (the number of close contacts per day per infected) as $c\lambda$. Then we could divide β by the population size N to get the coefficient α that appears in the differential equations. Thus, while the model is primarily "qualitative" in its description of epidemic behavior, public health officials can use it retrospectively to get a quantitative analysis of a particular epidemic.

For the remainder of the chapter, we studied price evolution in a single-product market, given very simple assumptions about supply and demand functions (both linear as functions of price) and a requirement that prices adjust so supply equals demand. We had two purposes in doing this: (a) to illustrate the very beginnings of a calculus-based study of economic theory, and (b) to show that discrete and continuous models based on the same underlying assumptions can lead to very different conclusions — even when each "approximates" the other.

In the continuous case, we modeled the rate of change of the price function as being proportional to "excess demand." We found that this model led to an initial value problem that we have seen many times now, starting with Newton's Law of Cooling in Chapter 3. You used the scaling-the-dependent-variable technique to solve the equation (Exercise 4 in Section 5.3) and to show that the price evolution is an exponential decay toward the equilibrium price, no matter what the starting price is.

What made the discrete model significantly different from the continuous one was (sensibly) allowing the producers to react to price at a *previous* trading session rather than at the *current* session. (This possibility did not arise in the continuous model, because when trading times are continuous, there is no "immediately preceding" trading session.) When the suppliers' reaction was "moderate," we saw behavior similar to that of the continuous model, namely, convergence at an exponential rate toward the equilibrium price — but with prices bouncing above and below equilibrium rather than staying on one side. In a laboratory exercise (but not in the text), you saw that more extreme reaction by the suppliers could lead to *instability*, that is, to wild oscillations away from the equilibrium price.

This behavior is not unlike some actually seen in real markets. For example, automobile manufacturers may produce a glut of new cars as a reaction to a previous season of strong sales at good prices (for the suppliers). This leads to supply far in excess of demand, and dealers are forced to offer deep discounts and other incentives to get sales moving again. If this reaction (lowering prices) is extreme, sales will pick up with a vengeance, and shortages may develop. That's an incentive for the manufacturers to both produce more (to meet demand) and raise prices (to benefit more from the increased demand). The cycle starts over at this point — but the swings may become more extreme, at least until somebody goes broke.

The process of solving the difference equation (discrete initial value problem) that constituted the discrete model bore some similarity to the process we used at the end of Chapter 2 to solve the discrete natural growth problem. But there was a difference: Here we got not just powers of a growth factor but *sums* of those powers. Thus, we found it necessary to digress briefly to work out a formula for calculating geometric sums:

$$1 + \beta + \beta^2 + \cdots + \beta^{k-1} \; = \; \frac{1 - \beta^k}{1 - \beta}. \tag{40R}$$

In Problem 6 we ask you to explore what happens to this expression as k becomes large. Geometric sums turn up in a lot of different places; knowing how to calculate them is a necessary tool to have in your toolbox.

Practice with Calculations

*Exercises in this category include (a) calculations that can be done by machines and (b) practice on important topics from courses that precede calculus. You need to develop **facility** with both categories — not because such calculations are a central feature of the course, but because they should not frustrate you or keep you from concentrating on the more important parts of the course by occupying a lot of your time or by leading to lots of mistakes. Even though routine calculations can be done quickly and accurately by computer or calculator, you need to develop judgment about when to use a machine and when not to, and you need to know how to tell when you might have pressed the wrong button. These skills are acquired and sharpened by **practice**.*

You should expect to see exercises like these as some portion of your homework assignments, quizzes, and tests.

Calculate each of the following derivatives:

1. $\dfrac{d}{dt} 2\ln t$ 2. $\dfrac{d}{dt}\ln(t^2)$ 3. $\dfrac{d}{dt}\ln 2t$

4. $\dfrac{d}{dt}\ln(t+2)$ 5. $\dfrac{d}{dt}(\ln t)^2$ 6. $\dfrac{d}{dt}e^{3t}\ln t$

7. $\dfrac{d}{dt}\sqrt{1+t^2}\,\ln 3t$

Calculate each of the following second derivatives:

8. $\dfrac{d^2}{dt^2} 2\ln t$ 9. $\dfrac{d^2}{dt^2}\ln(t^2)$ 10. $\dfrac{d^2}{dt^2}\ln 2t$

11. $\dfrac{d^2}{dt^2}\ln(t+2)$ 12. $\dfrac{d^2}{dt^2}(\ln t)^2$ 13. $\dfrac{d^2}{dt^2}e^{3t}\ln t$

Calculate each of the following sums exactly (i.e., do not compute decimal approximations):

14. $1+2+4+8+16$ 15. $1+\frac{1}{2}+\frac{1}{4}+\frac{1}{8}+\frac{1}{16}$

16. $1+\frac{1}{2}+\frac{1}{4}+\frac{1}{8}+\frac{1}{16}+\cdots+\frac{1}{1024}$ 17. $1+\frac{1}{2}+\frac{1}{4}+\frac{1}{8}+\frac{1}{16}+\cdots+\frac{1}{2^{12}}$

18. $\frac{1}{3^4}+\frac{1}{3^5}+\frac{1}{3^6}+\frac{1}{3^7}+\cdots+\frac{1}{3^{12}}$ 19. $\frac{7}{13}+\left(\frac{7}{13}\right)^2+\left(\frac{7}{13}\right)^3+\cdots+\left(\frac{7}{13}\right)^{20}$

20. $\frac{2}{3}+\left(\frac{2}{3}\right)^2+\left(\frac{2}{3}\right)^3+\cdots+\left(\frac{2}{3}\right)^{14}$ 21. $\frac{13}{7}+\left(\frac{13}{7}\right)^2+\left(\frac{13}{7}\right)^3+\cdots+\left(\frac{13}{7}\right)^{25}$

Conceptual Exercises

Exercises in this category use and/or further develop the concepts introduced in this and earlier chapters. The level of difficulty is similar to that of the exercises embedded in the text and at the ends of sections. While many of these exercises are presented in "realistic" contexts, they are not representative of real problems. Rather, these exercises are to help you get ready for tackling real problems.

You should expect to see exercises like these as a major portion of your homework assignments, quizzes, and tests.

1. In Exercise 2 of Section 5.1, you applied Euler's Method to the initial value problem $\frac{dP}{dt} = t^2$, $P(0) = 1$,   to find approximate values of $P(0.1)$, $P(0.2)$, and $P(0.3)$.

    (a) Find the *exact* solution of the initial value problem. [Hint: Determine all the functions that have derivative t^2, and let $P(t)$ be the one that satisfies the initial condition.]

    (b) Evaluate your solution function from part (a) at $t = 0.1$,   0.2,   and 0.3. Compare these exact values with the approximate values from Euler's Method. How good are the approximations?[16]

2.[17] Table 2 lists recorded contact numbers in the United States for several different diseases. Assuming 100% effectiveness of the vaccine in each case, what percentage of the population would have to be vaccinated against each disease to assure herd immunity?

Disease	Date	c
Whooping cough	1943	17.3
Chicken pox	1943	11.3
Scarlet fever	1908-1917	8.5
Mumps	1943	8.1
Diphtheria	1943-1947	7.4
Poliomyelitis	1955	4.9

Table 2. Historical contact numbers.

---

[16]We are not ready yet to take up exact solution of the initial value problem in Exercise 1 of Section 5.1; that will come in Chapter 8. As you have already seen in the lab, Euler's Method can generate numerical and graphical solutions even when you have no algebraic tools for finding a solution as a formula.

[17]Based on "Herd Immunity," a student project handout by A. L. Miller and K. D. Stroyan.

3.[17] Rubella (German measles) had a contact number of 7.7 in West Germany in 1972 and 7.0 in England and Wales in 1979. The rubella vaccine, like the measles vaccine is 95% effective. What fraction of the population must be inoculated to acquire herd immunity from rubella?

4.[17] Laws in the United States require children to have a rubella immunization prior to entering school. The typical vaccine is called "MMR," for measles-mumps-rubella; that is, it is a combined vaccine for all three. It is estimated that 98% of children entering school have been vaccinated. Is this fraction high enough to acquire herd immunity for any or all of the diseases?

5.[17] The contact numbers for measles in England and Wales (1956-1959) and in Nigeria (1960-1968) were 15.6 and 17.0, respectively. With a 95% effective vaccine, what percentage of each of these populations would have to be vaccinated to assure herd immunity?

6. Many important epidemics are not at all like flu epidemics; the most important one afflicting the world right now is the AIDS epidemic. In Exercise 8 of Section 5.2, you showed that a flu-like epidemic should grow exponentially at first. Table 3 shows the data from the Centers for Disease Control on reported cases for the first four years of the AIDS epidemic in the United States. Show that this data is *not* growing exponentially.

Date	Cumulative Cases Reported
Sep. 1981	129
Feb. 1982	257
July 1982	514
Jan. 1983	1,029
Aug. 1983	2,057
Apr. 1984	4,115
Feb. 1985	8,229
Jan. 1986	16,458

Table 3. The start of the AIDS epidemic.

7. Figure 6 is a representation of roughly the same data (the line labeled "TOTAL") as that shown in Table 3, with an added breakdown of the total by race. The graph is *not* a log-log or semilog graph; look at what the vertical axis represents. Explain carefully what the graph shows about the data. (See also Problem 11 in Chapter 2.)

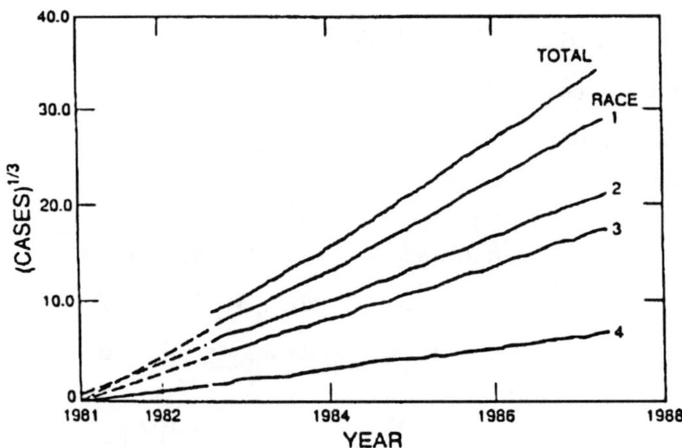

Figure 6. Data on AIDS cases by race:
1. White, 2. Black, 3. Hispanic, 4. Unknown.[18]

8. (a) Find and list the assumptions we made about the New York flu epidemic. How reasonable are these assumptions?

(b) What other assumptions might we make that would be appropriate for a flu epidemic? How would these change the system of differential equations?

(c) Think of some other infectious disease, and determine which of the assumptions in the flu model would not apply to it. How would you modify the assumptions to construct a model for the spread of your selected disease in the form of an initial value problem?

---

[18]The figure is reproduced from "Modeling the AIDS Epidemic," by Allyn Jackson, in the October 1989 issue of *Notices of the American Mathematical Society*. This is a highly readable source, with references, that reports on what is known about reasons for the non-exponential behavior of the epidemic.

9. The NCAA Championship basketball tournament is a single elimination tournament that starts with 64 teams. All teams play in the first round, and the winners progress to the second round. This pattern is repeated until only one team is left undefeated.

(a) Express the number of games played as a geometric sum, and calculate this number.

(b) Describe a simpler method for determining the number of games played in the tournament.

Problems and Projects

The problems presented here are not intended for individual homework assignments or for tests. These problems should be attempted by groups of two to four students sharing ideas, whether in the classroom or elsewhere. Some of the more extensive problems are suitable for projects with time frames ranging from a class period to a week.

1. There was no flu vaccine in 1968, but there is now. How would you incorporate the effect of a vigorous inoculation program into the flu epidemic model? (Assume the program is *not* complete before the first trace infection appears in the population, but that it continues as the disease is spreading.) How would you expect this change to affect the solution of the initial value problem?

2.[19] Before vigorous world-wide attempts to wipe out smallpox, its contact number in India was 5.2, relatively low among the contagious diseases considered in this chapter. Smallpox vaccinations began in 1958, but the disease persisted until 1977, almost 20 years. What explanations would you offer for this?

3.[19] (Start on this after you have done Exercise 13 in Section 5.2, Conceptual Exercises 2-5, and Problem 2.) In your view, why do we still have measles and rubella, but not polio or smallpox? About fifteen to twenty years ago, there were some problems with the measles vaccines; how are those problems related to recent outbreaks of measles and to vigorous re-vaccination campaigns on college campuses? If measles and rubella vaccines are given together and have the same level of effectiveness, how would you explain the facts that the incidence of rubella has been steadily declining in recent years, and the incidence of measles has not?

---

[19]Based on "Herd Immunity," a student project handout by A. L. Miller and K. D. Stroyan.

4. Use equations (8) and (11) to show that

$$\frac{ds}{dr} = -cs, \tag{41}$$

where $s = S/N$ is the fraction of the population that is susceptible, $r = R/N$ is the fraction of the population that is recovered, and c is the contact number. Add an appropriate initial condition, and solve the initial value problem to find the relationship between s and r. Check your answer against the data in Figure 2 at the end of the flu epidemic: Calculate c from the given values of β and λ, and estimate s_∞ and r_∞ from the Figure; then check that these numbers satisfy the relationship you have derived from the initial value problem.

5.[20] (a) Suppose the supply and demand for a given product are given by $Q = 2 + 9p$ and $D = 40 - 10p$, respectively, and that the market for the product fits the continuous price evolution model (26) with $\gamma = 0.17$. If the initial price is 10% above the equilibrium price, at what time will the price be 5% above equilibrium?

(b) Answer the same question as in part (a) if the market instead fits the discrete price evolution model (34).

6. Here again is the formula for calculating a geometric sum with k terms, each term being β times the previous term:

$$1 + \beta + \beta^2 + \cdots + \beta^{k-1} = \frac{1 - \beta^k}{1 - \beta}. \tag{40R}$$

(a) Explain why, if $|\beta| < 1$, the sum has a finite limiting value as k becomes infinitely large. Explain why that limiting value is $\frac{1}{1 - \beta}$.

(b) If $\beta > 1$ or $\beta < -1$, what is the limiting sum (if any) as k becomes infinitely large?

(c) If $\beta = 1$, what is the limiting sum (if any) as k becomes infinitely large?

(d) If $\beta = -1$, what is the limiting sum (if any) as k becomes infinitely large?

---

[20]From *An Introduction to Mathematical Models in Economic Dynamics* by David Clements (see Note 9).

7. (a) Use the formula from the part (a) of the preceding problem to explain why 0.3333... is the same number as $\frac{1}{3}$.

(b) What fraction is the same number as 0.353535...?

(c) What fraction is the same number as 0.199219921992...?

8. (a) Find the *exact* sum:

$$1+\frac{6}{7}+\left(\frac{6}{7}\right)^2+\left(\frac{6}{7}\right)^3+\cdots+\left(\frac{6}{7}\right)^{40}.$$

(b) Find a *decimal approximation* to the sum in part (a) with 7SD accuracy.

(c) Find the *exact* sum with infinitely many terms:

$$1+\frac{6}{7}+\left(\frac{6}{7}\right)^2+\left(\frac{6}{7}\right)^3+\cdots.$$

9. Zeno of Elea, a Greek philosopher of the fifth century B.C., constructed several paradoxes to show the impossibility of motion. Here is one of them:

> You cannot walk across the room because, to do so, you would first have to walk half-way across the room, then half the remaining distance, half of that distance, and so on. As you have to cover "half the remaining distance" an infinite number of times, you will never complete the trip. Therefore, motion is an illusion. (If you are having any doubts about the wisdom of this argument, get up and walk across the room.)

(a) Using the width of the room as a unit of distance, how far do you walk in the first step of this process? In the first two? The first three?

(b) Without any further adding of fractions (but using your formula for calculating a geometric sum), how far do you walk in the first 27 steps?

(c) What happens when the number of steps becomes infinite? And just how is it that you can complete an infinite number of these "steps" in a finite time?

10. Suppose that 70 cents of every dollar spent in the U. S. is spent *again* in the U. S. (Economists call this the "multiplier effect.") If the federal government pumps an extra billion dollars into the economy, how much total spending in the U. S. occurs as a result?

11. Suppose you drop a ball from a height of one meter, and the ball bounces to a height of three-quarters of a meter. (This is about the right "coefficient of restitution" for a Superball.) You let the ball continue to bounce until it comes to rest.

    (a) How high does it rise on the second bounce? On the third?

    (b) Estimate the total distance that the ball travels up and down.

    (c) Estimate the total time the ball is bouncing before it comes to rest.

    [If you have a reasonably "bouncy" ball handy, you can check to see how close the theoretical model is to reality. First, determine what number should replace "three-quarters," and solve the exercise with a coefficient of restitution that matches your real ball. Compare the time result with your observation of total time.]

12. In Chapter 1, we stated without explanation the following formula for the monthly payment on a new car:

$$p = \frac{(P - D)\, r\, (1 + r)^n}{(1 + r)^n - 1},$$

    where p is the monthly payment, P is the price of the car, D is the down payment, r is the monthly interest rate (as a decimal fraction), and n is the number of months required to pay back the loan. Explain the formula. [Hints: $P - D$ is the amount borrowed. The total amount paid back is np. The factor by which the outstanding balance grows each month (before a payment is made) is $1 + r$.]

13. A physician decides to give a patient an infusion of glucose at a rate of 10 grams per hour. The body of the patient simultaneously converts the glucose and removes it from the bloodstream at 3 grams per hour per gram of glucose present.

    (a) Explain why the amount $G = G(t)$ of glucose present at time t can be modeled by the differential equation $\frac{dG}{dt} = 10 - 3G$.

    (b) Given 2 grams of glucose in the bloodstream at time 0, find an explicit formula for the amount present at any time t.

    (c) At what time is the glucose level in the patient's bloodstream rising fastest? What is that fastest rate of increase?

(d) Suppose Euler's Method is used to generate approximate values of the glucose level, G_1, G_2, G_3, and so on, at times t_1, t_2, t_3, and so on, that are 15 minutes apart. What will the calculated approximate level of glucose be one hour after the infusion starts?

(e) Use your formula from part (b) to calculate the glucose level at the end of an hour. How does your approximation in part (d) compare with the glucose level calculated from the formula?

CHAPTER 6

Periodic Motion

Our investigations of population growth, cooling, falling bodies, electrical circuits, optics, epidemics, price evolution, and other phenomena have led to models based on exponential, logarithmic, and algebraic functions. We turn now to phenomena that exhibit *repetitive* behavior — for which none of the functions used so far is an especially good model. Exponential and logarithmic functions never repeat *any* of their values; some of the algebraic functions do — after passing through peaks or valleys — but not in a regular, repeating pattern.

Where do we see repetitive phenomena? Our hearts (if healthy) beat regularly, returning to the same state every second or so. The sun rises and sets in a regular cycle of about 24 hours. A long-play record spins at 33 revolutions per minute — that is, it returns to the same state 33 times every minute. The disks in our compact disk players also spin, but not at a *constant* speed. The electric current provided by our local utility (if properly regulated) alternates in direction 60 times every second. The planets, asteroids, and even some comets, travel around the Sun in repeating orbits. The hands of our analog clocks and watches traverse a complete circle every minute, hour, or half-day. (Which hand does which?) And all the operations in our computers are timed by an internal clock that cycles several million times a second.

To model such repetitive processes, we need to identify and study "basic" repetitive functions, the ones that will play a similar role to that of the power functions in the study of falling bodies and exponential functions in the study of biological growth. Do you already know some functions that repeat the same values over and over? If so, write their names here: __________________

6.1 Circular Functions: Sine and Cosine

We obtain the simplest repeating functions by considering the simplest repeating process we know: going around in circles.

Functions Generated by Circular Motion

Imagine a circular running track[1] of radius one hundred meters, and imagine an xy-coordinate system placed on the track with the center of the circle at the origin. (See Figure 1, which leaves only unessential details to your imagination.) Suppose you start at the point with coordinates $x = 100$ and $y = 0$ and run in a counterclockwise direction with a speed of five meters per second. How can we describe your position as a function of the angle θ made by your line to the origin and the x-axis? How can we describe your position as a function of time?

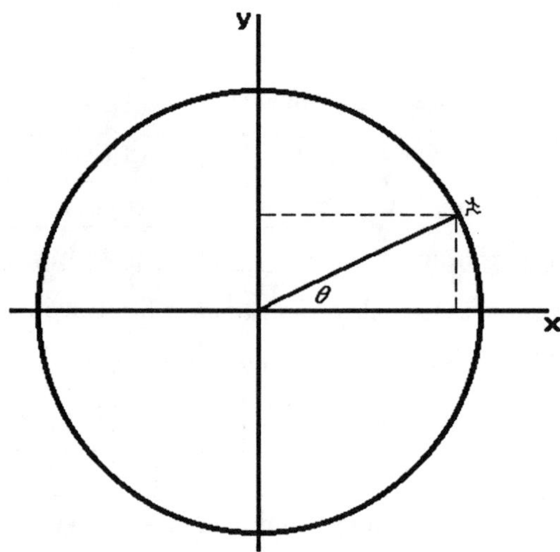

Figure 1. Position of runner on a track of radius 100 meters.

Consider first the question of describing your position in terms of the angle θ. We'll measure the angle in radians, so one lap corresponds to 2π radians, a quarter lap to $\frac{\pi}{2}$ radians, and so on. To locate your position in the plane, we need to specify both the x and y coordinates. The functions that do this should be old friends — or at least passing

---

[1]Thought question: Why are our running tracks never circular? The answer may have more to do with economics than with athletics.

acquaintances. If you have traveled through an angle of θ radians, then you are located at the point with x-coordinate $100 \cos \theta$ meters and y-coordinate $100 \sin \theta$ meters. That is,

$$x = 100 \cos \theta, \tag{1}$$

and

$$y = 100 \sin \theta. \tag{2}$$

Exercise 1. *(a) Label the distances in Figure 1 that give the x- and y-coordinates of the runner and the* 100 *meter radius.*

(b) Explain why x *and* y *are related to* θ *by cosine and sine functions, respectively. (Look at an appropriate right triangle in the Figure.)*

(c) Explain why, as you continue running laps, both your x-coordinate and your y-coordinate oscillate between 100 *meters and* -100 *meters.*

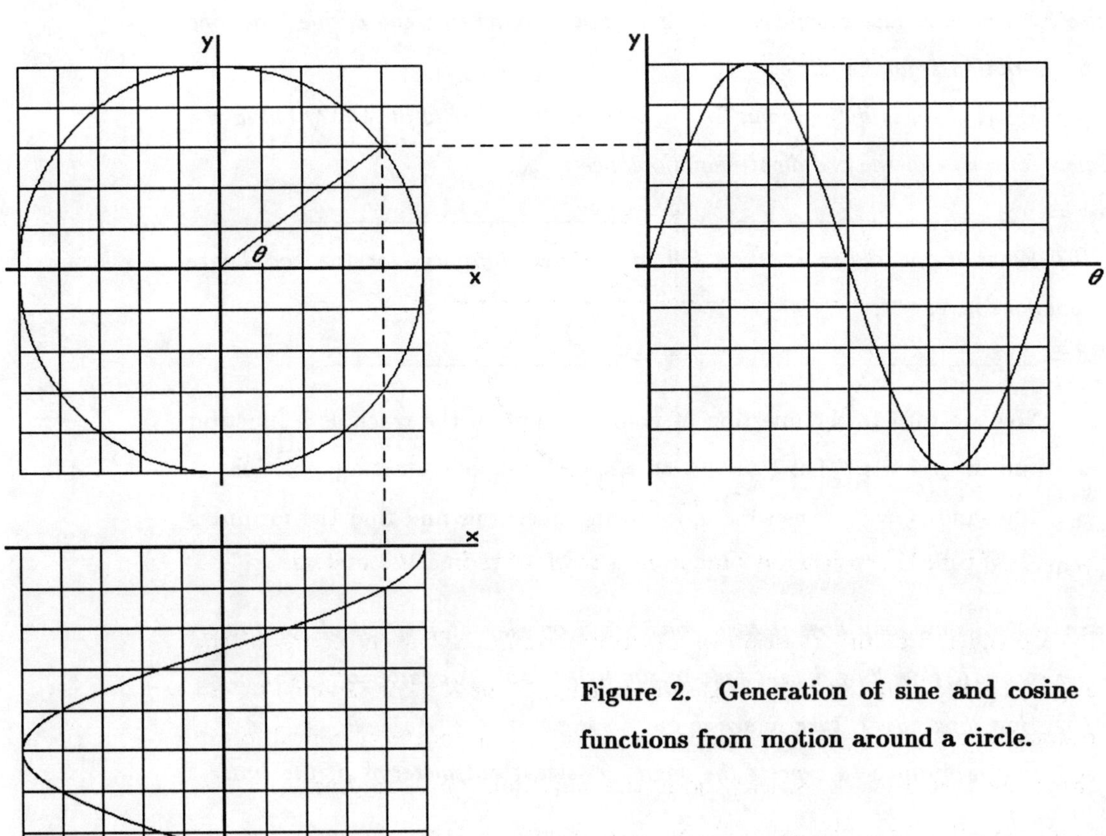

Figure 2.   Generation of sine and cosine functions from motion around a circle.

In Figure 2, we show another circle of radius 100 with separate graphs of the x- and y-coordinates that result from motion around the circle. Put the index finger of your left hand on the circle at the point $(100, 0)$ and the index finger of your right hand on the graph of the y-coordinate at the point $(0, 0)$. As you move your left finger along the circle, let your right finger trace out the graph of y as a function of θ.

Notice that the axes for the x-coordinate graph are rotated $\frac{\pi}{2}$ radians ($90°$) clockwise from the usual position; turn the page counterclockwise to be sure you understand what that part of the Figure shows. Now place your left index figure at the point $(0, 100)$ on the x-coordinate graph and your right index finger on the circle at the point $(100, 0)$. As you move your left finger along the circle, let your right finger trace out the graph of x as a function of θ.

Exercise 2. *You have just experienced the generation of the sine and cosine functions from motion around a circle.*

(a) What changes if the radius of the circle is 27 instead of 100? Fill in the new formulas for the coordinate functions here:

$x =$ _________________________ , $y =$ _________________________

(b) What if the radius is 1? Fill in the new formulas for the coordinate functions here:

$x =$ _________________________ , $y =$ _________________________

Now we turn to the question of your location on the track as a function of time in seconds. Let $t = 0$ correspond to your starting position at $x = 100$ and $y = 0$. Answer the following questions and find the formulas requested.

Exercise 3. (a) How long does it take you to run one lap at a speed of 5 meters per second? How long does it take to run half a lap? A quarter of a lap?

(b) How long does it take to sweep out an angle of $\frac{\pi}{2}$? An angle of $\frac{\pi}{3}$?

(c) Find a formula to express the angle θ swept out in terms of the time t elapsed.

(d) Now find formulas for the x-coordinate and the y-coordinate as functions of time.

Exercise 4. (a) *What circle is traveled by a moving object with coordinate functions*

$x = \cos\theta$ *and* $y = \sin\theta$?

(b) *Use your answer to part (a) to explain the following facts about sine and cosine:*

(i) *For every angle* θ, $\cos(\theta + 2\pi) = \cos\theta$ *and* $\sin(\theta + 2\pi) = \sin\theta$.

(ii) *For every angle* θ, $\sin^2\theta + \cos^2\theta = 1$.[2]

(iii) *For every angle* θ, $|\sin\theta| \leq 1$ *and* $|\cos\theta| \leq 1$.

We can interpret the property described in Exercise 4 (b) (i) in the following way: For both the sine function and the cosine function, no matter where you start with θ, when you increase θ by 2π, you will return to the same function value as when you started. In Figure 2, we have graphed each of these functions on the θ-interval from 0 to 2π; the "repeating" property tells us that we could construct the rest of the graph for each of these functions by copying over and over the part we have done already. Try it here:

Exercise 5. *On the axes below, copy from Figure 2 the graph of* $y = \sin\theta$ *from* 0 *to* 2π. *Then replicate this graph as many times as necessary to construct the graph of* $y = \sin\theta$ *from* -3π *to* 7π. *On the second set of axes, repeat for the graph of* $x = \cos\theta$.

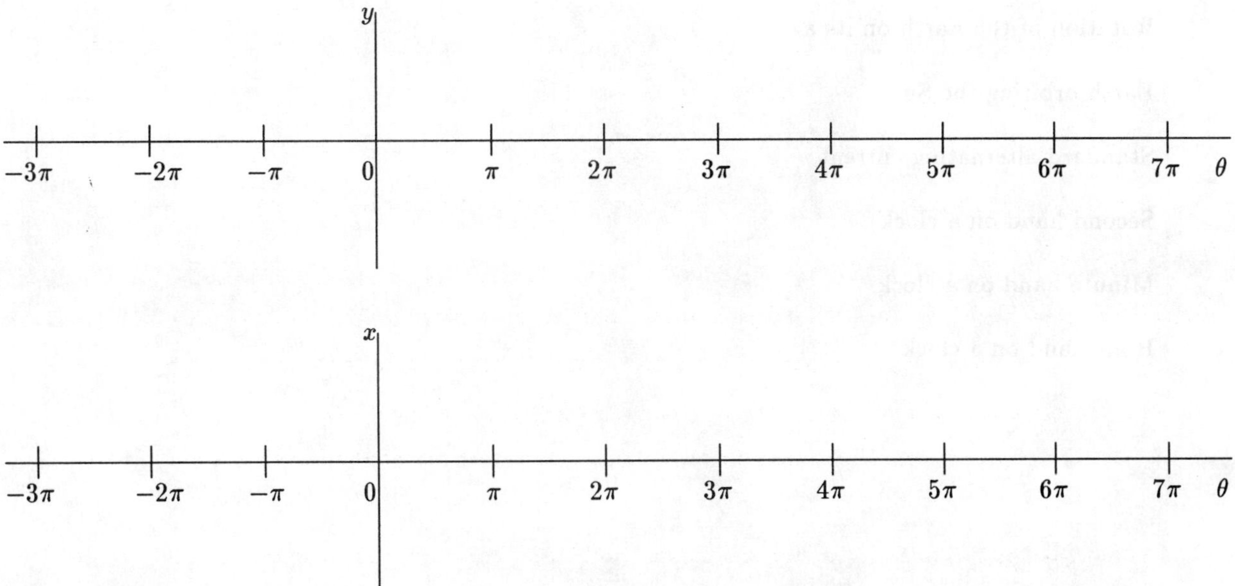

----

[2]Recall that $\sin^2\theta$ is a conventional notation that means the same thing as $(\sin\theta)^2$.

Period and Frequency

A function that repeats over and over, as do sine and cosine, is called
"periodic," and the horizontal length of the shortest pattern that repeats is
called the **"period."**   We can also think of the period as being the *time*
required to complete a single "cycle" (if time is the independent variable).
Thus, sine and cosine are periodic functions with period 2π.

A concept closely related to period is that of **"frequency,"** which is the
rate at which periods are being completed.   Thus, if the unit of time is
seconds, and a periodic function has period 5 seconds, then its frequency is
$\frac{1}{5}$ cycles per second. In general, frequency and period are reciprocals of each
other. The unit for frequency is "cycles per unit of time," and the unit for
period is "units of time per cycle."

Exercise 6. *Read the second paragraph of this chapter again. Next to each of the*
following repeating phenomena, write the period and frequency (in compatible
time units) of whatever function describes the phenomenon.

Phenomenon	Period	Frequency
Normal heart beat		
LP record		
Rotation of the Earth on its axis		
Earth orbiting the Sun		
Standard alternating current		
Second hand on a clock		
Minute hand on a clock		
Hour hand on a clock		

Derivatives of Sine and Cosine

In order to use sine and cosine functions to model phenomena that change in repeating patterns, we need to know the rates of change of these two functions. Thus, it is time to calculate those derivatives.

We start with some pictorial evidence of what to look for. In Figure 3, we show a graph of the sine function and in Figure 4 a graph of an *approximate* derivative, calculated as

$$\text{Approximate derivative of } \sin t \;=\; \frac{\sin (t + 0.001) - \sin t}{0.001}. \qquad (3)$$

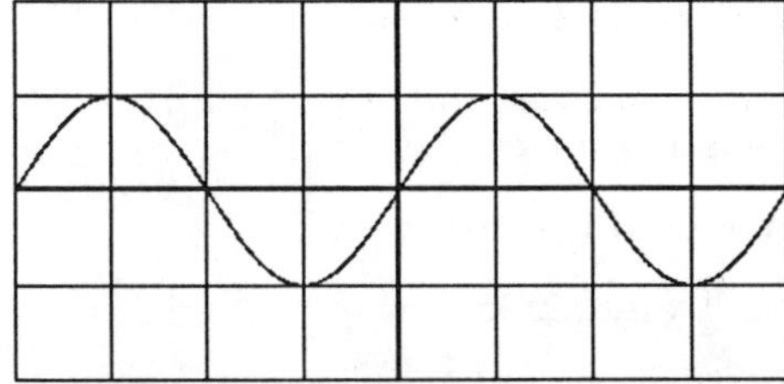

Figure 3. The sine function.

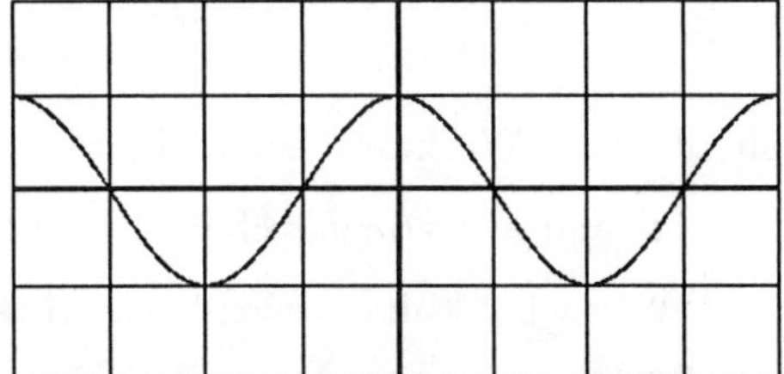

Figure 4. An approximation to the derivative of $\sin t$.

Exercise 7. *(a) Label the axes and indicate the horizontal and vertical units on the left-hand graph in Figure 3.*

(b) The scales are the same for Figure 4; label it also.

(c) Why does the expression on the right in (3) approximate the derivative?

(d) What do you see in the right-hand graph? What function do you think is the derivative of $\sin t$? Write your answer here: _______________

In Figures 5 and 6, we do the same thing with the cosine function; the picture on the right is the graph of

$$\text{Approximate derivative of } \cos t \;=\; \frac{\cos (t + 0.001) - \cos t}{0.001}. \qquad (4)$$

Exercise 8. *(a) Label the axes and indicate the units in Figures 5 and 6. (The units are the same as in Figures 3 and 4.)*

(b) What function do you think is the derivative of the cosine? Write your answer here: _____________

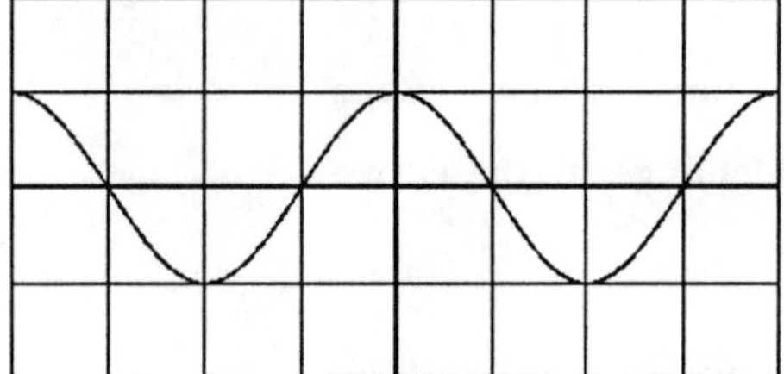

Figure 5. The cosine function.

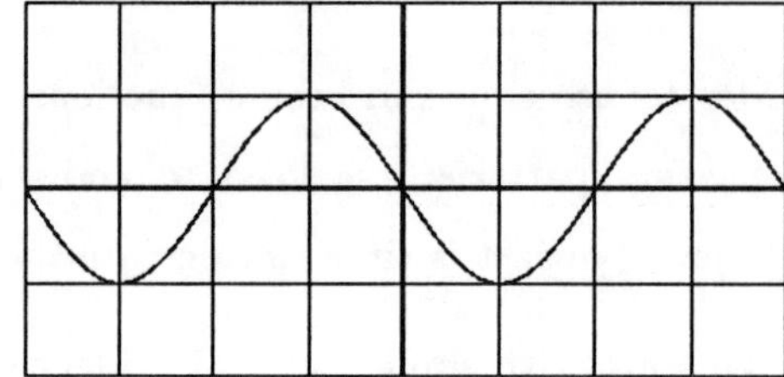

Figure 6. An approximation to the derivative of cos t.

To actually calculate the derivative of $\sin t$, we need to examine algebraically the difference quotient

$$\frac{\sin (t + \Delta t) - \sin t}{\Delta t}$$

for small Δt. We know from Figure 3 what this expression looks like for varying t and a *particular* choice of Δt, and you have used this evidence to conjecture what the derivative is. However, there isn't any obvious way to use algebra to verify the conjecture, i.e., to find a factor of Δt in the numerator to cancel with the one in the denominator.

There is, however, a way to make progress with algebra, if we write the numerator in some other form. For this purpose, we need to recall the trigonometric identity for "sine of a sum":

$$\sin (\alpha + \beta) \;=\; \sin \alpha \cos \beta + \cos \alpha \sin \beta, \tag{5}$$

Letting $\alpha = t$ and $\beta = \Delta t$, we may write $\sin(t + \Delta t) = \sin t \cos \Delta t + \cos t \sin \Delta t$. When we substitute this expression in the numerator of the difference quotient, we obtain

$$\frac{\sin (t + \Delta t) - \sin t}{\Delta t} = \frac{\sin t \cos \Delta t + \cos t \sin \Delta t - \sin t}{\Delta t}.$$

Now we regroup the terms on the right (check the details!) to obtain

$$\frac{\sin (t + \Delta t) - \sin t}{\Delta t} = \frac{\sin \Delta t}{\Delta t} \cos t + \sin t \frac{\cos \Delta t - 1}{\Delta t}. \tag{6}$$

This algebraic manipulation has allowed us to rewrite the difference quotient [left side of (6)] in a form that effectively separates "t" from "Δt" [right side of (6)]. Now, in order to determine what happens to the difference quotient as Δt shrinks to zero, we only need to find the limiting values of $\frac{\sin \Delta t}{\Delta t}$ and $\frac{\cos \Delta t - 1}{\Delta t}$. In Figures 7 and 8, we show graphs of these two functions of Δt; what do you think the limiting values are as Δt approaches 0?

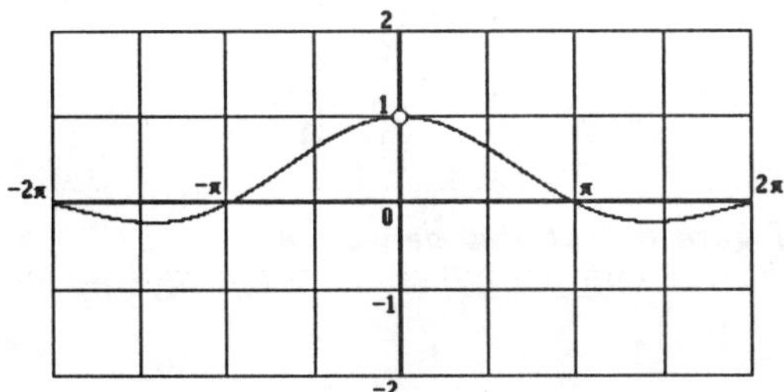

Figure 7. Graph of $y = \dfrac{\sin \Delta t}{\Delta t}$.

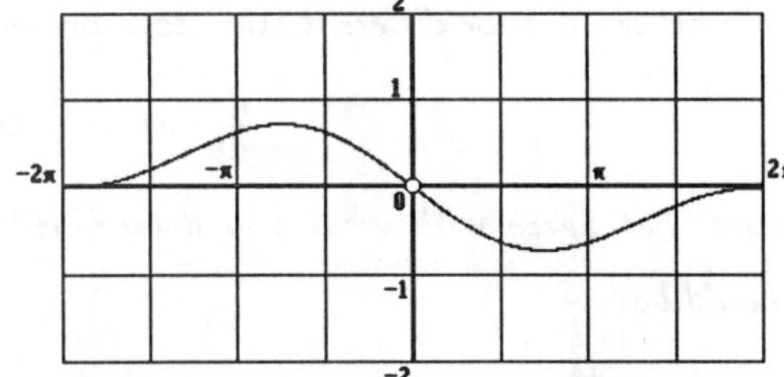

Figure 8. Graph of $y = \dfrac{\cos \Delta t - 1}{\Delta t}$.

Exercise 9. *(a)* *Check your perception by using your calculator to fill in the missing entries in the following table:*

Δt	$\sin \Delta t$	$\dfrac{\sin \Delta t}{\Delta t}$	$\cos \Delta t - 1$	$\dfrac{\cos \Delta t - 1}{\Delta t}$
1.0	0.84147	0.84147	-0.45970	-0.45970
0.1	0.099833	0.99833	-0.0049958	-0.049958
0.01				
0.001				
0.0001				
0.00001				

(b) *What is the limiting value of* $\dfrac{\sin \Delta t}{\Delta t}$ *as* Δt *approaches* 0?

(c) *What is the limiting value of* $\dfrac{\cos \Delta t - 1}{\Delta t}$ *as* Δt *approaches* 0?

(d) *Use your answers to (b) and (c), together with equation (6), to show that the derivative of the sine function is what you already knew it had to be.*

We could find the derivative of $\cos t$ in much the same way (see Problem 1 at the end of the chapter), but it is easier to use what we already know about the derivative of $\sin t$; the two functions are related by this pair of identities:

$$\cos \alpha = \sin\left(\frac{\pi}{2} - \alpha\right) \tag{7}$$

and

$$\cos\left(\frac{\pi}{2} - \alpha\right) = \sin \alpha. \tag{8}$$

Exercises

10. *Use identities (7) and (8), your answer to the previous exercise, and your other rules for differentiation to show that*

$$\frac{d}{dt}\cos t = -\sin t. \tag{9}$$

Does that agree with what you determined from Figure 6 that this derivative should be?

11. *(a) Find $\dfrac{d^2}{dt^2}\sin t$.*

(b) Find $\dfrac{d^2}{dt^2}\cos t$.

6.2 Spring Motion

The Spring-Mass System

We shall see that the sine and cosine functions are the keys to describing all repetitive or periodic behavior. For our first illustration of this fact, we consider the motion of a mass bobbing up and down on the end of a spring. That may not sound like an important problem (unless you have an old car in which the shock absorbers are really shot), but it serves nicely as a "prototypical" oscillatory behavior, one that we can easily observe and measure. It's much harder to observe the crystals in our watches, the electrical current in our walls, or the silicon circuits in our computers. And study of the bouncing mass on a spring is the beginning of analyses of phenomena such as the vibrations of a bridge caused by traffic and wind or the flutter of an airplane wing caused by air flowing past and by vibrations of the engine. In all of these cases we are interested in knowing such things as the frequency and amplitude[3] of the vibrations.

In Problem 6 of Chapter 1, we asked you to express "Hooke's Law" by a simple formula. This physical "law" asserts that the force required to stretch a spring a distance s from its natural length is proportional to s, i.e., the force function has the form $F(s) = ks$ for some constant k, which is called the "spring constant" for that spring. To find the spring constant for a given spring, we can hang a known weight[4] on the spring and measure the amount of displacement. For example, if a five-pound weight stretches a spring six inches, then (in English units of feet and pounds), $5 = k \cdot \frac{1}{2}$, so the spring constant k is 10 lbs/ft.

In Figure 9 we show two views of the same spring, one at its natural length (no force being applied) and the other stretched by hanging an object

---

[3]Frequency is defined in the previous section; "**amplitude**" is the maximum amount of displacement from zero in either direction. Thus, the functions $A \sin \theta$ and $A \cos \theta$ both have amplitude A.

[4]"Weight" means the force exerted on an object by gravity. By Newton's Second Law, this force is the *mass* of the object times the acceleration g due to gravity. Thus, weight and mass are related by $w = mg$.

of mass m on it. We assume the spring-mass system is in equilibrium in the second view, i.e., the object is not moving. Thus, the force the spring exerts on the object exactly balances the (gravitational) force the object exerts on the spring, namely, mg. By Hooke's Law, this force is also ks_0, so the mass, spring constant, and equilibrium displacement are related by $mg = ks_0$.

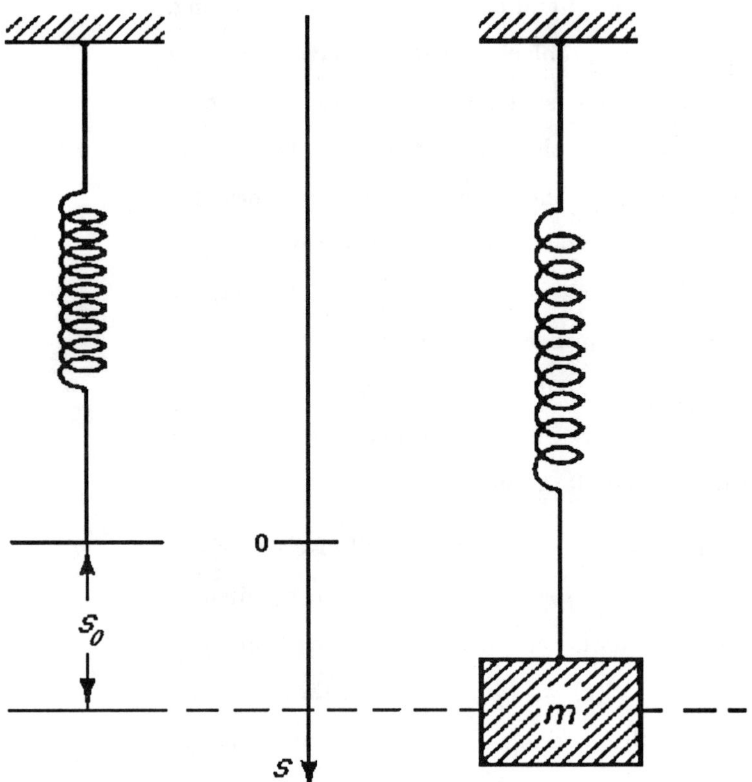

Figure 9. Stretching a spring by hanging a mass on it.

Now we consider what happens if we stretch the spring an additional distance x, as in Figure 10. (Think of rescaling the vertical axis additively so that the new origin $x = 0$ is at $s = s_0$.) The system consisting of just the mass and spring is no longer in equilibrium; if we let go, there will be an imbalance of forces that will cause the object to move. On the one hand, there is the gravitational force mg (positive because our positive direction is downward); on the other hand, there is the "restoring" force of the spring, which, according to Hooke's Law, is $-ks = -k(x + s_0)$ (negative because this force opposes increase in s). Thus, the total

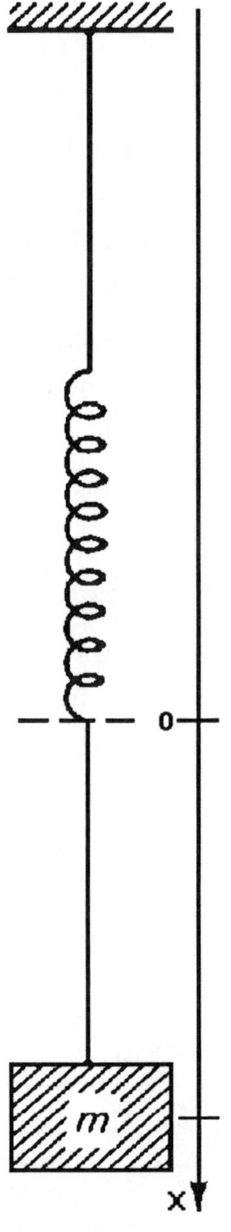

Figure 10. Stretching a spring a distance x.

force on the object is $mg - kx - ks_0$, which simplifies to just $-kx$, because $mg = ks_0$.

The Initial Value Problem

By Newton's Second Law of Motion, the total force on the moving object of mass m is also its mass times its acceleration. Thus, we have a relationship between the displacement x and its second derivative:

$$m \frac{d^2 x}{dt^2} = -kx. \tag{10}$$

Some features of equation (10) should look familiar and some should appear new. First, under the influence of gravity, our bouncing mass is a "falling body" — but with a second force acting on it (the spring). When the falling body had no other force acting on it, our differential equation model expressed the second derivative of the position function as a *constant*; here the second derivative is proportional to the position function.

Second, as was the case with the falling body, we should expect to "undo" differentiation twice to find x as a function of t. But, unlike the falling body model, it is not at all clear how to do that. We might also expect that each "undifferentiation" step will involve an initial condition in some way, probably an initial velocity and an initial position.

Finally, note that "proportional to the unknown function" has appeared before — in natural population growth. What's different here is that it is the *second* derivative, not the first, that is proportional to the unknown function. Of course, an exponential function, say, $f(t) = e^{rt}$, has a second derivative that is proportional to the function itself, but it seems unlikely that such a function could tell us anything about bouncing up and down on a spring. Let's rule out such functions as solutions to (10) right away:

Exercise 1. (a) Calculate the second derivative of $f(t) = e^{rt}$.

(b) Show that f'' is proportional to f and that the proportionality constant must be positive, no matter what sign r has.

(c) Explain why f cannot be a solution of equation (10).

We have just eliminated a promising candidate (promising mathematically, but not physically) for a solution to the differential equation (10). We have made some assertions (with little supporting evidence) that sines and cosines should have something to do with the oscillatory behavior of the spring-mass system. Look again at your results in Exercise 11 at the end of Section 6.1: The second derivatives of the sine and cosine functions *are* proportional to those functions, with proportionality constant -1 — at least the *sign* is right. Before we pursue the questions of (a) how get the *proportionality constant* right and (b) whether to use *sine* or *cosine* or some combination of the two, we will study the differential equation from a numerical point of view first.

A Numerical Solution: Euler's Method

If we can formulate our problem as an initial value problem, then we have some hope that we can generate a numerical or graphical solution, which may in turn provide some evidence about the possible role to be played by sine and cosine functions.

In addition to the equation of motion,

$$m \frac{d^2x}{dt^2} = -kx, \tag{10R}$$

we know something about how the motion starts. In particular, if we pull the mass to an initial displacement x_0 before we let go, then we know that $x(0) = x_0$. And if we simply *let go*, that is, we don't "throw" the mass either up or down, then the initial velocity, $x'(0)$, must be zero. Thus, we can state our initial value problem [after solving (10) for the second derivative] in the following way:

$$\frac{d^2x}{dt^2} = -\frac{k}{m}\,x, \quad \text{with} \quad x(0) = x_0 \quad \text{and} \quad x'(0) = 0. \tag{11}$$

Euler's Method, as we have used it so far, seems to apply only to "first-order" equations, that is, to equations in which the *first* derivative is expressed by a formula involving the independent or dependent variables. But, as we have seen with the flu epidemic model in the previous chapter, the method can be used with a *system* of first-order differential equations, provided we have an initial condition for *each* of the unknown functions. We can use that knowledge here if we can write a single second-order differential equation as a system of two first-order equations.

We have two strong hints in our previous work about where to look for the second unknown function. First, we already have two initial conditions, one for the function x we are trying to find, the other for its derivative x'. Perhaps the second function should be the velocity! That would fit with the second hint, which comes from the falling body problem, in which our actual approach to solving the equation $\dfrac{d^2 s}{dt^2} = g$ was to solve a sequence of two first-order problems, one for finding velocity from acceleration, the other for finding position from velocity.[5]

Suppose $x(t)$ is the desired function for position of the bouncing mass at time t, i.e., the solution of the initial value problem (11). If we write $v(t) = \dfrac{dx}{dt}$, then x and v must satisfy the system of *first-order* equations

$$x' = v, \tag{12}$$

$$v' = -\frac{k}{m}\, x, \tag{13}$$

with the initial conditions

$$x(0) = x_0, \tag{14}$$

$$v(0) = 0. \tag{15}$$

(Be sure you know where each of these four equations comes from.)

Recall the basic idea of Euler's Method: We step forward in time in equal steps of length Δt, so $t_0 = 0$, $t_1 = \Delta t$, $t_2 = 2(\Delta t)$, and so on. Initially, we know starting values for x and v from (14) and (15), and we can calculate starting slopes for the two functions by substituting these values in (12) and (13). Given slopes, we can calculate a rise for each function as "slope $\times$ run"; adding each rise to the corresponding starting value gives the next value for each function. And then we start over: From

---

[5]If you have completed the solution of the falling-body-with-air-resistance problem (Problem 7 in Chapter 3), you followed exactly the same route: The physical assumptions led to a differential equation for the velocity function; its solution then gave a differential equation for the position function, which you could solve by guessing the form of the solution and calculating the constants.

known function values, we use (12) and (13) to calculate slopes, use slopes to calculate rises, and add rises to find the next values.

Of course, as we saw in section 5.1, the rises we are calculating are really dx and dv (rises to tangent lines) instead of what we really want to know, Δx and Δv (rises to the unknown curves). Euler's Method is only as good as the approximations (see Section 4.7) $dx \approx \Delta x$ and $dv \approx \Delta v$.

Verify that the discussion in the last two paragraphs leads to the forward-stepping formulas:

$$x_{k+1} = x_k + v_k \cdot \Delta t, \tag{16}$$

$$v_{k+1} = v_k - \frac{k}{m} \cdot x_k \cdot \Delta t, \tag{17}$$

with

$$x_0 = \text{the starting displacement,} \tag{18}$$

$$v_0 = 0. \tag{19}$$

In Figures 11 and 12 we show solutions for x and v generated from formulas (16)–(19) for a starting displacement x_0 of one unit and $k/m = 1$.

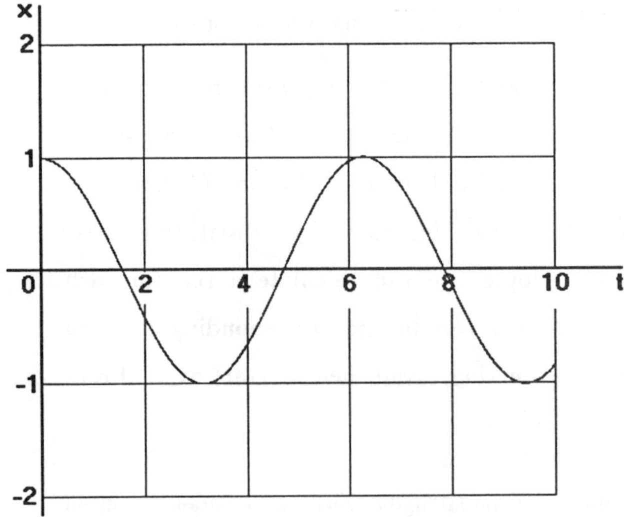

Figure 11. Numerical solution for position $x(t)$

of the mass bouncing on the spring.

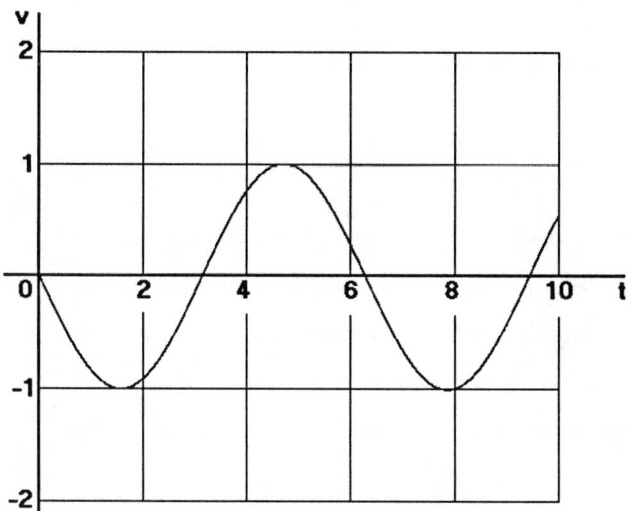

Figure 12. Numerical solution for velocity $v(t)$

of the mass bouncing on the spring.

Exercise 2. (a) *Examine Figures 11 and 12 carefully. Do the curves behave the way you would expect from what you know about the physics of the spring-mass system?*

(b) Are the curves believable as a solution to the initial value problem

$$x' = v, \quad \text{with} \quad x(0) = 1,$$

$$v' = -x, \quad \text{with} \quad v(0) = 0?$$

Why or why not?

(c) How do you think the numerically generated graphs of $x = x(t)$ and $v = v(t)$ are related to the sine and/or cosine functions? Does the relationship fit with what you know about derivatives of these functions?

A Symbolic Solution

We are ready now to find symbolic solutions of the differential equation,

$$\frac{d^2x}{dt^2} = -\frac{k}{m}\,x, \tag{10R}$$

where k and m are arbitrary positive constants. Then we can use the initial values for x and x' to complete our search for formulas for the solutions we generated graphically in Figures 11 and 12. As usual, we attack this problem by first solving an easier *inverse* problem. That is, we calculate derivatives of functions that would appear to have the right form to satisfy (10).

Exercise 3.[6] *(a) Show that $\dfrac{d^2}{dt^2}\sin \omega t = -\omega^2 \sin \omega t$.*

(b) Show that $\dfrac{d^2}{dt^2}\cos \omega t = -\omega^2 \cos \omega t$.

Exercise 4. (a) Show that $x(t) = \sin \omega t$ and $x(t) = \cos \omega t$ are both solutions of the differential equation

$$\frac{d^2x}{dt^2} = -\omega^2 x. \tag{20}$$

(b) Show that any function of the form $x(t) = A \sin \omega t + B \cos \omega t$ is a solution of the same differential equation.[7]

Our original differential equation (10) has precisely the form of equation (20) if we identify ω^2 with $\frac{k}{m}$, that is, if we set $\omega = \sqrt{\frac{k}{m}}$. And we can do that, because we know that both k and m are positive. Thus, according to Exercise 4, any function of this form is a solution of (10):

$$x(t) = A \sin \sqrt{\tfrac{k}{m}}\, t + B \cos \sqrt{\tfrac{k}{m}}\, t\ . \tag{21}$$

---

[6]The symbol ω is a lower case Greek omega; it is a conventional symbol for a period-altering scale factor in trigonometric functions.

[7]Notice the parallel here with what we did in Chapter 2. There we looked for functions to satisfy an equation of the form "first derivative is proportional to the function" and found that the solutions were all of the form Ae^{kt}. In particular, the scale factor for the independent variable was the proportionality constant in the differential equation. Now we are solving "second derivative is *negatively* proportional to the function" and finding a similar result with sines and cosines. In particular, the scale factor for the independent variable is the square root of the absolute value of the proportionality constant. "Second derivative *positively* proportional to the function" is a different, but also similar, situation — see Problems 4-6 at the end of the chapter.

Now we need to determine what combination of coefficients A and B (if any) will give us *one* function that also satisfies the initial conditions for displacement and velocity. We start by substituting $t = 0$ and $x = x_0$ in (21):

$$x_0 \;=\; A \sin 0 + B \cos 0. \tag{22}$$

Since $\sin 0 = 0$ and $\cos 0 = 1$, equation (22) tells us that B must be the starting displacement x_0.

Since A dropped out of the equation when we substituted $t = 0$, we know nothing yet about the value of A. But we have another initial condition! In order to use it, we need a formula for $v(t)$, which you can provide by differentiating the formula for $x(t)$ in equation (21).

Exercise 5. (a) Show that

$$v(t) = A \sqrt{\tfrac{k}{m}} \; \cos \sqrt{\tfrac{k}{m}}\, t \;-\; B \sqrt{\tfrac{k}{m}} \; \sin \sqrt{\tfrac{k}{m}}\, t \;. \tag{23}$$

(b) Substitute the initial values for t and v to show that A must be zero.

Conclusion: The unique solution of the spring-mass initial value problem,

$$\frac{d^2 x}{dt^2} \;=\; -\,\frac{k}{m}\, x, \quad \text{with} \quad x(0) = x_0 \quad \text{and} \quad x'(0) = 0, \tag{11R}$$

is

$$x(t) \;=\; x_0 \cos \sqrt{\tfrac{k}{m}}\, t. \tag{24}$$

In order to generate a numerical solution for Figures 11 and 12, we had to give numerical values for the starting displacement and for the ratio of spring constant to mass. We chose both of those numbers to be 1, so the solution we generated was just $x = \cos t$; the velocity function was then the derivative of $\cos t$, that is, $v = -\sin t$. Were those your answers to Exercise 2 (c)?

6.3 Pendulum Motion[8]

We don't make much use of pendulum clocks (sometimes called "grandfather" clocks) anymore, but pendulum motion provides another example of a repetitive phenomenon that we can easily observe and measure. We will see it has some features that are similar to spring motion and some that are different.

Experiments and Questions

Stop reading at the end of this paragraph, and make yourself a pendulum; string and a heavy object (such as a doorknob) work well. If the only doorknob you have is still attached to your door, substitute a coffee mug (empty) or your key case. If nothing else is at hand, take off your shoe and use your shoelace and the shoe. Hold the top of the string fixed, and pull the object at the bottom (called the "bob") to one side at an angle of approximately 30°, hold it still, then let it go. The pendulum will swing back and forth, with the maximum angle from the vertical decreasing with time until the swinging stops.

Thought question: What causes the pendulum to stop?[9]

Experimental question: The length of time it takes the pendulum to swing out and back is the "period" of the pendulum. Does the period seem to depend on the initial angle? Try various starting angles to see.

Another experimental question: The length of the string from your hand to the bob is called the "length" of the pendulum. Move your hand up and down the string to vary the length. How does the *period* of the pendulum seem to depend on the *length* of the pendulum?[10]

---

[8]At the discretion of the instructor, the rest of this chapter (except the Summary) may be omitted without loss of continuity in the course.

[9]The spring-mass system stops too, although our theoretical solution in equation (24) says it never does. There was a hidden assumption in our treatment of the spring, but we can't hide it any longer. That's what we're asking you to think about.

An Initial Value Problem

We will construct a model to describe how the *angle* of the pendulum varies as a function of time; then we will use this model to describe how the *period* of the pendulum varies as a function of the length of the pendulum. In particular, suppose a bob of mass m (say, 50 grams) is attached to the end of a string of length L (say, one meter). Assume the bob is pulled to one side so that the string makes an angle of 30 degrees with the vertical. The bob is held still and released.

To obtain a reasonably simple model for pendulum motion, we make a number of simplifying assumptions, namely:

(i) Relative to the mass m of the bob, the mass of the string is negligible.

(ii) Compared to the length of the string, the diameter of the bob is negligible.

(iii) At the stable point of swing (the pivot), friction and other retarding forces are negligible.

(iv) The retarding force of air resistance is negligible.[11]

You may well question some or all of these assumptions — particularly if your pendulum is your shoe on your shoestring. Compare these assumptions to your answer to the Thought Question in the "Experiments and Questions" section. Under these assumptions, is there anything that can make the pendulum come to rest?

Holding our skepticism in abeyance, let's see where these assumptions lead us. We can use Newton's Second Law of Motion to model the motion of the bob. Let s (a function of time t) be the distance along the arc from the lowest point to the current position of the bob, with displacement to the right considered positive. (Label s in Figure 13.) By Newton's Law, the force on the bob *in the direction of motion* is equal to $m\dfrac{d^2s}{dt^2}$.

---

[10]We could have asked a similar question about the spring-mass system — the period depends on both the mass and the spring constant, and equation (24) tells you exactly how. Unless you have access to a variety of weights and springs, it might be difficult to actually observe this variation. But with the pendulum, it is at least easy to see that there is a variation, even if it is not so easy to describe it by a formula.

[11]We made similar assumptions about the spring-mass system, but we did not list them explicitly. What did we assume that is analogous to each item in this list?

The force on the bob due to gravity is *downward* with a value of mg, where g is the acceleration due to gravity. As gravity is the principal (only?) cause of motion here, these two forces must be related somehow.

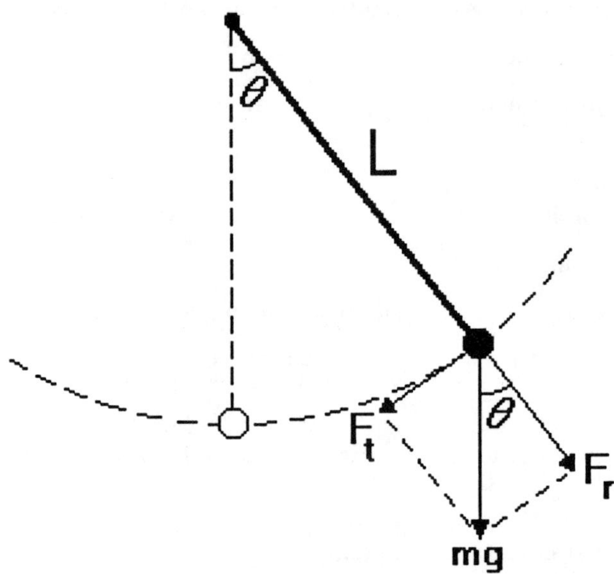

Figure 13. Tangential and radial components of gravitational force on the pendulum bob.

We have a small problem: The direction of motion is along (i.e., tangent to) the circle of motion, and this direction is definitely not downward.   To get around this difficulty, we employ a useful trick.   We (actually our industrious ancestors) have learned by experience that a single force may be considered to be the result of two different forces acting at different angles.   For example, if you want to push a heavy chair from behind, you could obtain the same result if you and a friend each pushed with less force from different angles behind the chair.

In the present situation, the gravitational force may be considered to be the result of a "tangential force" F_t and a "radial force" F_r.   The magnitudes of these forces should be such that they fit together as sides of a rectangle of which the diagonal is the gravitational force (see Figure 13); this is the way forces at different angles of application add together.[12,13]

Exercise 1. Use Figure 13 to explain why $F_t = -mg \sin \theta$. *Where does the factor of* $\sin \theta$ *come from? Where does the minus sign come from?*

Now we have two expressions for the tangential force, one from the application of Newton's Second Law to the motion in the tangential direction, the other from Exercise 1. If we set the two expressions equal, we see that s obeys the differential equation

$$m \frac{d^2 s}{dt^2} = -mg \sin \theta.$$

When we cancel the common factor m, we get

$$\frac{d^2 s}{dt^2} = -g \sin \theta. \tag{25}$$

There are two "features" of this differential equation that we need to think about — and then to do something about: (i) The variable s appears on the left side of the equation, while the variable θ appears on the right; each is an unknown function of t, and they are clearly related. (ii) The derivative in equation (25), as was the case with the spring-mass system, is a *second* derivative. The first of these features is easily dealt with, as we will see now; the second we will treat as we did with the spring. First things first.

Recall that s is distance along the circumference of a circle of radius L. For a particular pendulum, L is fixed and s varies with θ. We need to find the relationship between s and θ. Fill in the missing entries in the table below.

angle θ	distance s	
2π	$2\pi L$	(a full circle)
π		(a half circle)
$\dfrac{\pi}{2}$		

---

[12]The radial force is exactly balanced by the force of tension in the string; therefore, neither contributes anything to the motion of the pendulum, so we may safely ignore them. If the string were replaced by a *spring*, we would have a really interesting (and much tougher) problem to solve!

[13]We will discuss this decomposition of forces into perpendicular "component" forces (and other related matters) when we investigate vectors in Calculus III.

We see that $s = L\theta$; indeed, that's what it *means* to measure the angle in radians.[14] Hence $\dfrac{d^2s}{dt^2} = L\dfrac{d^2\theta}{dt^2}$ (why?). If we substitute this expression in (25) and divide through by L, we obtain

$$\frac{d^2\theta}{dt^2} = -\frac{g}{L}\sin\theta. \tag{26}$$

Now we have a differential equation with a single dependent variable θ.

Equation (26) is *similar* to the spring equation

$$\frac{d^2x}{dt^2} = -\frac{k}{m}x, \tag{10R}$$

with θ replacing x, g replacing k, and L replacing m. But an important *difference* between (26) and (10) is the presence of the sine finction in equation (26). Recall that, for the spring, trigonometric functions turned up only in the *solutions*.

For our assumed conditions, (starting at a $30°$ angle to vertical and letting the bob drop), our initial conditions are

$$\theta = \frac{\pi}{6} \quad \text{when} \quad t = 0 \tag{27}$$

and

$$\frac{d\theta}{dt} = 0 \quad \text{when} \quad t = 0. \tag{28}$$

Next to each of these conditions, write a description in words of its physical meaning.

In summary, the function we seek, in order to describe the motion of the pendulum, is the function $\theta(t)$ that satisfies the second-order initial value problem:

$$\frac{d^2\theta}{dt^2} = -\frac{g}{L}\sin\theta \;\; \text{with} \;\; \theta(0) = \frac{\pi}{6} \;\; \text{and} \;\; \theta'(0) = 0. \tag{26R-28R}$$

How can we describe the solution of an initial value problem such as this? One solution of equation (26) is easy: $\theta(t) = 0$. However, this "trivial" solution does not satisfy our initial conditions.

---

[14]One way to define radian measure is this: The radian measure of a central angle is the ratio of the length of any circular arc it subtends to the radius of the arc. In particular, if you consider only arcs of radius 1, the radian measure of the angle and the length of the arc are the same number.

Exercise 2. (a) What would the initial conditions have to be if they are to be satisfied by a trivial solution $\theta = 0$?

(b) What physical situation is described by this solution?[15]

Trivial solutions are not very interesting, because they describe only trivial physical situations that need no mathematics for their understanding. A few minutes of trial and error will show you that it is not easy to describe other functions satisfying the differential equation. In fact, we cannot give a simple formula to describe the function $\theta(t)$ that is the solution of the initial value problem (26)-(28). Instead, we will take two other approaches to describing the function — which we know must exist, because we see the pendulum move.[16]

A Numerical Solution: Euler's Method

Our first approach to solving the pendulum problem will be to find a numerical solution as we did for the spring-mass problem. We know now how to apply Euler's Method to a second-order initial value problem, and the fact that the differential equation is slightly different for the pendulum should have little effect on our ability to carry out the solution.

Exercise 3. Suppose $\theta(t)$ is the unknown solution function for position of the pendulum bob at time t. We write $v(t) = \dfrac{d\theta}{dt}$; explain why θ and v must satisfy the system of first-order equations

$$\theta' = v, \tag{29}$$

$$v' = -\frac{g}{L}\sin\theta, \tag{30}$$

with the initial conditions

$$\theta(0) = \frac{\pi}{6}, \tag{31}$$

$$v(0) = 0. \tag{32}$$

---

[15]The differential equation (10) for the spring has a similar trivial solution. What physical situation does *it* describe?

[16]Zeno of Elea notwithstanding — see Problem 9 in Chapter 5.

Exercise 4. *Explain why the forward-stepping formulas for Euler's Method are*

$$\theta_{k+1} = \theta_k + v_k \cdot \Delta t, \tag{33}$$

$$v_{k+1} = v_k - \frac{g}{L} \cdot \sin(\theta_k) \cdot \Delta t, \tag{34}$$

with

$$\theta_0 = \frac{\pi}{6}, \tag{35}$$

$$v_0 = 0. \tag{36}$$

In Figure 14 we show solutions for θ and v generated from formulas (33)–(36) for a pendulum of length one meter. In a lab associated with this chapter, you may experiment with Euler's Method to see what it tells us about the relationship between period and length of the pendulum.

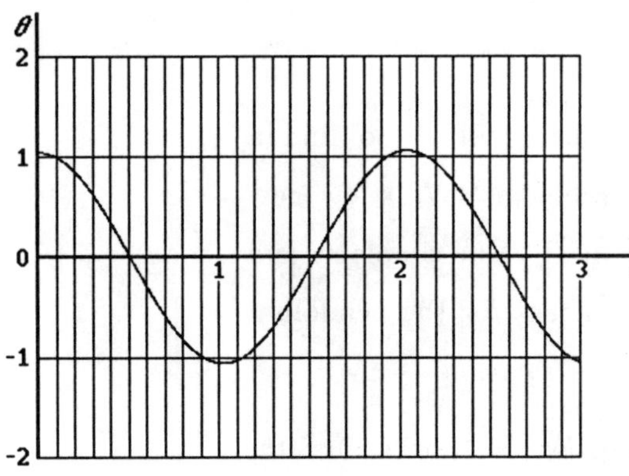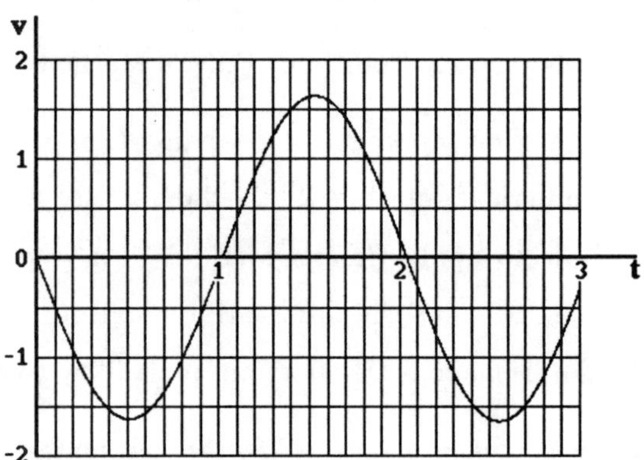

Figure 14. Numerical solutions for the position and velocity of the pendulum bob.

Exercise 5. *(a) Examine Figure 14 carefully. Do the curves behave the way you would expect from what you know about the physics of the pendulum?*

(b) Are the curves believable as a solution to the initial value problem

$$\theta' = v \quad \text{with} \quad \theta(0) = \frac{\pi}{6}, \tag{29R, 31R}$$

$$v' = -\frac{g}{L} \sin \theta \quad \text{with} \quad v(0) = 0? \tag{30R, 32R}$$

(c) Estimate the period of a one-meter pendulum from Figure 14. (Time is measured in seconds.)

(d) Why doesn't this period depend on the mass of the bob?

A Simpler Initial Value Problem: Symbolic Solution

Let's recall the problem we set out to solve: We are attempting to determine the relationship between the period of a pendulum and its length. We modeled the motion of the (idealized) pendulum by the second-order differential equation

$$\frac{d^2\theta}{dt^2} = -\frac{g}{L}\sin\theta, \qquad (26\text{R})$$

where θ is the angle the pendulum makes with the vertical at time t. We saw that, for a particular choice of the length L, and for particular starting values of θ and θ', we could generate a numerical and graphical solution of the differential equation by using Euler's Method. However, that gives us no direct insight into the symbolic form of a solution. It *does* show us a "shape" for the solution (see Figure 14). And, if repeated with enough different values of L (as you may do in the laboratory), it can give us some insight into the relation we seek between L and the period.

Suppose, for a moment, that we simplify the equation by choosing a length L that is numerically equal to the gravitational constant g; then the equation reduces to

$$\frac{d^2\theta}{dt^2} = -\sin\theta.$$

Does either $\theta = \sin t$ or $\theta = \cos t$ satisfy this equation? Not quite! For example, if $\theta = \sin t$, then $\theta'' = -\sin t = -\theta$, *not* $-\sin\theta$.

But we're also not so far off, either. Indeed, you showed in Exercise 9 in Section 6.1 that, for small values of θ, $\sin\theta$ and θ are approximately equal. Thus, instead of making a direct assault on symbolic solution of equation (26),[17] we can replace the differential equation with one that is similar but has an easily described family of solutions. If we replace $\sin\theta$ by θ in equation (26),[18] we obtain

$$\frac{d^2\theta}{dt^2} = -\frac{g}{L}\theta. \qquad (37)$$

Can we justify this replacement? For $\theta = \frac{\pi}{6} \approx 0.5236$, we have $\sin\theta = 0.5$.

---

[17]The direct assault would be doomed to failure anyway. Even with all our simplifying assumptions, this simple physical situation is one that cannot be represented exactly by a formula that uses only so-called "elementary" functions. We chose the problem in part to make that point.

So the substitution of θ for $\sin\theta$ is not a great approximation for θ near $\frac{\pi}{6}$, but it becomes better for smaller values of θ (again, see your table in Section 6.1, Exercise 9).

If we shift our attention to the differential equation (37), what can we say about the family of solutions? We already know what they are! Indeed, equation (37) is just an "alias" for equation (10) — with the names of the variables changed to protect the innocent:

$$\frac{d^2x}{dt^2} = -\frac{k}{m}\,x. \tag{10R}$$

Specifically, in equation (37), x has been replaced by θ, k by g, and m by L:

$$\frac{d^2\theta}{dt^2} = -\frac{g}{L}\,\theta. \tag{37R}$$

In the following exercises, we ask you to recall what you already know about solutions of (37) and to answer the question about relationship between period and length of the pendulum.

Exercises

6. (a) *Show that every function in the family* $A\,\sin\sqrt{\frac{g}{L}}\,t + B\,\cos\sqrt{\frac{g}{L}}\,t$ *is a solution of the differential equation (37).*

(b) *For each such function, explain why* B *is the value of the function at* $t = 0$.

(c) *How is* A *related to the value of the first derivative at* $t = 0$?

---

[18]This replacement is called "linearization" — we are replacing the nonlinear expression $\sin\theta$ by one that is *linear* in θ, namely θ itself. There is a parallel here with our introduction of Newton's Method in Chapter 4. Because it is often impossible to solve a nonlinear algebraic equation algebraically, we replace it by an approximating linear equation and solve that instead. Also, it is often impossible to solve a nonlinear differential equation algebraically, so we replace it with an approximating linear differential equation and solve that instead. If you find it hard to see the parallel, keep in mind that when we solve algebraic equations, we are looking for *numbers* as solutions, and when we solve differential equations, we are looking for *functions* as solutions.

7. Find values for the constants A and B such that the function

$$\theta(t) \;=\; A \sin \sqrt{\tfrac{g}{L}}\, t + B \cos \sqrt{\tfrac{g}{L}}\, t \tag{38}$$

satisfies the initial value problem

$$\frac{d^2\theta}{dt^2} \;=\; -\frac{g}{L}\,\theta, \quad \text{with } \theta(0)=\tfrac{\pi}{6}, \quad \text{and} \quad \frac{d\theta}{dt}(0)=0. \quad \text{(26R-28R)}$$

8. We know that the period of $\sin t$ is 2π; that is, 2π is the smallest number
T such that $\sin t$ goes through a complete swing from 0 to 1 to -1 and
back to 0 again in the interval $[0,T]$.

(a) What is the period of $\sin 7t$? of $\sin 0.2t$?

(b) What is the frequency of each of the functions in part (a)?

9. (a) Find a formula for the period of the function $\sin \omega t$; for $\cos \omega t$.

(b) Find the period of the solution (38) to the initial value problem in Exercise
7, and simplify.

(c) How is the period related to the length L?

(d) Does your answer agree with Figure 15, in which we have plotted period as
a function of length?

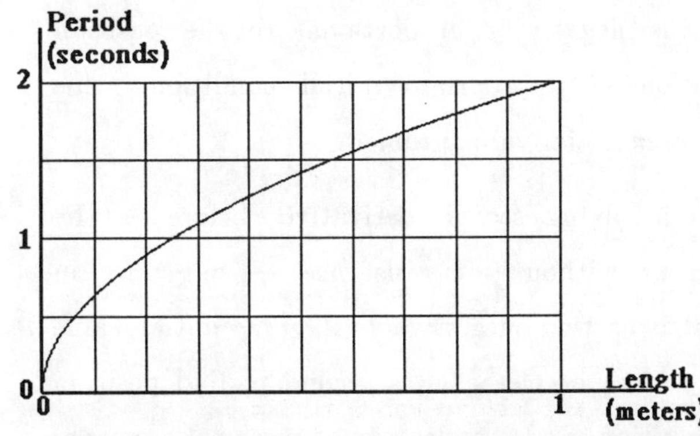

Figure 15. Graph of period T versus length L
for the approximate pendulum equation (37).

SUMMARY

This chapter began with a review of the most basic properties of the sine and cosine functions, with an emphasis on the origin of these functions in motion around a circle. We found, first graphically and then by studying limiting values of difference quotients, that sine and cosine have very simple derivative formulas:

$$\frac{d}{dt} \sin t \;=\; \cos t, \tag{39}$$

and

$$\frac{d}{dt} \cos t \;=\; -\sin t. \tag{40}$$

Next we turned to the first of two physical problems as a "prototype" for problems whose solutions are periodic: the spring-mass system that oscillates back and forth because the spring is always pulling in the direction opposite to the displacement from equilibrium. We saw that Newton's Second Law of Motion leads to a description of the spring-mass motion as "the acceleration is negatively proportional to the position function." With the addition of appropriate initial conditions, this statement becomes a *second order* initial value problem.

We have seen equations involving second derivatives before — the falling body problems, with or without air resistance — but only in situations that could be treated as two successive *first order* initial value problems, one to find velocity from acceleration, a second to find position from velocity. Here the position and acceleration functions appeared together in the same equation.

Our first approach to solving such a problem was to rewrite it as a system of two first order problems, using velocity as an "intermediate variable." That turned our problem into one that we could solve numerically, just as we solved the S-I-R epidemic model in Chapter 5, by using Euler's Method. The numerical solution looked a lot like a cosine function, which fit with our calculation of derivatives of sine and cosine.

By scaling the independent variable (i.e., adjusting the period or frequency by replacing t by ωt), using the Chain Rule, and calculating second derivatives, we saw that we could find solutions to the spring-mass equation of the form $A \sin \omega t + B \cos \omega t$ if we took ω to be $\sqrt{\frac{k}{m}}$, where k is the spring constant and m is the mass. Finally, we saw that initial conditions of the form $x(0) =$ initial displacement x_0 and initial velocity equal to zero required that B be x_0 and A be zero. Thus, the solutions were indeed cosine functions, with period determined by the spring constant and mass, and amplitude determined by the initial displacement.

We took up motion of a simple pendulum as a second problem for the study of repeating phenomena, one that is more challenging than the spring problem, but also one for which the solution of the spring problem provides important insights. In particular, we focused on the question of period as a function of length to see if we could describe this relationship. We found three different ways to address this question.

First, we can do physical measurements, as you did with your casual experiments. In the classroom you may have already done more careful measurements of periods and lengths to get a data table representing the physical relationship.

Second, we can model the pendulum with a second-order initial value problem. Even with simplifying assumptions, this problem turns out to be *nonlinear* because one term in the differential equation involves the sine of the unknown function. Nevertheless, by treating the first derivative as another unknown function (which of course it is), we can turn the single second-order differential equation into a system of two first-order equations. Then, as we did for the spring equation, we can use Euler's Method to generate numerical and graphical solutions. Euler's Method "doesn't care" whether the slope information is linear or not.

Third, we can deal with the nonlinearity by "linearizing": We can replace the sine of the unknown function by the unknown function itself. This approximating differential equation has solutions we can easily discover the same way we found solutions of the spring equation. Those solutions should *approximate* the still-unknown solutions of the nonlinear differential equation — although we have no way of knowing at this time how good the approximations are. Nevertheless, the solutions of the linear equation have the attractive feature that their period is easy to calculate as a function of the length of the pendulum, thus giving at least an approximate answer to our primary question.

In a laboratory or classroom project, you may be asked to discuss the relationships among these three "solutions": physical data, subject to measurement errors; numerical approximation of the solution of an exact (but idealized) differential equation; and exact solution of an approximating differential equation. Each contributes important information to our understanding of the pendulum; taken together, they still leave us without anything we can call an "exact" answer. And that's the way the world works.

Derivative Formulas

Derivatives of exponential functions (Chapter 2):

$$\frac{d}{dt}\, e^t \;=\; e^t, \quad \text{where } e \text{ is the natural base, } 2.71828\ldots. \tag{41}$$

$$\frac{d}{dt}\, e^{kt} \;=\; k\, e^{kt}, \quad \text{for any constant } k. \tag{42}$$

$$\frac{d}{dt}\, b^t \;=\; (\ln b)\, b^t, \quad \text{for any constant exponential base } b. \tag{43}$$

Derivatives of polynomial functions (Chapter 2):

Use the Sum Rule to differentiate term-by-term.

Use the Constant Multiple Rule and Power Rule on each term:

$$\frac{d}{dt}\, ct^n \;=\; cnt^{n-1}, \quad \text{for any power } n \text{ and any constant } c. \tag{44}$$

Derivatives of power functions:

Power Rule: $\quad \dfrac{d}{dt}\, t^r \;=\; r\, t^{r-1}, \quad$ for any real constant r. $\tag{45}$

Derivative of the natural logarithm function (Chapter 4):

$$\frac{d}{dt} \ln t = \frac{1}{t}. \tag{46}$$

Derivatives of trigonometric functions (this chapter):

$$\frac{d}{dt} \sin t = \cos t. \tag{47}$$

$$\frac{d}{dt} \cos t = -\sin t. \tag{48}$$

Scaling formulas for independent and dependent variables (Chapters 2 and 4):

If $T = kt$, where k is constant, then

$$\frac{d}{dt} f(kt) = k \frac{d}{dT} f(T). \tag{49}$$

CHAIN RULE: $\quad \dfrac{dy}{dt} = \dfrac{dy}{du}\dfrac{du}{dt}. \tag{50}$

Constant Factor Rule:

$$\frac{d}{dt} Af(t) = A \frac{d}{dt} f(t), \quad \text{for any constant } A. \tag{51}$$

Formulas that apply to combinations of functions (Chapters 2, 4, and 6):

Sum Rule: $\quad \dfrac{d}{dt}\left[f(t) + g(t)\right] = \dfrac{d}{dt} f(t) + \dfrac{d}{dt} g(t). \tag{52}$

Product Rule: $\quad \dfrac{d}{dt}\left[g(t) \cdot h(t)\right] = g(t) \cdot \dfrac{d}{dt} h(t) + h(t) \cdot \dfrac{d}{dt} g(t). \tag{53}$

Power Rule: $\quad \dfrac{d}{dt} f(t)^r = r\, f(t)^{r-1} \dfrac{d}{dt} f(t), \quad \text{for any real constant } r. \tag{54}$

Exponential Rule: $\dfrac{d}{dt} e^{f(t)} = e^{f(t)} \dfrac{d}{dt} f(t). \tag{55}$

Logarithmic Rule: $\dfrac{d}{dt} \ln f(t) = \dfrac{1}{f(t)} \dfrac{d}{dt} f(t). \tag{56}$

Sine Rule: $\quad \dfrac{d}{dt} \sin f(t) = \cos f(t) \dfrac{d}{dt} f(t). \tag{57}$

Cosine Rule: $\quad \dfrac{d}{dt} \cos f(t) = -\sin f(t) \dfrac{d}{dt} f(t). \tag{58}$

Examples of Derivative Calculations

(a) $\dfrac{d}{dx}(7x^5 - 4x^3 + 2) = 35x^4 - 12x^2$ (Sum Rule, Constant Multiple Rule, Power Rule)

(b) $\dfrac{d}{du}\sqrt{5u^4 - u + 1} = \dfrac{1}{2}\dfrac{1}{\sqrt{5u^4 - u + 1}}\dfrac{d}{du}(5u^4 - u + 1)$ (Power Rule, Chain Rule)

$$= \dfrac{20u^4 - 1}{2\sqrt{5u^4 - u + 1}}$$

(c) $\dfrac{d}{dt}(\ln t)^4 = 4(\ln t)^3\dfrac{1}{t}$ (Power Rule, Chain Rule, Logarithmic Rule)

(d) $\dfrac{d}{dt}\left(e^{2t}\cos 3t\right) = e^{2t}\dfrac{d}{dt}\cos 3t + \cos 3t\dfrac{d}{dt}e^{2t}$ (Product Rule)

$$= 3\,e^{2t}(-\sin 3t) + 2\,e^{2t}\cos 3t$$ (Exponential Rule, Cosine Rule, Chain Rule)

(e) $\dfrac{d}{d\theta}\left(\dfrac{e^{2\theta}}{\cos 3\theta}\right) = \dfrac{d}{d\theta}\left(e^{2\theta}\dfrac{1}{\cos 3\theta}\right)$

$$= e^{2\theta}\dfrac{d}{d\theta}\left(\dfrac{1}{\cos 3\theta}\right) + \dfrac{d}{d\theta}\left(e^{2\theta}\right)\dfrac{1}{\cos 3\theta}$$ (Product Rule)

$$= e^{2\theta}(-1)(\cos 3\theta)^{-2}\dfrac{d}{d\theta}\cos 3\theta + 2e^{2\theta}\dfrac{1}{\cos 3\theta}$$ (Chain Rule, Power Rule)

$$= e^{2\theta}(-1)(\cos 3\theta)^{-2}\,3(-\sin 3\theta) + 2e^{2\theta}\dfrac{1}{\cos 3\theta}$$ (Chain Rule)

$$= \dfrac{e^{2\theta}}{\cos 3\theta}\left(3\dfrac{\sin 3\theta}{\cos 3\theta} + 2\right)$$

(f) $\dfrac{d}{ds}\ln(5s + 6) = \dfrac{1}{5s + 6}\dfrac{d}{ds}(5s + 6) = \dfrac{5}{5s + 6}$ (Logarithmic Rule, Chain Rule)

Practice with Calculations

*Exercises in this category include (a) calculations that can be done by machines and (b) practice on important topics from courses that precede calculus. You need to develop **facility** with both categories — not because such calculations are a central feature of the course, but because they should not frustrate you or keep you from concentrating on the more important parts of the course by occupying a lot of your time or by leading to lots of mistakes. Even though routine calculations can be done quickly and accurately by computer or calculator, you need to develop judgment about when to use a machine and when not to, and you need to know how to tell when you might have pressed the wrong button. These skills are acquired and sharpened by **practice**.*

You should expect to see exercises like these as some portion of your homework assignments, quizzes, and tests.

Calculate each of the following derivatives:

1. $\dfrac{d}{dt}\, e^{2t} \sin t$ 2. $\dfrac{d}{dt}\, e^{t} \cos 2t$ 3. $\dfrac{d}{dt}\, \ln t \sin 2t$

4. $\dfrac{d^2}{dt^2}\, \sin 2t$ 5. $\dfrac{d^2}{dt^2}\, \sin^2 t$ 6. $\dfrac{d^2}{dt^2}\, e^{2t} \sin t$

7. $\dfrac{d}{dt}\,(10\,t^6 - 4\,t^3 + \pi)$ 8. $\dfrac{d}{dx}\, \dfrac{\sin x}{x}$ 9. $\dfrac{d}{dx}\,(\cos x)^3$

10. $\dfrac{d}{dt}\, \ln\sqrt{3\,t - 9}$ 11. $\dfrac{d}{d\theta}\, \dfrac{\sin \theta}{\cos 2\theta}$ 12. $\dfrac{d}{dw}\, \dfrac{1}{w^3}$

13. $\dfrac{d}{dt}\, t^{7/5}$ 14. $\dfrac{d}{dt}\, t\,e^{-t}$ 15. $\dfrac{d^2}{dt^2}\, t\,e^{-t}$

Find the derivative of each of the following functions:

16. $y = \sin 2t - 3 \cos t$ 17. $y = \cos^2 t$ 18. $y = \cos\left(t^2\right)$

19. $y = (\sin 3t)(\cos 2t)$ 20. $z = \dfrac{2}{t-1}$ 21. $z = \dfrac{1}{1+7t}$

22. $y = x^2 + 3x + 6$ 23. $y = 5^x$ 24. $y = x^3 \sin(2x)$

25. $y = \ln\left((x^{10} - 3x)^{1/3}\right)$ 26. $y = y(x)$, where $y^3 + x^2 = 3x + 7$

27. $y = \dfrac{\sin x}{\cos x + 1}$ 28. $y = \ln(e^x)$ 29. $f(x) = 3\,e^{-4x}$

30. $f(x) = \sin(x^2)$ 31. $f(x) = \dfrac{x^2}{1+2x}$ 32. $y = \dfrac{\cos x + x^2}{x^3}$

Find all <u>anti</u>derivatives of each of the following functions:

33. e^{2t}

34. t^n, where n is a positive integer

35. $\dfrac{1}{t^2}$

36. $\cos t$

37. $\dfrac{1}{t}$

38. $\sin 2t$

39. $\dfrac{2}{t-1}$

40. $\dfrac{1}{1+7t}$

41. Given that $\dfrac{dP}{dt} = 2 - \cos(\pi P)$ and $P(0) = 1$, use Euler's Method with $\Delta t = 0.2$ to approximate $P(1)$.

42. Find the maximum and minimum values of

$$y = \sin x + \cos x$$

on the interval $[0,\pi]$.

Conceptual Exercises

Exercises in this category use and/or further develop the concepts introduced in this and earlier chapters. The level of difficulty is similar to that of the exercises embedded in the text and at the ends of sections. While many of these exercises are presented in "realistic" contexts, they are not representative of real problems. Rather, these exercises are to help you get ready for tackling real problems.

You should expect to see exercises like these as a major portion of your homework assignments, quizzes, and tests.

1. (a) Sketch the graph of $\cos 2t$ for $0 \le t \le 2\pi$.

 (b) Sketch the graph of $\cos^2 t$ for $0 \le t \le 2\pi$.

 (c) Use the identities $\cos 2t = \cos^2 t - \sin^2 t$ and $\cos^2 t + \sin^2 t = 1$ to show that

$$\cos^2 t \;=\; \frac{1 + \cos 2t}{2} \quad \text{for all } t.$$

 (d) Find all the points of inflection on the curve $y = \cos^2 t$, and mark them on your graph in part (b). Why must these points also be points of inflection for the curve $y = \sin^2 t$?

 (e) Sketch the graph of $\sin^2 t$ for $0 \le t \le 2\pi$.

 (f) Show that

$$\sin^2 t \;=\; \frac{1 - \cos 2t}{2} \quad \text{for all } t.$$

2. (a) Show that sine is an odd function and cosine is an even function.

 (b) How do the symmetries of these functions fit with the information in Problem 1 in Chapter 4?

3.[19] Give a reason why $\cos x$ is not a polynomial.

4.[19] Determine the smallest positive number x for which the function

$$f(x) \;=\; -4 \sin\left(4x + \frac{\pi}{6}\right)$$

has

 (a) the value zero;

 (b) the maximum value of $f(x)$;

 (c) the minimum value of $f(x)$.

---

[19]From *Calculus Problems for a New Century*, edited by Robert Fraga, MAA Notes Number 28, 1993.

5.   If the mass attached to the spring is "thrown" (vertically up or down) instead of let go, all that changes is that the initial velocity is no longer zero.

(a) The initial velocity can be either positive or negative; which sign corresponds to throwing the mass "up" and which "down"?

(b) Find values for the constants A and B such that the function

$$x(t) \;=\; A \sin \sqrt{\tfrac{k}{m}}\, t + B \cos \sqrt{\tfrac{k}{m}}\, t$$

satisfies the initial value problem

$$\frac{d^2 x}{dt^2} \;=\; -\,\frac{k}{m}\, x, \quad x(0) = 1, \quad \text{and} \quad \frac{dx}{dt}(0) = -0.5.$$

6.   If the pendulum bob is "thrown" (along the circle of motion) instead of being dropped, all that changes is that the initial velocity is no longer zero.

(a) The initial velocity can be either positive or negative; which sign corresponds to throwing the bob "up" and which "down"?

(b) Find values for the constants A and B such that the function

$$\theta(t) \;=\; A \sin \sqrt{\tfrac{g}{L}}\, t + B \cos \sqrt{\tfrac{g}{L}}\, t$$

satisfies the initial value problem

$$\frac{d^2 \theta}{dt^2} \;=\; -\,\frac{g}{L}\, \theta, \quad \theta(0) = \frac{\pi}{6}, \quad \text{and} \quad \frac{d\theta}{dt}(0) = -0.5.$$

7.   (Cf. Problem 8 in Chapter 4) Recall that the tangent and secant functions are defined by

$$\tan t = \frac{\sin t}{\cos t} \quad \text{and} \quad \sec t = \frac{1}{\cos t}.$$

(a) Show that

$$\frac{d}{dt} \tan t \;=\; \sec^2 t.$$

(b) Show that

$$\frac{d}{dt} \sec t \;=\; \sec t \tan t.$$

8. Recall that the cotangent and cosecant functions are defined by

$$\cot t = \frac{\cos t}{\sin t} \quad \text{and} \quad \csc t = \frac{1}{\sin t}.$$

Find formulas for the derivatives of

(a) the cotangent function;

(b) the cosecant function.

9.[20] Define a function f by $f(x) = \frac{x + \sin x}{\cos x}$ for $-\frac{\pi}{2} < x < \frac{\pi}{2}$.

(a) Is $f(x)$ an even function, an odd function, or neither? Justify your answer.

(b) Find $f'(x)$.

(c) Find an equation of the line tangent to the graph of f at the point where $x = 0$.

10. A mass of 4 kilograms is suspended from a spring attached to the ceiling that has a spring constant of 9 kg/sec^2. The mass is pulled downward one meter below its equilibrium position and then released from this point with an initial velocity (upward) of 1.5 meters/sec.

(a) Write down the differential equation and initial conditions satisfied by the function $x(t)$ that gives displacement of the mass from equilibrium as a function of time.

(b) Solve the initial value problem to find $x(t)$.

(c) At what times will the mass be closest to the ceiling?

11. If we begin with $x_0 = 0$, then Newton's Method fails to find a root for one of the following functions. Which function is it?

(i) $f(x) = \sin x$ (ii) $f(x) = \cos x$

(iii) $f(x) = 2e^x - 1$ (iv) $f(x) = e^{-x} - x$

---

[20]From *Calculus Problems for a New Century*, edited by Robert Fraga, MAA Notes Number 28, 1993.

Problems and Projects

The problems presented here are not intended for individual homework assignments or for tests. These problems should be attempted by groups of two to four students sharing ideas, whether in the classroom or elsewhere. Some of the more extensive problems are suitable for projects with time frames ranging from a class period to a week.

1. Use a trigonometric identity for the cosine of a sum to rewrite the difference quotient $\dfrac{\cos{(t+\Delta t)} - \cos t}{\Delta t}$ as in equation (6). Use what you already know about limiting values to show (another way) that $\dfrac{d}{dt}\cos t = -\sin t$.

2. Figure 16 shows the solution of Conceptual Exercise 6 for a pendulum of length one meter. The figure suggests that the solution is just a sinusoidal function, shifted horizontally (i.e., with an additive scaling of the independent variable). Show that every function of the form

$$\theta(t) = A \sin \omega t + B \cos \omega t$$

can also be written in the form

$$\theta(t) = C \sin(\omega t + \delta)$$

for some constants C and δ (called, respectively, the "amplitude" and "phase shift"). You may find it simpler to work this problem "backwards," that is, to expand the second form, using the formula for sine of a sum, and then match coefficients with the first form. That will show how C and δ must be related to A and B.

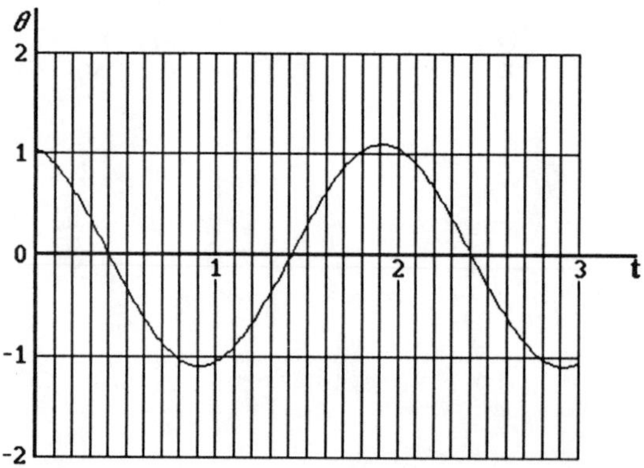

Figure 16. Graph of the solution of Conceptual Exercise 6.

3. In Figure 17, we show the long-term average monthly temperatures (in degrees Fahrenheit) in Durham, NC. Each average temperature is plotted as a dot. We have superimposed on the plotted data the graph of an approximating function of the form $f(t) = b + C \sin(\omega t + \delta)$, where time t is measured in months. The constant term (b) is an additive scaling of the vertical axis to "center" the data around an average value.

(a) Estimate from the graph the values of b (average temperature for the year), C (amplitude), ω (related to the period), and δ (phase shift).

(b) Check your answers by substituting into your function several integers between 0 and 12. If you don't get values close to those in Figure 17, make the necessary adjustments in your function until you do.

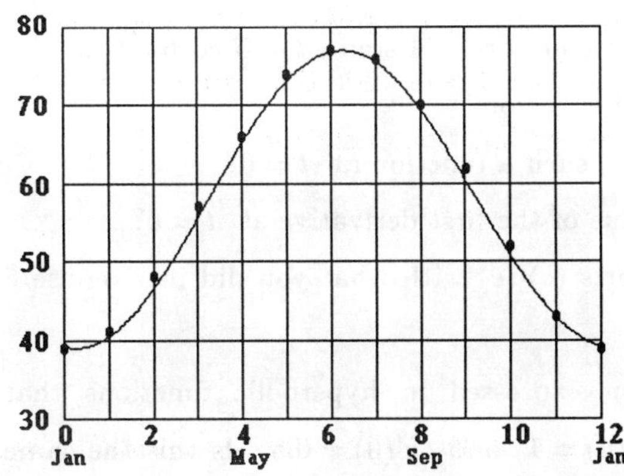

Figure 17. Long-term average monthly temperatures,

with a superimposed sine function.

4. (Cf. Note 7) We have found (in Chapter 2) that solutions of $\dfrac{dy}{dt} = ky$ are all of the form $y = Ae^{kt}$,   and we have found (in this chapter) that solutions of $\dfrac{d^2y}{dt^2} = -\omega^2 y$ are all of the form $A \sin \omega t + B \cos \omega t$. Now we seek solutions of $\dfrac{d^2y}{dt^2} = \omega^2 y$.

(a) Show that $e^{\omega t}$ and $e^{-\omega t}$ are both solutions of $\dfrac{d^2y}{dt^2} = \omega^2 y$.

(b) Show that all functions of the form $Ae^{\omega t} + Be^{-\omega t}$ are solutions.

(c) If $\omega = 2$, find a solution that satisfies the initial conditions $y(0) = 1$ and $y'(0) = 0.5$.

5.   The solutions of $\dfrac{d^2y}{dt^2} = \omega^2 y$ can be written in another form that is more suggestive of the parallel with trigonometric functions (also called "circular" functions). This other form uses so-called "hyperbolic" functions, the hyperbolic sine (abbreviated "sinh") and hyperbolic cosine (abbreviated "cosh"). These functions may be defined from exponential functions by the following formulas:

$$\sinh \omega t = \tfrac{1}{2}\left(e^{\omega t} - e^{-\omega t}\right); \quad \cosh \omega t = \tfrac{1}{2}\left(e^{\omega t} + e^{-\omega t}\right).$$

(a) How are the first derivatives of these hyperbolic functions related to the functions themselves?

(b) Explain why you already know that $\sinh \omega t$ and $\cosh \omega t$ are solutions of $\dfrac{d^2y}{dt^2} = \omega^2 y$.

(c) Explain why every function of the form $A \sinh \omega t + B \cosh \omega t$ is also a solution of the same differential equation.

(d) Why must B be the value of such a function at $t = 0$?

(e) How is A related to the value of the first derivative at $t = 0$?

(f) Compare what you did in parts (c)-(e) with what you did in Exercise 6 of Section 6.3.

(g) If $\omega = 2$, find a solution expressed in hyperbolic functions that satisfies the initial conditions $y(0) = 1$ and $y'(0) = 0.5$. Is this the same solution you found in the previous problem? Why or why not?

6.   You showed in Exercise 4 of Section 6.1 that the circular functions

$$x = \cos \omega t \quad \text{and} \quad y = \sin \omega t$$

satisfy the equation

$$x^2 + y^2 = 1$$

— the equation of a circle — which is not surprising, given the origins of these function in a circle.

(a) Find a similar equation satisfied by the hyperbolic functions

$$x = \cosh \omega t \quad \text{and} \quad y = \sinh \omega t.$$

(Hint: Calculate the squares of the hyperbolic sine and the hyperbolic cosine directly from their definitions in the preceding problem.)

(b) Why are these functions called "hyperbolic"?

7. (a) Make a sketch, on a single set of coordinate axes, of the graphs of $y = e^x$ and $y = e^{-x}$.

(b) The function $y = \cosh x$ is the average of the two functions whose graphs you just sketched. Sketch the average of the two graphs to find the graph of $y = \cosh x$.

(c) Construct the graph of $y = \sinh x$ in a similar manner, as the average of two exponential graphs.

(d) Check your work on a computer or graphing calculator.

8. (a) Show that, for a given spring, the period of a spring-mass system is determined by the mass in approximately the same way that the period of a pendulum is determined by the length. Why is the analogy only approximate?

(b) What feature of the pendulum problem is analogous to "spring constant"? Why didn't we study the dependence of period on this feature?

9. Suppose the same mass m is attached to a variety of different springs with different spring constants. How do the periods of the different spring-mass systems depend on the spring constants?

10. An airplane is flying directly overhead at an altitude of $20,000$ feet. One minute later, the line of sight to the airplane makes an angle of $49°$ with the ground. What is the approximate speed of the airplane in miles per hour?

11.[21] (Of special interest to outfielders and quarterbacks) For reasons we will see in the next chapter, the range R of a projectile (if air resistance is negligible) is given by

$$R = \frac{v^2 \sin 2\alpha}{g},$$

where v is the muzzle velocity, α is the angle of elevation, and g is the acceleration of gravity. What angle of elevation gives the maximum range of the projectile?

---

[21]Adapted from *Calculus Problems for a New Century*, edited by Robert Fraga, MAA Notes Number 28, 1993.

12. What is the cone of largest volume that can be formed by rotating a right triangle of fixed hypotenuse h around one of its legs? What is the volume of that cone? [Note: This was also Problem 20 in Chapter 4. At that point you might have used the length of one leg of the triangle as your independent variable. Try the problem again using one angle of the triangle as your independent variable. Do you get the same answer both ways? Which choice leads to an easier calculation?]

13.[21] (Cf. Conceptual Exercise 8) According to Poiseuille's Law, if blood flows into a straight blood vessel of radius r branching off another straight blood vessel of radius R at an angle α, the total resistance T of the blood in the branching vessel is given by

$$T \;=\; C\left(\frac{a - b\,\cot\alpha}{R^4} + \frac{b\,\csc\alpha}{r^4}\right),$$

where a, b, C, r, and R are constants, and $r < R$. Show that the total resistance is minimized when $\cos\alpha = \left(\dfrac{r}{R}\right)^4$.

14.[22] Suppose you walk counterclockwise around the perimeter of the square with corners at $(\pm 1, \pm 1)$, starting at the point $(1,0)$. Let $(x(t), y(t))$ be your position on the square after you have walked t units.

(a) Find formulas for $x(t)$ and $y(t)$ as functions of t.

(b) Sketch the graph of $x(t)$ as a function of t.

(c) Sketch the graph of $y(t)$ as a function of t.

(d) Are the functions $x(t)$ and $y(t)$ periodic? If so, what is the period of each? Explain.

---

[22]Thanks to John Frampton, Northeastern University.